PLANT BIOCHEMISTRY
AND
MOLECULAR BIOLOGY

It is with great sadness that I learned of the untimely death of my ex-colleague, Dr. Janice Forde, of Rothamsted Experimental Station, on 21st November 1992. I would like to dedicate this volume to her memory.

P.J. Lea

PLANT BIOCHEMISTRY
AND
MOLECULAR BIOLOGY

Edited by

PETER J. LEA

Division of Biological Sciences, Lancaster University, UK

RICHARD C. LEEGOOD

Department of Animal and Plant Sciences, University of Sheffield, UK

JOHN WILEY & SONS

Chichester · New York · Brisbane · Toronto · Singapore

Copyright © 1993 by John Wiley & Sons Ltd,
Baffins Lane, Chichester,
West Sussex PO19 1UD, England

Other Wiley Editorial Offices

John Wiley & Sons, Inc., 605 Third Avenue,
New York, NY 10158-0012, USA

Jacaranda Wiley Ltd, G.P.O. Box 859, Brisbane,
Queensland 4001, Australia

John Wiley & Sons (Canada) Ltd, 22 Worcester Road,
Rexdale, Ontario M9W 1L1, Canada

John Wiley & Sons (SEA) Pte Ltd, 37 Jalan Pemimpin #05-04
Block B, Union Industrial Building, Singapore 2057

Library of Congress Cataloging-in-Publication Data

Plant biochemistry and molecular biology/edited by Peter J. Lea and
    Richard C. Leegood.
        p. cm.
      Includes bibliographical references and index.
      ISBN 0 471 93313 9 (pbk)   0 471 93895 5 (cloth)
      1. Botanical chemistry.   2. Plant molecular biology.   I. Lea,
    Peter J.   II. Leegood, Richard C.
    QK861.P52   1993
    581.19′2—dc20   92-22064
                                                     CIP

British Library Cataloguing in Publication Data

A catalogue record for this book is available from the British Library

ISBN 0 471 93313 9 (pbk)   0 471 93895 5 (cloth)

Typeset in Times 10/12 pt by Acorn Bookwork, Salisbury, Wiltshire.
Printed and bound in Great Britain by The Bath Press, Avon.

CONTENTS

CONTRIBUTORS

J.A. BRYANT — Department of Biological Sciences, University of Exeter, Exeter EX4 4QG, UK

J.H. BRYCE — Department of Biological Sciences, Heriot-Watt University, Riccarton, Edinburgh EH14 4AS, UK

A.C. CUMING — Department of Genetics, University of Leeds, Leeds LS2 9JT, UK

J.R. GALLON — School of Biological Sciences, University of Wales, Swansea SA2 8PP, UK

J.A. GATEHOUSE — Department of Biological Sciences, University of Durham, Durham DH1 3LE, UK

G.A.F. HENDRY — NERC Unit of Comparative Plant Ecology, Department of Animal and Plant Sciences, University of Sheffield, Sheffield S10 2TN, UK

S.A. HILL — Department of Biological Sciences, Heriot-Watt University, Riccarton, Edinburgh EH14 4AS, UK

P.J. LEA — Division of Biological Sciences, Lancaster University, Lancaster LA1 4YQ, UK

R.C. LEEGOOD — Department of Animal and Plant Sciences, Robert Hill Institute, University of Sheffield, Sheffield S10 2UQ, UK

D.J. MURPHY — John Innes Centre for Plant Science Research, Norwich NR4 7UH, UK

N.J. ROBINSON — Department of Biological Sciences, University of Durham, Durham DH1 3LE, UK

A.H. SHIRSAT — Department of Biological Sciences, University of Durham, Durham DH1 3LE, UK

C.J. SMITH — School of Biological Sciences, University of Wales, Swansea SA2 8PP, UK

R.J. SMITH — Institute of Environmental and Biological Sciences, Lancaster University, Lancaster LA1 4YQ, UK

R. WALDEN   Max-Planck-Institut für Züchtungsforschung, D-5000 Köln 30, Germany

M.D. WATSON   Department of Biological Sciences, University of Durham, Durham DH1
                     3LE, UK

PREFACE

Although our knowledge of plant metabolism has improved greatly in the last few years, it remains a neglected field, with considerable ignorance about the structure and regulation of important pathways, let alone the synthesis of the myriad of plant products. It is, more than ever, quite obvious that plant cells cannot simply be regarded as animal cells with chloroplasts. Plant cells possess many unique pathways and they even regulate those pathways which they share with other organisms in distinctive ways. Equally, their molecular biology has many unique features. This book is designed to give an overview of plant metabolism and molecular biology to second-year undergraduates, but will also be useful to third-year and postgraduate students.

The book opens with the mechanisms by which energy is generated both within the chloroplasts and mitochondria, using either the photosynthetic light reactions or the metabolism of carbohydrates via glycolysis and the Krebs cycle. Photosynthetic carbon metabolism is considered in two chapters, first the process of CO_2 assimilation in the Calvin cycle, then the photosynthetic adaptations which have occurred in C_4 and CAM plants (and in aquatic organisms) in response to particular environmental constraints. The way in which this carbon is utilized in the synthesis of various products such as carbohydrates, lipids and pigments is then considered. Photosynthesis is also the driving force for the reduction of N_2 and assimilation of nitrate, and subsequent incorporation into amino acids required for protein synthesis.

The latter part of the book deals with the structure of the genome in nucleus, chloroplasts and mitochondria. This is followed by the regulation of gene expression and the control of development. Finally, the technology of genetic manipulation of plants is described. This blend of plant biochemistry and molecular biology is justified because, increasingly, they are coming together in such areas as the regulation of gene expression by metabolism and the control of development. In these areas, transgenic plants are beginning to have a large impact. The use of the powerful techniques of molecular biology to generate transgenic plants is also beginning to be used for the manipulation of selected enzymes in plants so that control of metabolism can be studied. It is now becoming essential for the plant biochemist to become well versed in molecular biology and *vice versa*. We hope that this book will go some way towards achieving that blend.

P.J. Lea
R.C. Leegood
Sheffield, July 1992

Abbreviations

In the majority of instances in this volume an abbreviation is defined on the first usage in each chapter. However, there are a number of abbreviations that are used universally throughout all aspects of Biochemistry. These are defined below:

Abbreviation	Full name	Abbreviation	Full name
A	adenosine	$NADP^+$	nicotinamide adenine dinucleotide phosphate, oxidized (NADPH, reduced form)
ACP	acyl carrier protein		
ADP	adenosine 5′-diphosphate		
AMP	adenosine 5′-monophosphate		
ATP	adenosine 5′-triphosphate	P_i	inorganic phosphate
C	cytosine	PP_i	inorganic pyrophosphate
CDP	cytidine 5′-diphosphate	RNA	ribonucleic acid (mRNA, messenger; rRNA, ribosomal; tRNA, transfer)
CMP	cytidine 5′-monophosphate		
CoA	coenzyme A		
CTP	cytidine 5′-triphosphate	T	thymine
DNA	deoxyribonucleic acid	TCA	tricarboxylic acid
FAD	flavine adenine dinucleotide	T-DNA	transforming DNA
G	guanine	TPP	thiamine pyrophospahte
GDP	guanosine 5′-diphosphate	U	uracil
GMP	guanosine 5′-monophosphate	UDP	uridine 5′-diphosphate
GTP	guanosine 5′-triphosphate	UMP	uridine 5′-monophosphate
kDa	kilodalton, (unit of molecular weight)	UTP	uridine 5′-triphosphate
K_m	Michaelis–Menten constant		
NAD^+	nicotinamide adenine dinucleotide, oxidized (NADH reduced form)		

1

ENERGY PRODUCTION IN PLANT CELLS

J.H. Bryce and S.A. Hill

Department of Biological Sciences, Heriot-Watt University, Edinburgh, UK

THE IMPORTANCE OF PHOTOSYNTHESIS AND RESPIRATION

Higher plants, in particular angiosperms, provide the major supply of food (energy) for animal and microbial life on earth. In the seas, the microscopic phytoplankton are the food source for zooplankton and fish. The phytoplankton are a source of food because they assimilate carbon dioxide into organic compounds which can be oxidized either by themselves or by animals that feed on them. The oxidation is carried out by the respiratory pathways and provides energy, reducing power and intermediate compounds for biosynthesis and growth. These organic compounds are produced using energy from sunlight

Plant Biochemistry and Molecular Biology. Edited by P.J. Lea and R.C. Leegood

by a process known as photosynthesis. These are autotrophic organisms and require only light, carbon dioxide, water and mineral ions to produce their respiratory substrates. Heterotrophs, in contrast, have a requirement for organic compounds produced by autotrophs.

A major feature of most photosynthetic organisms is that they release oxygen from water. The life that we know today is therefore due not only to the supply of food that is produced from photosynthesis, but also relies upon the oxygen that is formed. The maximum potential energy from the respiration of organic compounds will only be obtained if their oxidation is coupled to the reduction of oxygen back to water. It is only because oxygen, a product of photosynthesis, is a gas, that heterotrophic organisms have ready access to the molecule that enables them to obtain the full benefit of the energy available in organic compounds. However, some microorganisms and animals are capable of sustained life in the absence of oxygen (anaerobic conditions) and many plants can survive fairly long periods with their roots in flooded anoxic soils where the level of oxygen, if it is present, may be very low. Under anaerobic conditions, the energy obtained from respiration is greatly restricted and the rate of growth is much reduced.

In this chapter, we will describe the major processes of photosynthesis and respiration that are going on around us on this planet.

ENERGY FOR BIOSYNTHESIS

Biosynthesis in cells occurs by a series of reactions that will reduce the entropy or randomness of the molecules involved. Energy is therefore required to promote these biosyntheses. Without an input of energy the equilibrium position of the biosynthetic reactions would favour the starting materials which, in the case of plants, would be carbon dioxide, water and various mineral nutrients. The growth and biosynthesis of living organisms therefore requires that unfavourable reactions are coupled to favourable ones so that there is an overall decrease in the free energy of the coupled reac-

tion. In cells the hydrolysis of adenosine triphosphate (ATP) is coupled to biosynthesis:

$$ATP + H_2O \rightarrow ADP + P_i$$

At pH 7.0, in the presence of 10 mM Mg^{2+}, the equilibrium constant

$$K_{eq} = [ADP][P_i]/[ATP] = 10^5$$

The free energy change (ΔG) that occurs when a reaction takes place is related to enthalpy and entropy changes by the equation

$$\Delta G = \Delta H - T\Delta S$$

where ΔH is the enthalpy change, T is the absolute temperature and ΔS is the entropy change. The free energy change is dependent on the temperature and concentration of reactants and products. In order to compare reactions, a standard free energy change $\Delta G^{0\prime}$ has been defined which is the energy required to convert 1 mol of reactants to 1 mol of products at 25 °C when all concentrations are initially 1 M, except H^+ which is 10^{-7} M (pH 7.0). The free energy is related to the equilibrium constant by the equation

$$\Delta G^{0\prime} = 2.3RT\log K_{eq}$$

where R is the universal gas constant. By definition, the concentration of water is taken as unity rather than 55 M and is therefore omitted from equilibrium equations. The K_{eq} for the hydrolysis of ATP is 10^5. In the presence of 10 mM P_i and 10 mM ADP, which are in the cytosolic range, the equilibrium concentration of ATP would be only 10^{-9} M. However, it has been shown that isolated mitochondria can maintain a mass action ratio as low as 10^{-5} M. If this ratio occurred *in vivo* with the above concentrations of P_i and ADP then hydrolysis of 1 mol of ATP would yield 13.6 kcal[1]. The hydrolysis of ATP to ADP and P_i is complicated because these compounds are partially ionized at physiological pH and also because Mg^{2+} chelates ATP and ADP with different affinities. At pH 7 the following forms of ATP can exist:

$$ATP^{4-}, ATP^{3-}, ATPMg^{2-}, ATPMg^{1-}$$

Since ATP is involved in biosynthesis it may be extensively and differentially bound to membranes and proteins so that in cytosol, mitochondria and chloroplasts measured ratios of ATP/ADP may bear very little relationship to free ATP and ADP. It is therefore not possible to say exactly what the free energy of hydrolysis of ATP is *in vivo*. However, it is the free energy released upon hydrolysis of ATP that enables biosynthesis to occur. For further information see ref. 1.

REDUCING POWER FOR BIOSYNTHESIS

The reducing power of a compound is dependent upon its affinity for electrons and the concentration of its reduced and oxidized forms. This affinity can be defined by its redox potential. For convenience, all redox potentials are compared to that of hydrogen gas at 1 atmosphere of pressure bubbling over a platinum electrode in a solution at pH 0:

$$H_2 \rightarrow 2H^+ + 2e^-$$

This couple is arbitrarily given a redox potential of zero volts. At pH 7.0 the reaction is pulled to the right because $[H^+]$ is low, and the redox potential therefore drops to -4.2 V. At equilibrium the difference in redox potential E_0 between two couples is related by the equation:

$$\Delta E_0 = 2.3RT/nF \log K_{eq}$$

where F is the Faraday constant and n is the number of electrons transferred. In practice, if each redox couple initially has equal concentrations of oxidized and reduced forms, then electrons will move from the couple with the more negative potential to that with the more positive potential.

There are two particularly important molecules in oxidation/reduction reactions within cells.

These are the cofactors nicotinamide adenine dinucleotide (NAD^+) and nicotinamide adenine dinucleotide phosphate ($NADP^+$). The redox potential at pH 7.0 of both $NAD^+/NADH + H^+$ and $NADP^+/NADPH + H^+$ is -0.32 V. It is believed that in cells NAD^+ is approximately 90% oxidized and $NADP^+$ 90% reduced. These values will, of course, vary with the particular metabolic state of cells but would mean that their redox potentials were respectively -0.29 V and -0.35 V. The redox potential of $\frac{1}{2}O_2 + 2H^+/H_2O$ is 0.82 V. Photosynthesis couples the oxidation of water to the reduction of $NADP^+$ using the electron transport components in the thylakoid membrane of chloroplasts. This is a redox potential difference of 1.15 V. The energy for this reduction comes from the absorption of light. NADPH is used for the assimilation of carbon dioxide into sugars (see chapter 2) and for many other biosynthetic reactions. The respiration of sugars to carbon dioxide is associated with NAD^+ reduction. The subsequent oxidation of NADH in mitochondria is linked to the reduction of oxygen, a reversal of photosynthesis.

ATP SYNTHESIS

Chloroplasts and mitochondria are the two organelles which are involved in ATP synthesis. Chloroplasts mediate the synthesis using energy from sunlight, and mitochondria utilize metabolites supplied by the degradation of carbohydrates, lipids and proteins. It is these organelles that maintain the high ATP/ADP ratio necessary for a large release of free energy that can be coupled to biosynthesis.

The key to the synthesis of ATP in chloroplasts and mitochondria is an intact membrane across which an electrochemical gradient can be established. The gradient is produced by electron transport which leads to the vectorial translocation of protons (Figure 1.1). The synthesis of ATP is then mediated by flux of protons back through what is known as an F-type ATPase. The ATPase of mitochondria and chloroplasts consists of two oligomeric protein complexes. The F_0 com-

4

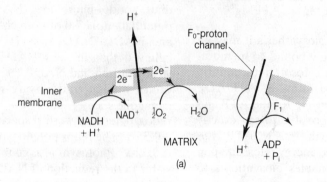

(a)

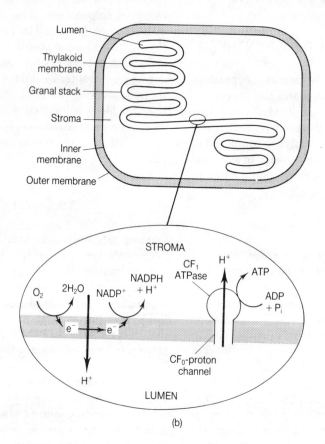

(b)

Figure 1.1. The flow of electrons along the electron transport chain components of the inner mitochondrial membrane (a) and the thylakoid membrane (b) of chloroplasts is associated with vectorial transport of protons. This produces an electrochemical gradient across the membrane. The energy stored in the gradient is coupled to ATP synthesis when protons return through the F_0F_1 ATPase.

plex is located within the membrane and acts as a proton channel. Attached to this complex is the F_1 moiety, which protrudes into the stroma of the chloroplast or matrix of the mitochondria and consists of a complex of five distinct polypeptides. There are three copies of the subunit β and it is to these subunits that ADP and P_i binding occurs. The exact mechanism by which the H^+ gradient is used for the synthesis of ATP in F-type ATPases is not known. However, there is a subunit of F_0 that is thought to be involved in regulating the flow of protons through the complex. The antibiotic oligomycin binds to this subunit and inhibits the use of the proton gradient for ATP synthesis. It is believed that the gradient increases the affinity of the F_1 complex for P_i. The formation of ATP then occurs within a hydrophobic catalytic site in the enzyme. Within such a site, very little free energy would be required for the dehydration of ADP and synthesis of ATP; the energy from the proton gradient would be used to promote the release of ATP from this hydrophobic site. The ATPase is also inhibited by dicyclohexylcarbodiimide (DCCD) or diethylstilboestrol (DES) owing to their binding to the F_0 complex. ATP synthesis in mitochondria and chloroplasts is thus indirectly coupled to an electrochemical proton gradient.

OVERVIEW OF PHOTOSYNTHESIS

In order for plants to grow they must be able to convert the energy from the sun into a useful form. Humans produce pigments in their skin that protect them from harmful effects of solar radiation; in contrast, exposure of plants to light stimulates them to produce pigments that absorb and utilize light energy. In higher plants the final products of photosynthesis are sucrose, and starch or fructans. Sucrose is the major form into which carbon from carbon dioxide is assimilated for transport throughout the plant, and starch and fructans are the major forms in which carbon is stored. In algae a wide variety of sugar alcohols and glucose polymers have been found to be the products of photosynthesis. Photosynthetic carbon metabolism will be discussed in detail in

chapter 2; in this chapter we will concentrate on the compounds that are formed as a direct result of light absorption—the compounds that are required to drive the reactions of photosynthetic carbon metabolism, ATP and NADPH.

PHOTOSYNTHETIC FORMATION OF ATP AND NADPH

Light Energy and Photosynthesis

The energy for the formation of ATP and NADPH comes from visible radiation of wavelength 400–700 nm. The reduction of $NADP^+$ is catalysed by ferredoxin–NADP reductase on the stromal face of the thylakoid membrane. This is a two-electron reduction and occurs as a result of the transfer of electrons from water via the electron transport chains in the thylakoid membrane. As already discussed, the difference in redox potential between water and NADPH is at least 1.15 V. ATP synthesis occurs due to electron transport being coupled to the production of an electrochemical proton gradient across the thylakoid membrane.

Light interacts with matter in discrete packets or photons. The energy per photon is dependent upon its frequency:

$$E = h\nu$$

where ν is the frequency and h is Planck's constant (1.58×10^{-34} cal s^{-1}). The frequency (ν) is related to the wavelength of light by

$$\nu = c/\lambda$$

where c equals the velocity of light and λ is the wavelength.

The energy of blue light is therefore greater than that of red light. Light with a wavelength of 400 nm has an energy of 71 kcal mol^{-1} of photons compared to that at 680 nm, which has an energy of 42 kcal mol^{-1} of photons[2]. Blue light excites electrons in chlorophyll to a higher energy state than red light (Figure 1.2). However, the additio-

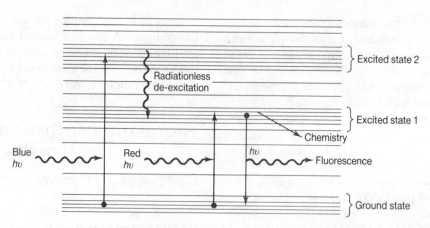

Figure 1.2. Excitation of chlorophyll. The parallel lines represent energy sub-state or electronic orbitals. Thus the energy delivered by the absorption of a blue photon (left) is sufficient to raise an electron to excited state 'two' from where it rapidly returns by a process of radiationless de-excitation, 'cascading' through sub-states, to excited state 'one'. A photon of red light (centre) only has enough energy to raise an electron to excited state 'one' but this excited state is sufficiently stable to permit useful chemical work and is, in effect, the starting point of all other events in photosynthesis. Modified from D. Walker, 1979, *Energy, Plants and Man*, Packard Publishing, Chichester, with permission.

nal energy is lost when an electron falls back from state 2 to state 1—the level to which red light excites electrons. Therefore, the useful energy for photosynthesis that can be obtained from blue and red light is the same. The energy to move 1 mol of electrons through a redox potential difference of −1.1 V is 26 kcal. Therefore, the efficiency of energy conservation would have to be extremely high (>62%) if one photon were to be able to provide the energy for one electron to move through this potential difference. In practice, promotion of an electron from a redox potential of +0.82 V to that sufficient to reduce $NADP^+$ (at least −0.32 V) involves two photons and occurs in two stages using two photosystems. As the $NADPH/NADP^+$ ratio rises, its redox potential will decrease towards −0.35 V. The successive photosystems, first photosystem II and then photosystem I, absorb photons and increase the energy of the electrons to the required level. The transport of an excited electron from photosystem II to photosystem I results in a decrease in its energy, but this is coupled to the transport of protons from the stroma to the thylakoid lumen.

The process of photosynthesis involves the transfer of two electrons from water to $NADP^+$

Since the formation of oxygen releases four electrons, the oxidation of 2 mol of water is coupled with the formation of 2 mol of NADPH. Figure 1.3 shows the pathway of electron transfer that occurs within the thylakoid membrane of chloroplasts[3]. This is known as the 'Z-scheme' and shows the mid-point redox potentials of the different components. However, very little is known of the extent to which these components are oxidized or reduced *in vivo* and hence of their true steady-state redox potentials.

The Pigments that Absorb Light

Light energy is absorbed by pigments and the energy used to excite electrons to a higher energy level. The pigments are associated with proteins in the core complexes of the two photosystems—photosystem II (P680) and photosystem I (P700)—as well as in light-harvesting complexes that are associated with each photosystem. The reaction centre of photosystems II and I are chlorophyll a–protein complexes that absorb light at 680 nm and 700 nm respectively. The pigments in the light-harvesting complexes are in very close asso-

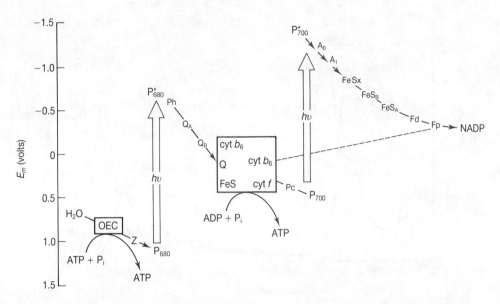

Figure 1.3. The Z scheme for oxygenic photosynthesis. The locations of two 'phosphorylation sites' associated with electron transfer are shown by bold curved arrows. The electron transport carriers are arranged vertically according to measured (if available) or estimated midpoint redox potentials (at pH 7.0). No relationship between the reactants or products of ATP synthesis and the energetic scale is intended. OEC, oxygen-evolving complex; Z, donor to photosystem II (PS II); P_{680}, reaction centre chlorophyll of PS II; Ph, phaeophytin acceptor of PS II; Q, plastoquinone; cyt, cytochrome; FeS, Rieske iron–sulphur protein; Pc, plastocyanin; P_{700}, reaction centre chlorophyll of PS I; A_0 and A_1, early acceptors of PS I; FeS_x, FeS_B and FeS_A, bound iron–sulphur protein acceptors of PS I (FeS_B and FeS_A may operate in parallel); Fd, ferredoxin; Fp, flavoprotein (ferredoxin–NADP reductase). The dashed line indicates cyclic electron flow around PS I. Reproduced with permission from D. R. Ort and N. E. Good, 1988, *Trends Biochem. Sci.*, **13**, 467–469.

ciation (<10 nm between chlorophylls) so that the energy from excited electrons can be passed via a number of chlorophyll molecules to the chlorophyll in the associated core complex with minimum loss of energy. Photosystem II predominates in the appressed regions of the thylakoid membrane and photosystem I in the unstacked open regions of the thylakoid membrane (see Figure 1.1). Photosystem I produces a strong reductant that donates electrons to $NADP^+$. It receives its electrons from photosystem II via a number of electron transport chain components ending with the mobile electron carrier plastocyanin. Photosystem II generates a weaker reductant than photosystem I, but has the ability to take electrons from water.

There are three classes of photosynthetic pigments in photosynthetic organisms: the chlor-ophylls, the phycobilins and the carotenoids (see chapter 8). The chlorophyll molecule consists of a porphyrin ring with a dense cloud of π-electrons. The absorption of a photon of light promotes the molecule to an excited electronic state (see Figure 1.2). The long-chain lipophilic terpenoid side-group, unique to chlorophyll a and b, may provide hydrophobic bonding components for associations within the thylakoid membrane or with proteins.

Chlorophyll a is the only chlorophyll found in all photosynthetic groups of plants and is also the only chlorophyll found in the core complex of photosystems. The wavelengths of light absorption of chlorophyll a and b are shown in Figure 1.4. The binding of chlorophyll to proteins alters the precise wavelengths of absorption; however, this figure shows that there is a window from 500

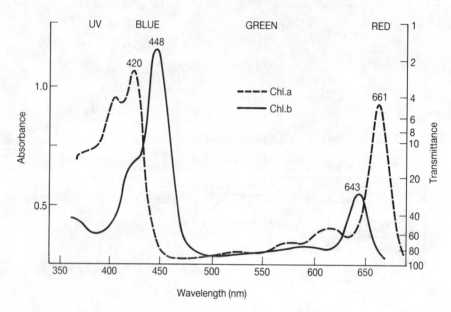

Figure 1.4. Absorption spectra of chlorophylls a and b showing absorption in the red and blue and transmission in the green. Modified from D. Walker, 1979, *Energy, Plants and Man*, Packard Publishing, Chichester, with permission.

to 600 nm where there is very little chlorophyll absorption. In higher plants the useful wavelengths of photosynthetically active radiation can be extended by the carotenoids. These pigments can absorb light at certain wavelengths in the 'window' region, particularly just above 500 nm, and pass the energy from their excited electronic state to chlorophyll. Carotenoids can also react with oxygen and help deactivate the chlorophyll triplet state. They can, therefore, play a role in protecting chlorophyll from excess photo-oxidation.

In aquatic plants the accessory pigments are crucial in the absorption of light, particularly for those plants growing at the greatest depths of water where the light intensity is low and, owing to differential absorption of light, where green light may be the predominant wavelength available. The red algae contain two phycobilins—phyco-erythrin, which is a red pigment, and phycocyanin, which is a blue pigment—that allow them to utilize the light of low intensity which is available. This is one factor in determining the ability of red algae to grow at the greatest depths of water.

Photosystems and Electron Transport

The reduction of $NADP^+$ and the formation of an electrochemical proton gradient used to synthesize ATP involves electron transport. This requires three discrete multi-enzyme complexes: the photosystem II complex, the cytochrome b/f complex and the photosystem I complex. The link between these complexes is through the lateral mobility of the plastoquinone pool (PQ) and plastocyanin. The order in which these complexes react is shown in Figure 1.3.

The photosystem II core complex contains about 50–60 light-harvesting chlorophyll a molecules and is associated tightly with an inner light-harvesting complex and a more loosely associating peripheral complex. Over half of the chlorophyll a molecules and most of the chlorophyll b and carotenoids are associated with light-harvesting complex II. Energy absorbed by chlorophyll in these complexes passes energy to Chl_2, which acts as the primary photosystem II donor and transfers an electron via another chlorophyll to phaeophytin, which is a 'chlorophyll' molecule without a chelated Mg atom (Figure 1.5). This

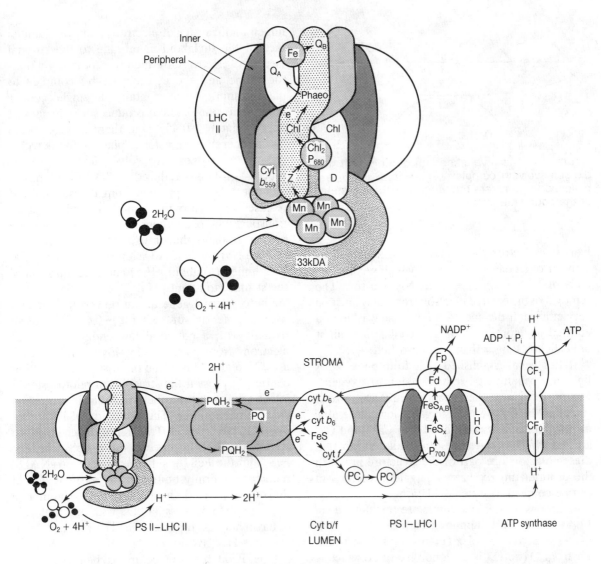

Figure 1.5. Organization of the higher plant thylakoid membrane components participating in light-harvesting and photosynthetic electron transfer from water to $NADP^+$ and proton translocation. The continuous network of interconnected thylakoid membranes is structurally differentiated into oppressed and non-appressed regions. The photosystem I complexes (PS I) and ATP synthase are located only in non-appressed membranes, most photosystem II complexes (PS II) and their associated light-harvesting complexes (LHC II) are located in appressed membranes, and the cytochrome b/f complexes (Cyt b/f) are randomly distributed. This figure shows a schematic representation of the polypeptides of the PS II reaction centre and the associated water-splitting apparatus. Q_A and Q_B are plastoquinone molecules which act in series as electron acceptors and are associated with a ferrous ion, Fe. Phaeo are phaeophytin molecules, one of which acts as a primary electron acceptor. Chl_2 is speculated to be a special pair acting as the primary donor P680. Chl represents each of the monomeric chlorophyll molecules, one or both of which may help to facilitate electron transfer to Phaeo. Z is an electron donor to P680 and the 33 kDa protein is responsible for stabilizing the manganese cluster necessary for water oxidation. D is a component similar to Z (same EPR signal but different kinetics) which does not seem to be involved in electron transfer from water to $P680^+$. Cyt b is cytochrome b_{559} of unknown function. Arrows show possible routes for electron transport. Modified from J. Barber, *Trends Biochem. Sci.*, **12**, 321–326; J.M. Anderson and B. Anderson, 1988, *Trends Biochem. Sci.*, **13**, 351–355, by permission of W.H. Freeman and Company.

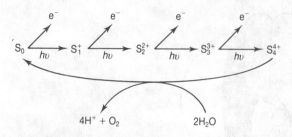

$$S_0 \xrightarrow{h\upsilon} S_1^+ \xrightarrow{h\upsilon} S_2^{2+} \xrightarrow{h\upsilon} S_3^{3+} \xrightarrow{h\upsilon} S_4^{4+}$$

$4H^+ + O_2 \qquad\qquad 2H_2O$

Figure 1.6. The five states through which the oxygen-evolving complex can cycle. S_1 to S_4 are believed to represent the various oxidation states of the bound Mn ions.

leaves an electron 'hole' in Chl_2. This is filled by transfer of an electron from Z, which receives its electrons from water via an Mn cluster. The removal of the excited electron from Chl_2 must be very efficient if the energy is not to be wasted and emitted as fluorescence at a wavelength of light only slightly longer than that absorbed.

It has been established that four photons of light are required to reduce water to oxygen[4]. Figure 1.6 shows the reaction sequence, which requires Mn, Ca^{2+} and Cl^- ligated in an ensemble of polypeptides. It is believed that Ca^{2+} and Cl^- are not directly related to Mn oxidation but that these cofactors are part of the structure in which the accumulating oxidizing equivalents on the Mn can be coupled to water oxidation.

Electrons from phaeophytin are transferred via Q_A and Q_B and mobile plastoquinone to the cytochrome b/f complex (Figure 1.5). The transfer from Q_A to Q_B is via a non-haem Fe which is regulated by the tight binding of bicarbonate between Q_A and Q_B. The transfer of electrons to the cytochrome b/f complex therefore depends upon the presence of carbon dioxide. The Q_B site is able to bind a wide range of photosystem II herbicides, including the triazines and substituted ureas. These herbicides prevent the flow of electrons out of photosystem II which causes the electron acceptor molecules in the photosystem to become reduced. This leads to enhanced fluorescence from the photosystem[5].

The excitation of electrons in photosystem II leads to deposition of protons from water in the lumen and transfer of electrons from the luminal face of the thylakoid membrane to the stromal face. The electrons are then transported via plastoquinone and the cytochrome b/f complex to plastocyanin, which contains a single copper atom. This photosystem produces a reductant of approximately -0.45 V but there is a significant loss of energy on transfer to plastoquinone with a redox potential of zero volts. However, in the process of being reduced and oxidized pastoquinone takes up protons from the stroma and releases them into the lumen. Hence, electron transport via plastoquinone contributes to the proton gradient that is used for ATP synthesis (Figure 1.5) and leads to acidification of the thylakoid lumen (to about pH 5) and alkalinization of the stroma (to about pH 8).

Photosystem I generates a powerful reductant with a potential of at least -0.60 V and an oxidant with a potential low enough to accept electrons from plastocyanin (Figure 1.3). There are 12 Fe and 12 S in the photosystem I reaction centre. At present the structure of photosystem I is not known but it is believed that it contains a chlorophyll a dimer. The excitation of P700 produces $[P700^+ - A_1^-]$. Transfer of the electrons to ferredoxin then occurs via three FeS_4 clusters, FeS_x and the $FeS_{A,B}$ system. Ferredoxin–NADP reductase is firmly bound to the stromal side of the non-appressed thylakoids. The protein is a single peptide of mass 34 kDa with one molecule of flavin adenine dinucleotide. The reduced flavin transfers H^- (hydride) to $NADP^+$.

One P700 reaction centre is believed to be associated with about 40 accessory chlorophyll a molecules which pass their absorbed energy directly to that centre. The exact number of chlorophyll molecules associated with the photosystem I light-harvesting complex is unknown, but is almost certainly variable. However, it is probably chlorophyll a which predominates in this particular light-harvesting complex.

The flow of electrons from the oxygen-evolving complex of photosystem II to $NADP^+$ via photosystem I is known as non-cyclic electron transport. This pathway may be supplemented by cyclic electron transport in which ferredoxin transfers its electrons to cytochrome b_6 in the cytochrome b/f complex (Figure 1.3). Electron transport back to P700 then occurs via plastoquinone and cytochrome f. This flux of electrons is coupled to ATP synthesis via the production of an electrochemical proton gradient due to plastoquinone reduction/oxidation (Figure 1.5) and is of particular importance in systems such as heterocysts (chapter 6) and the bundle-sheath of C_4 plants such as maize (chapter 3).

Non-cyclic flow of electrons will take precedence over the cyclic process if ferredoxin preferentially reduces $NADP^+$ rather than cytochrome b_6. In theory the ratio of ATP to NADPH production by non-cyclic electron transport is 3:2. This is appropriate for the fixation of one molecule of carbon dioxide into a hexose sugar, but may not be appropriate for synthesis of more complex compounds. An increase in cyclic electron transport would increase the synthesis of ATP. However, it must be emphasized that the precise stoichiometry of proton extrusion during electron transport is not known and that the precise number of protons per ATP synthesized is not known. The proton leak through the membrane may also vary with the electrochemical gradient. Therefore, non-cyclic electron transport may not provide ATP and NADPH precisely in a ratio of 3:2 and cyclic electron flow will allow a flexibility in the ratio of NADPH to ATP production.

Regulatory mechanisms can alter the balance between electron transport in the two photosystems. It is known that if plastoquinone is overreduced by photosystem II, because of an excess of light exciting photosystem II, then a membrane-bound kinase is activated. This kinase rapidly phosphorylates a protein of peripheral light-harvesting complex II and results in an increase in repelling negative charges which forces some of these peripheral complexes to dissociate from photosystem II and migrate laterally to the non-appressed regions of the thylakoid membrane, where photosystem I is located. This alters the distribution of light excitation energy between the two photosystems, greatly reducing the energy passed to photosystem II.

Increased excitation of photosystem I may then favour cyclic electron flow. These changes can occur in as little as 20 s and are known as state transitions[6].

Within plants there is a tremendous flexibility in the absorption and utilization of light. Shade leaves have more chloroplasts with greater amounts of thylakoid membrane and large granal stacks compared to sun leaves. Shade leaves are also thicker, larger and have more photosynthetic pigments so that they can maximize their light-harvesting capacity. At high irradiance photosynthetic activities are increased by increased synthesis of the cytochrome b/f complex, plastoquinone, ferredoxin and the ATPase in comparison to the light-harvesting components.

CARBOHYDRATES

Basis for Control of Substrate Oxidation

As already stated, the final products of photosynthesis are sucrose, and starch or fructans. Starch is formed in chloroplasts and amyloplasts, sucrose in the cytoplasm, and where fructans are the storage compound in leaves of certain grasses their synthesis takes place in the vacuole. The respiration of these compounds produces energy and carbon skeletons for biosynthesis.

The products of photosynthesis are converted to pyruvate or malate, which can enter an organelle known as a mitochondrion for the final stages of oxidation to carbon dioxide via the tricarboxylic acid cycle. At any stage during their oxidation, intermediate compounds may be withdrawn and enter pathways of protein, lipid and cell wall synthesis. The oxidation of photosynthate involves the complex interaction of metabolites moving between a variety of organelles. This adds greatly to the difficulty of studying metabolism in plants in comparison to that in animals because operation of a particular pathway may occur simultaneously in two or more organelles within the same cell.

All reactions in any sequence must make some contribution to control since control must be regarded as the property of the pathway as a whole. Nonetheless, the contribution of individual reactions may be very different. For the major pathways where we have a knowledge of regulation it appears that the bulk of control is vested in a minority of the reactions. These were termed the regulatory reactions[7]. The role that these reactions could play in regulation would depend on four factors: first, the extent to which the enzyme is saturated by substrate (if any enzyme is not saturated by substrate then its rate will be determined by the rate at which it receives its substrate); second, the position of the reaction in the pathway; third, its ability to communicate a change in its activity with other reactions in the pathway; and finally it must be directly or indirectly responsive to regulation by compounds other than pathway substrates. It was the *non-equilibrium* reaction in a pathway that met these criteria that was regarded as the flux-generating step. Reactions close to equilibrium would therefore not be catalysed by regulatory enzymes.

Our knowledge of the extent to which particular enzymes in a metabolic pathway regulate flux is generally very limited. This is unfortunate, because the techniques of genetic engineering have now become sufficiently sophisticated to enable the production and activity of specific enzymes to be manipulated. The question that will have to be addressed over the next decade is what the enzymes are in any particular pathway whose activity must be altered if the flux through that pathway is to be manipulated. We believe that over the next few years it will be the techniques of metabolic control analysis as developed by Kacser that will provide the answers. This is because the method seeks to establish a quantitative measure of the extent to which an enzyme is exerting control without the need to know how that control is manifested.

The extent of control by any enzyme can be defined by its flux control coefficient[8]. The extent to which any enzyme exerts control can be determined by measuring the effect of altering its activity on the flux through the pathway. In practice this can be achieved through the use of specific enzyme inhibitors or the use of plants with specific enzyme mutations. Also, work is underway to determine the effect of genetic manipulation of enzymes on the activity of pathway fluxes. Initial experiments have provided rather unexpected results so it is possible to anticipate some very interesting findings over the coming years. In practice the relationship between 'cause' and 'effect' is measured. In the simplest case, where there is a linear relationship between the inhibitor concentration and the velocity of an individual enzyme, the control coefficient can be defined as:

$$C_j = (dJ/J)/(dI/I_{max})$$

where J is the flux in the absence of an inhibitor and I_{max} is the minimum amount of an irreversible inhibitor required to give maximum inhibition. If, for example, 10% inhibition of an enzyme reduced the flux through the pathway by 5%, then the flux control coefficient for that particular step in the pathway would be 5%/10% = 0.5. The flux summation property of a pathway defines the sum of all flux control coefficients for any one flux as equal to unity. It is therefore possible to measure the extent to which any enzyme in a pathway regulates the flux through that pathway without the need to determine how the enzyme is regulated (for further detail and references as to how this can be applied see ref. 8).

Gluconeogenesis

Gluconeogenesis is the process by which plant cells synthesize sucrose and reducing sugars either from storage reserves in non-photosynthetic cells (to provide energy and carbon skeletons for the germinating seed), or directly from carbon dioxide in photosynthetic tissues. There is no good evidence that this is an important pathway in starch-storing tissues, therefore gluconeogenesis will only be considered in lipid-storing tissues. The carbon metabolism of leaves and the pathways of sucrose synthesis from hexose phosphates will be dealt with in chapters 3 and 4, respectively, so that in this chapter attention will be focused on the conversion of triose phosphates to hexose phosphates in the cytosol of non-photosynthetic plant tissues. The discussion will consider three areas: first, the evidence that there is net conversion of triose phosphates into hexose phosphates in some plant tissues; secondly, the pathway of this conversion; and thirdly, the regulation of this pathway.

A large number of seeds store lipid as their major reserve (e.g. castor bean, cucumber and marrow). Mobilization of this lipid on germination involves the oxidation of fatty acids to produce acetyl-CoA, the synthesis of malate from acetyl-CoA via the glyoxylate cycle (see chapter 5) and the subsequent synthesis of sucrose within the cytosol from this malate. The distribution of ^{14}C label following feeding of [^{14}C]acetate to both castor bean endosperm and marrow cotyledons provides conclusive evidence that the conversion of malate into hexose phosphates proceeds via triose phosphate intermediates. The cytosolic enzymes malate dehydrogenase and phosphoenolpyruvate carboxykinase (the latter is found only in lipid-mobilizing tissues) catalyse the conversion of malate into phosphoenolpyruvate (PEP) via oxaloacetate. Since the reactions of glycolysis from triose phosphates to PEP are freely reversible, this PEP can equilibrate with the cystosolic triose phosphate pool (Figure 1.7). The only confirmed occurrence of net conversion of triose to hexose phosphates is thus in lipid-mobilizing tissues. Therefore, the remainder of this section will focus on the pathway of cytosolic conversion of triose to hexose phosphates, and its regulation in lipid-mobilizing tissues.

The Pathway of Gluconeogenesis and its Control

The triose phosphates dihydroxyacetone phosphate and glyceraldehyde 3-phosphate are in equilibrium with one another and can combine by means of the aldolase reaction to form fructose 1,6-bisphosphate (F1,6BP). There are three enzymes in plant cells involved in the interconversion of F1,6BP and the hexose phosphate, fructose 6-phosphate (F6P) (see Figure 1.7):

1. ATP–F6P 1-phosphotransferase (phosphofructokinase, PFK), which catalyses the irreversible conversion of F6P into F1,6BP.
2. PP$_i$–F6P 1-phosphotransferase (PFP), which has been shown to be near equilibrium *in vivo* in a number of plant tissues.
3. F1,6BP 1-phosphatase (F1,6BPase), which catalyses the irreversible conversion of F1,6BP into F6P.

Since PFK and F1,6BPase catalyse irreversible reactions their roles are relatively well established. Evidence that F1,6BPase mediates the conversion of triose phosphates to hexose phosphates in lipid-mobilizing tissues comes from studies of the developmental variation in the maximum catalytic activity of this enzyme. In both castor bean endosperm and marrow cotyledons the activity increases in parallel with the rate of lipid breakdown and sucrose synthesis. The role of PFP, which catalyses a freely reversible reac-

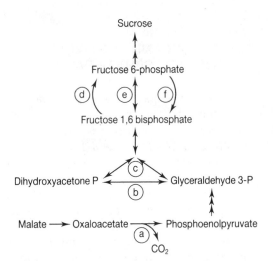

Figure 1.7. The reactions involved in the conversion of C$_4$ acids to hexose phosphates during gluconeogenesis: (a) phosphoenolpyruvate carboxykinase; (b) triose phosphate isomerase; (c) aldolase; (d) fructose 1,6-bisphosphatase; (e) pyrophosphate fructose 6-phosphate 1-phosphotransferase; (f) phosphofructokinase.

tion, is less clear. In castor bean endosperm the activity of PFP increases at the same time as F1,6BPase, and the enzymes are present at comparable levels, providing support for a gluconeogenic role for PFP. However, the regulatory properties of PFP suggest that its activity is antagonistic to F1,6BPase, and that it could operate in the glycolytic direction. This controversy is, as yet, unresolved, but there is now considerable evidence from a number of plant tissues which suggests that maintenance of pyrophosphate levels is the primary role of PFP.

Measurement of the levels of F1,6BP, F6P and phosphate (the substrates and products of the F1,6BPase reaction) confirms that this reaction is far from equilibrium *in vivo* and is therefore a potential site of control. Moreover, application of the quantitative methods of metabolic control theory also suggest that the reactions converting F1,6BP into F6P are important sites in the control of gluconeogenesis. F1,6BPase from castor bean endosperm is inhibited by the regulatory metabolite fructose 2,6-bisphosphate (F2,6BP) and by adenosine monophosphate (AMP), this inhibition being increased by high levels of phosphate. There is good evidence that F2,6BP plays an important role in the regulation of gluconeogenesis *in vivo*, since appreciable amounts of this metabolite are present in castor bean endosperm. Moreover anoxia or treatment with 3-mercaptopicolinic acid, both of which reduce the rate of gluconeogenesis, leads to significant increases in the level of F2,6BP. The enzymes responsible for the synthesis and degradation of F2,6BP respond so that high levels of F6P and phosphate cause an increase in the amount of F2,6BP, whereas high levels of glyceraldehyde 3-phosphate cause a decrease [9]. The flux through F1,6BPase is therefore dependent on the relative levels of triose phosphates and hexose phosphates. An additional complication to the control of the flux between triose and hexose phosphates occurs due to PFP. This enzyme also responds to F2,6BP levels, but in the opposite manner to F1,6BPase. Therefore an increase in the level of F2,6BP causes not only a reduction in the activity of F1,6BPase, but also an increase in the activity

of PFP. It has been proposed that these enzymes constitute a substrate cycle, the net activity of which determines the flux between triose and hexose phosphates. This is an attractive hypothesis, but the following must be borne in mind. First, in most substrate cycles so far described the activity of the enzyme catalysing the reverse reaction is generally lower than that of the forward reaction. However, the activity of PFP and F1,6BPase are comparable in castor bean endosperm. Secondly, both reactions in substrate cycles are usually irreversible, but PFP catalyses a freely reversible reaction. Thirdly, lipid-mobilizing tissues also contain appreciable activities of PFK, which does not respond to the level of F2,6BP. Finally, there is no direct evidence confirming that PFP operates in the glycolytic direction in lipid-mobilizing tissues, and in other tissues the available evidence would seem to indicate that it does not. In conclusion the flux between triose and hexose phosphates is likely to be controlled by the relative levels of these sugars, but the precise mechanism of this regulation is unclear.

GLYCOLYSIS AND THE OXIDATIVE PENTOSE PHOSPHATE PATHWAY

There are two major pathways by which carbohydrate is converted to pyruvate and malate, namely glycolysis and the oxidative pentose phosphate pathway (Figures 1.8 and 1.9). The common substrate for these pathways is the hexose phosphate, glucose 6-phosphate. The regulation of hexose phosphate metabolism is important in plants because not only do they act as substrates for respiration, but they are key intermediates in the interconversion of sucrose (transport carbohydrate) and starch (storage carbohydrate). The formation of hexose phosphates from sucrose and starch will be outlined in chapter 4.

The initial stages of glycolysis produce glyceraldehyde 3-phosphate via F6P and F1,6BP. F6P and glyceraldehyde 3-phosphate may also be formed via ribulose 5-phosphate through activity

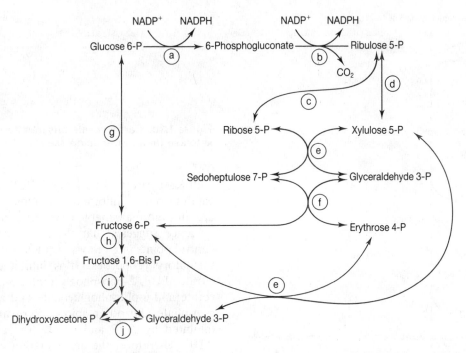

Figure 1.8. The association between glycolysis and the oxidative pentose phosphate pathway: (a) glucose 6-phosphate dehydrogenase and 6-phosphogluconolactonase; (b) 6-phosphogluconate dehydrogenase; (c) ribose phosphate isomerase; (d) ribulose phosphate 3-epimerase; (e) transketolase; (f) transaldolase; (g) hexose phosphate isomerase; (h) phosphofructokinase; (i) aldolase; (j) triose phosphate isomerase.

of the oxidative pentose phosphate pathway. The unique product of the pentose phosphate pathway is reduced NADPH formed through the reactions catalysed by glucose 6-phosphate dehydrogenase and 6-phosphogluconate dehydrogenase. The K_i values of NADPH for glucose 6-phosphate dehydrogenase and 6-phosphogluconate dehydrogenase have been reported as 11 μM and 20 μM respectively. There is therefore good reason to propose that the activity of the oxidative pentose phosphate pathway will be regulated by the ratio of $NADP^+/NADPH$. This is supported by the discovery that treatment of plant tissues with compounds such as methylene blue and nitrite, which preferentially accept electrons from NADPH, immediately stimulates flux through the oxidative pentose phosphate pathway. The further reactions of ribulose 5-phosphate to F6P and glyceraldehyde 3-phosphate in the pentose phosphate pathway, catalysed by ribose phosphate isomerase,

ribulose phosphate 4-epimerase, transketolase and transaldolase, are close to equilibrium[10].

Transketolase transfers a two-carbon unit whereas transaldolase transfers a three-carbon unit (Figure 1.10). The donor sugar is always a ketose whereas the recipient is always an aldose. These reactions create a reversible link between glycolysis and the initial reactions of the pentose phosphate pathway. The formation of ribose 5-phosphate for nucleic acid biosynthesis and erythrose 4-phosphate, a substrate for the shikimate pathway leading to the synthesis of aromatic amino acids, are not therefore dependent on the operation of the pentose phosphate pathway. However, these pathways may be dependent on the reduction of $NADP^+$.

The pathway of glycolysis to pyruvate as outlined in Figures 1.8 and 1.9 is well established. Also there is good evidence from a variety of both photosynthetic and non-photosynthetic tissues in

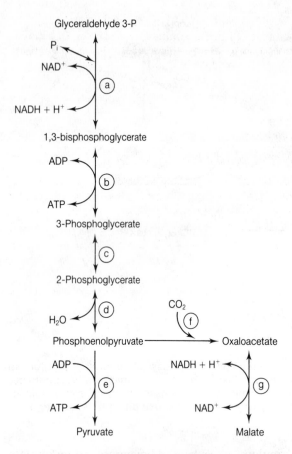

Figure 1.9. The oxidative and ATP-generating reactions of glycolysis: (a) glyceraldehyde 3-phosphate dehydrogenase; (b) 3-phosphoglycerate kinase; (c) phosphoglyceromutase; (d) enolase; (e) pyruvate kinase; (f) PEP carboxylase; (g) malate dehydrogenase.

several species that, as well as being converted to pyruvate, PEP may be carboxylated to oxaloacetate and reduced to malate. This has been established by demonstrating that in roots of peas, plantains and in the sterile appendage of the spadix of the thermogenic Araceae (see Figure 1. 11), $^{14}CO_2$ is assimilated into oxaloacetate, malate and related amino acids in the dark[11]. This is probably a widespread feature of plant respiration. Both malate and pyruvate are respired by mitochondria and malate may also be stored in the vacuole.

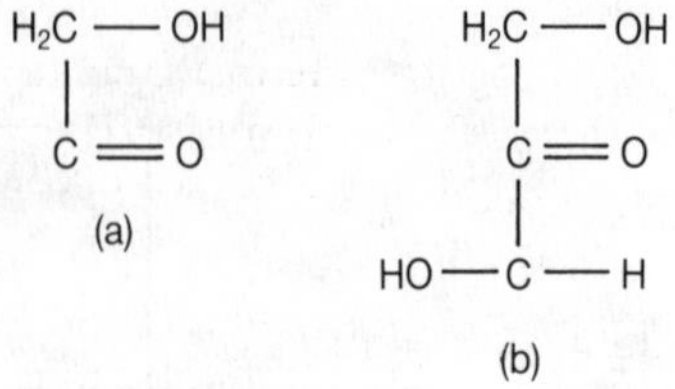

Figure 1.10. Carbon skeletons transferred by transketolase (a) and transaldolase (b).

PFK, pyruvate kinase and PEP carboxylase catalyse non-equilibrium reactions. These steps are thus more amenable to control, and available evidence suggests that these will be the main control points in glycolysis[12]. PFK shows complex regulatory properties. It is inhibited by ATP, citrate, PEP, 2-phosphoglycerate, 3-phosphoglycerate and 6-phosphogluconate and activated by phosphate and other anions. Pyruvate kinase is inhibited by ATP and citrate and activated by ADP. Therefore, the accumulation of ATP and citrate will not only reduce flux from glucose 6-phosphate to F1,6BP directly, but will also inhibit pyruvate kinase, thus leading to the accumulation of glycolytic intermediates which also inhibit PFK. PEP carboxylase is inhibited by oxaloacetate and malate, and phosphorylation of

Figure 1.11. The open spadix of *Arum maculatum* at thermogenesis. The unfurling of the green sheath reveals the sterile appendage known as the 'club'. When this occurs, there is a rapid degradation of stored starch in the 'club', its temperature rises and volatile amines are released which attract pollinating insects.

this enzyme increases its K_m for PEP tenfold, thus effectively reducing the rate of PEP carboxylation very substantially. Reduction of PEP carboxylase activity will also lead to an accumulation of glycolytic intermediates that inhibit PFK. Considerable work still remains to be done to establish that *in vivo* the changes in metabolic intermediates necessary to alter enzyme activities do in fact occur.

Many of the glycolytic and pentose phosphate pathway reactions occur in plastids as well as in the cytoplasm. This greatly increases the complexity of studying regulation, but it does mean that, within a single cell, pathways can be operating in opposite directions to meet the cell's metabolic demands.

PP$_i$–Fructose 6-Phosphate 1-Phosphotransferase

The discovery of PFP, which catalyses the phosphorylation of F6P to F1,6BP using pyrophosphate (PP$_i$) rather than ATP as the phosphate donor, has raised the question as to whether or not PFK can be bypassed in the glycolytic pathway. If this were so, then regulation of glycolysis by PFK as has been proposed would be considerably undermined. PFP is activated by low levels of F2,6BP ($0.01–0.5$ μM). Its concentration *in vivo* is controlled by phosphofructo-2-kinase and F2,6BPase; these enzymes respectively phosphorylate F6P and dephosphorylate F2,6BP. Upon illumination, stomatal guard cell protoplasts rapidly convert starch to malate. This breakdown is associated with a rise in F2,6BP and a fall in glucose 6-phosphate, implying an activation of the conversion of hexose phosphates to triose phosphates. The K_m of PFP from various plant tissues for PP$_i$ ranges from 0.004 to 0.02 mM. As yet, the exact distribution of pyrophosphate between the cytosol and different organelles remains largely unresolved; however, it is considered that sufficient pyrophosphate is available in the cytosol for PFP to function glycolytically. In non-photosynthetic tissues present data suggest that the concentration of F2,6BP will vary

between 0.1 and 6 μM. This would be sufficient to activate PFP. In bananas the recovery of F2,6BP which was added during their tissue extractions increased from $55–60\%$ in unripe bananas up to $87–92\%$ in ripe bananas[13]. Therefore, there was a marked difference in interference in its extraction at different stages of banana development. This illustrates the importance of recovery determinations at different stages of development even with one tissue.

The question of the role of PFP in glycolysis has been addressed in thermogenic tissues of the Araceae. The thermogenic aroids produce a spadix which contains at its base the floral organs (anthers and ovaries). Above this floral region is a sterile appendage known as the club (Figure 1.11). During development of the spadix, the club grows and accumulates starch. At the stage known as thermogenesis, the spadix unfurls and the club hydrolyses and respires its stored starch within a few hours. The heat generated by the respiration of this starch may raise the temperature of the club by up to $10\,°C$. This increase in temperature volatilizes amines which attract insects for pollination.

Clubs from the spadices of *Arum maculatum* at thermogenesis may degrade starch at rates of respiration up to $10\,000–30\,000$ μl oxygen h^{-1} g^{-1} fresh weight. Table 1.1 shows that the measured rates of respiration in these clubs increased from 0.11 μmol hexose min^{-1} g^{-1} fresh weight in the young clubs up to $7–10$ μmol hexose min^{-1} g^{-1} fresh weight in thermogenic clubs[14]. The amounts of the metabolites PP$_i$ and F2,6BP showed only minor changes during development. Likewise PFP activity remained between 4.02 and 5.80 μmol min^{-1} g^{-1} fresh weight. In contrast, PFK activity increased from 1.15 to 30.01 μmol min^{-1} g^{-1} fresh weight. Other enzymes closely associated with the rapid respiratory flux in club tissue—endoamylase, hexokinase, aldolase, glyceraldehyde 3-phosphate dehydrogenase and PEP carboxylase—also show marked increases in maximum catalytic activity during development of the spadix. At thermogenesis the activity of PFP would have been insufficient to mediate flux, whereas PFK appeared to have increased to an

Table 1.1. Respiration and activities of PFK and PFP and content of PP_i and fructose 2,6-bisphosphate during the development of the clubs of *Arum maculatum*.

Stage of development	Club wt (mg)	Activity (μmol hexose min^{-1} g^{-1} fr. wt)			Metabolite content (μmol g^{-1} fr. wt)	
		Respiration	PFP	PFK	PP_i	F2,6BP
α	239	0.11	5.60	1.15	21.1	0.56
β	616	0.24	5.24	3.76	36.9	0.36
Pre-thermogenesis	1346	0.31	4.02	18.15	27.5	0.31
Thermogenesis	1205	7–10	4.26	30.01	36.2	0.46

Modified from ap Rees *et al*, 1985, *Biochemical Journal*, **227**, 299–304, by permission of The Biochemical Society.

activity sufficient to have enabled it to mediate the thermogenic glycolytic flux. This suggests that, in this particular tissue during thermogenesis, the major glycolytic role is played by PFK and not PFP. The properties of PFK are such that it could regulate the entry of carbon into glycolysis. This, however, does not appear to be so for PFP. At present there is not sufficient evidence to attribute a major role in the glycolytic pathway to PFP.

MITOCHONDRIA

The respiration of carbon compounds by mitochondria involves three major processes: uptake of metabolites, oxidative decarboxylation and electron transport. The uptake of metabolites is necessary because mitochondria are surrounded by two membranes. The outer membrane contains large pores which do not restrict the entry of metabolites. It is the inner membrane whose surface area is increased by invaginations that form cristae which acts as a barrier to most of the metabolic exchanges with the cytosol. These exchanges are mediated by proteins embedded in the membrane which act as transporters. Compounds that are taken up include succinate from lipid breakdown, glycine from photorespiration and pyruvate and malate from glycolysis. Furthermore the supply of ADP and P_i for ATP synthesis is provided respectively by the adenine nucleotide

and phosphate translocators located in the inner membrane. The second process is the oxidation of carbon compounds by enzymes of the tricarboxylic acid cycle, malate by malic enzyme (as well as malate dehydrogenase) and glycine by glycine decarboxylase and serine hydroxymethyltransferase. These oxidations lead to the reduction of NAD^+ (or flavine adenine dinucleotide (FAD) if succinate is the substrate). NADH and $FADH_2$ are oxidized by the third major process: electron transport within the inner mitochondrial membrane. Electron transport leads to the vectorial translocation of protons from the mitochondrial matrix, thus generating a membrane potential which is used to drive ATP synthesis and transport processes.

Mitochondrial Transporters

The basis for uptake of metabolites and ions into the matrix of plant mitochondria is the electrochemical proton gradient generated across the inner mitochondrial membrane. In plant mitochondria H^+ may exchange with K^+ from the matrix, so that a high proportion of the energy is stored as an electrical potential rather than the proton gradient. In respiratory terms, the four key translocators and those common to all plant mitochondria transport protons (F_0F_1 ATPase), phosphate, adenine nucleotides and pyruvate. Other transporters may play specific metabolic functions in particular tissues. Glycine transport is associated specifically with mitochondria of photosynthetic tissues that photorespire, and glutamate/aspartate transport is associated with the transfer of reducing equivalents in, for example, gluconeogenesic tissues[15].

Phosphate is taken up rapidly by mitochondria during oxidative phosphorylation through cotransport with H^+. This transport is electroneutral and sensitive to thiol reagents such as mersalyl and N-ethylmaleimide. The adenine nucleotide translocator is specific for ADP and ATP and is electrogenic, exchanging ADP^{3-} for ATP^{4-}. There are three specific inhibitors of this translocator: atractyloside and carboxyatractyloside, which bind to protein on the outer side of the mitochondrial inner membrane; and bongkrekic acid (an antibiotic formed by a mould in decaying coconut meals), which binds from the inside.

Pyruvate can diffuse across the inner mitochondrial membrane at high 'non-physiological' concentrations, but at concentrations of less than 2 mM it is transported by a specific translocator that is sensitive to α-cyano-4-hydroxycinnamic acid. Doubts have been expressed as to whether the activity of this translocator is sufficient to account for observed rates of respiration. However, with mitochondria isolated from clubs of the spadices of *Arum maculatum*, the inhibition of pyruvate (plus malate) oxidation by α-cyano-4-hydroxycinnamic acid indicated that pyruvate transport could sustain a rate of respiration equivalent to 30 600 µl oxygen h^{-1} g^{-1}. This would be more than sufficient to account for the normal thermogenic rates of respiration in this tissue. Pyruvate oxidation is measured in the presence of malate to ensure a rapid turnover of CoA. This occurs with pyruvate because oxaloacetate formed from malate dehydrogenase enables acetyl-CoA to condense with oxaloacetate to form citrate, and hence release CoA.

There is recent evidence from work with plant mitochondria that the net import of cofactors may be important in determining the rate of respiration. It has been known for some time that the addition and uptake of thiamine pyrophosphate (TPP) is essential for the oxidation of pyruvate and 2-oxoglutarate by plant mitochondria. It is assumed that TPP is lost during isolation of the organelles, but this has yet to be shown. More recently, it has been established that there are separate carriers for the uptake of NAD^+ and CoA. NAD malic enzyme activity requires CoA, and it has been shown that the malate oxidation of freshly isolated potato tuber mitochondria by malic enzyme is enhanced by this coenzyme. The rates of respiration of NAD-linked substrates by mitochondria isolated from certain tissues show a very marked stimulation by the addition of NAD. The indications are that in dormant or storage tissues such as potato tubers, and in senescent tissue such as soya bean cotyledons, the rate of *in vivo* mitochondrial respiration may be limited by their internal level of NAD. Mitochondria isolated from such tissues also show low levels of adenine nucleotides, and it has been shown that there is a specific energy-dependent uptake of adenine nucleotides into mitochondria that is insensitive to carboxyatractyloside, one of the potent inhibitors of the adenine nucleotide translocator. It therefore appears that there is a distinct transporter for the one-way uptake of adenine nucleotides.

Oxidation of Carbon Compounds

Figure 1.12 shows the oxygen uptake of mitochondria respiring malate at pH 7.2. State 3 respiration is the rate in the presence of substrate plus ADP; state 4 is the rate when ADP is exhausted. The ratio of state 3 : state 4 is known as the respiratory control ratio (RC). An increase in the rate of respiration upon adding ADP indicates that the membrane has a low permeability to protons and that proton flux through the F_0F_1 ATPase is coupled to the synthesis of ATP. Because of this proton flux through the ATPase, the electrochemical proton gradient drops and this decrease enables electron transport to accelerate. State 4 rates of oxygen uptake can also be accelerated by adding uncouplers such as dinitrophenol (DNP) or carbonylcyanide

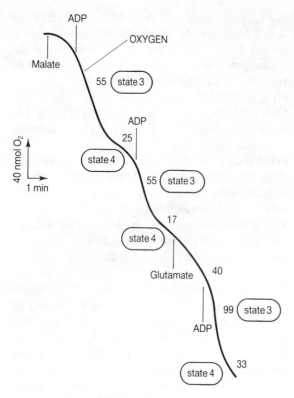

Figure 1.12. Oxygen uptake of soya bean cotyledon mitochondria oxidizing malate at pH 7.5. Additions as indicated were 10 mM malate, 10 mM glutamate and 200 nmol ADP. Numbers on traces are oxygen uptake rates in nmol min^{-1} mg^{-1} protein. Modified from J.H. Bryce and D.A. Day, 1990, *J. Exp. Bot.*, **41**, 961–967 and reproduced by permission of Oxford University Press.

p-trifluoromethoxyphenylhydrazone (FCCP) that increase the permeability of the membrane to protons. By convention, the ratio of the ADP consumed during state 3 to the consumption of oxygen (ADP:O) is calculated using the total oxygen uptake during state 3. It is assumed that during state 4 the high electrochemical gradient will lead to some proton leak through the membrane and that on addition of ADP the drop in the electrochemical gradient will lead to a cessation or marked reduction of this leak. Therefore, the state 4 rate of respiration is not subtracted from state 3 respiration in the calculation of the ADP:O ratio. The ATPase can be inhibited by

oligomycin, which binds to a polypeptide in the F_0 portion of the ATPase and inhibits the movement of protons.

There is a unique contrast between the malate oxidation of mitochondria isolated from animals and those from plants. The equilibrium for malate oxidation to oxaloacetate by malate dehydrogenase does not favour the formation of oxaloacetate. Therefore, in order to obtain rapid rates of malate oxidation by this enzyme in mitochondria it is necessary to remove the oxaloacetate. This can be done by adding glutamate so that oxaloacetate is transaminated to aspartate or through condensation of oxaloacetate with acetyl-CoA to form citrate (Figure 1.12)[16]. However, plant mitochondria have the ability to oxidize malate at a rapid rate without the removal of oxaloacetate. This is due to NAD–malic enzyme located within the matrix of the mitochondria and is particularly evident at pHs below 7, where malic enzyme is most active. The entry of carbon into mitochondria from glycolysis may either be as malate through activity of PEP carboxylase and cytosolic malate dehydrogenase or as pyruvate through activity of pyruvate kinase. It is not at present known what proportion of glycolytic flux is via PEP carboxylase rather than via pyruvate kinase. NAD–malic enzyme would enable the tricarboxylic acid (TCA) cycle to function in the absence of cytosolically produced pyruvate—for example, during lipid mobilization or if malate were supplied from a store in the vacuole. Malic enzyme is also involved in decarboxylating malate in the bundle-sheath cells of NAD–malic enzyme type C_4 species and certain crassulacean acid metabolism (CAM) plants (see chapter 3).

Plant mitochondria oxidize metabolites via the TCA cycle as outlined in Figure 1.13. For each turn of the cycle, this series of reactions releases two molecules of carbon dioxide, reduces three molecules of NAD, one of FAD and directly phosphorylates one molecule of ADP to ATP. Succinate dehydrogenase is bound to the inner surface of the inner mitochondrial membrane. All the other enzymes are soluble matrix enzymes. The regulatory properties of the enzymes of the TCA cycle have been studied in great detail[17].

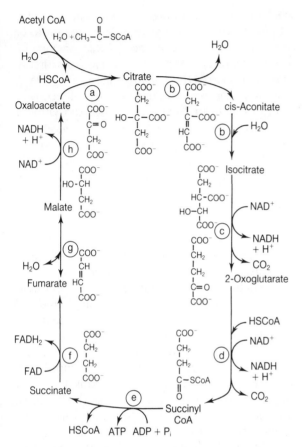

Figure 1.13. Reactions of the tricarboxylic acid cycle: (a) citrate synthase; (b) aconitase; (c) isocitrate dehydrogenase; (d) 2-oxoglutarate dehydrogenase; (e) succinyl-CoA ligase; (f) succinate dehydrogenase; (g) fumarase; (h) malate dehydrogenase.

However, these enzymes in the mitochondrial matrix exist in a viscous gel of 500 mg protein ml^{-1} (a 50% protein solution). This is far removed from the protein concentration of about 1 mg ml^{-1} in which biochemists generally study the properties of enzymes. At present, there is a debate as to whether or not all 'soluble' TCA cycle enzymes exist together in a close association known as a 'metabolon' with one enzyme passing its product on directly to act as the substrate for the next enzyme, such as occurs in fatty acid synthesis. If such a metabolon existed it would not only

increase the efficiency with which the cycle could operate, but it would also have implications for the exchange of metabolites between mitochondria and the cytosol.

The entry of carbon into the TCA cycle is tightly controlled by pyruvate dehydrogenase. The activity of this enzyme is controlled by reversible phosphorylation. It is activated by phosphorylation and inactivated by a phosphatase. Pyruvate in the presence of TPP is a potent inhibitor of inactivation and therefore pyruvate acts to maintain or increase the activity of pyruvate dehydrogenase. As well as its activity being modified by phosphorylation, the products of pyruvate dehydrogenase—NADH and acetyl-CoA—both inhibit its activity directly. Therefore, if the oxidation of NADH by the electron transport chains is limited, for example, by a lack of ADP, or if the availability of oxaloacetate to condense with acetyl-CoA is restricted, then pyruvate dehydrogenase will be inhibited. The properties of pyruvate dehydrogenase thus allow entry of carbon into the TCA cycle to be regulated.

Electron Transport and Terminal Oxidases

The major pathway of electron transport in the inner mitochondrial membrane of plants is shown in Figure 1.14. It is active following the oxidation of NADH formed by the mitochondrial dehydrogenases that oxidize malate (by malate dehydrogenase and malic enzyme), pyruvate, isocitrate and 2-oxoglutarate. The oxidation of NADH to NAD$^+$ provides two protons and two electrons to complex I. The mechanism of this oxidation in plants is not fully established but it involves flavine mononucleotide (FMN)-mediated proton transfer and an iron–sulphur centre. Iron–sulphur centres are able to mediate the flux of electrons through oxidation/reduction of Fe. Complex I transfers electrons to ubiquinone and results in the transfer of four H$^+$ ions from the matrix to the intermembrane space for every NADH oxidized. This transfer is inhibited by rotenone. However, in plant mitochondria there is a rotenone-insensitive bypass with a K_m for

NADH of approximately 80 μM, compared to 8 μM for complex I. This bypass is not coupled to the vectorial transfer of protons across the membrane and therefore the bypass is not associated with ATP synthesis. Its presence in plants is widespread and it can be engaged by high NADH:NAD^+ ratios, which can occur when malate is oxidized by NAD–malic enzyme or if electron transport is restricted by a high electrochemical proton gradient when, for example, the availability of ADP is restricted.

Succinate oxidation is mediated by complex II, which consists of FAD, several non-haem iron centres and two large polypeptides. It is bound to the inner surface of the inner membrane and is not involved in proton translocation. The enzyme is inhibited by oxaloacetate and malonate. Succinate, like complex I and the retenone-insensitive bypass, reduces ubiquinone (Q) to QH_2. Ubiquinone is also reduced through the oxidation of cytosolic NADH by a dehydrogenase located on the outer surface of the inner membrane. The exact nature of the interaction of this dehydrogenase with ubiquinone is not known, but it is insensitive to inhibitors of complex I and has an ADP:O ratio of approximately 2. The enzyme can be inhibited by platanetin, a 3,5,7,8-tetrahydroxy-6-isoprenyl flavone isolated from bud scales of the plane tree, or by chelation of Ca^{2+} with EGTA. Only micromolar levels of Ca^{2+} are required to activate this dehydrogenase. The direct oxidation of cytosolic NADH may be involved in the regulation of glyceraldehyde 3-phosphate dehydrogenase in glycolysis and the regulation of β-oxidation via 3-hydroxyacyl-CoA dehydrogenase (see chapter 5). There is evidence from work with isolated mitochondria that activity of the external NADH dehydrogenase is restricted by the oxidation of NADH from NAD-linked substrates in the matrix. Whenever substrate supply to mitochondria is restricted the external dehydrogenase could therefore play a role in stimulating glycolysis and lipid breakdown.

NADPH oxidation is also a feature of plant mitochondria. It is coupled via ubiquinone to two sites of ADP phosphorylation, is sensitive to chelators and has a lower pH optimum than that of the NADH dehydrogenase. It is thought to be located on the outer surface of the inner membrane and to be a separate enzyme from that of the external NADH dehydrogenase. The role of the enzyme is unclear, but the cytosolic conversion of citrate to 2-oxoglutarate for biosynthesis involves NADP–isocitrate dehydrogenase and aconitase. If the biosynthetic requirement for carbon skeletons from the pentose phosphate pathway were more important than the need for NADPH, then the external NADPH dehydrogenase could provide a means of regenerating $NADP^+$.

Ubiquinone passes electrons to complex III (the cytochrome b/c complex). Figure 1.15 shows what is termed the Q cycle, which enables two protons to be translocated across the membrane for every electron that is transferred to cytochrome c via an iron–sulphur protein and cytochrome c_1. The complex is inhibited by antimycin A. Cytochrome c is a mobile electron carrier that transfers electrons to complex IV (cytochrome c oxidase), which spans the inner mitochondrial membrane. This is the terminal complex of the electron transport chain and reduces oxygen to water on the matrix side of the membrane. The complex contains two cytochromes, cytochrome a and a_3, and two copper atoms. Cytochrome a receives electrons from cytochrome c and passes them to cytochrome a_3, which reacts with oxygen. Azide, cyanide and carbon monoxide inhibit activity of this complex. The reduction of oxygen by cytochrome c oxidase is the only irreversible step in oxidative phosphorylation. There is evidence from the stoichiometry of proton translocation by isolated complexes in vesicles that, as well as proton translocation associated with the redox loops, there may also be proton pumps. It is hypothesized that these could provide for the overall translocation of four protons per complex for each NADH oxidized. This would allow for up to three protons to be involved in the synthesis of ATP by the ATPase, with a proton available for coupling to the uptake of phosphate[18]. With chloroplasts there is much debate over the precise

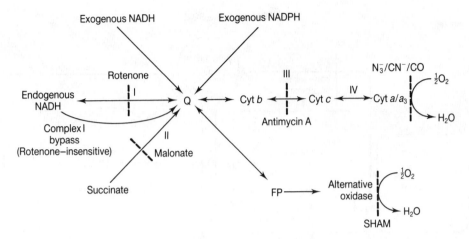

Figure 1.14. Pathways of electron transport in plant mitochondria. Complexes are denoted by roman numerals, Q is a mobile pool of ubiquinone which transfers electrons to complex III (the cytochrome b–c complex) and bars indicate sites of inhibitor action.

stoichiometry of vectorial proton translocation and the stoichiometry of its coupling to ATP synthesis.

Non-phosphorylating Electron Transport

Although an ADP : O ratio of 3 is predicted for oxidation of NAD-linked substrates, this is rarely achieved with plant mitochondria. In part this may be due to the presence in mitochondria from plants of electron transport pathways that are not coupled to proton translocation. These are the rotenone-insensitive bypass of complex I and the alternative pathway which branches from the main pathway at ubiquinone (Figure 1.14). Respiration via this pathway is insensitive to the inhibitor cyanide and is therefore known as cyanide-insensitive respiration. It is believed that the oxidase for this alternative pathway is a ubiquinol : oxygen oxidoreductase that catalyses the four-electron reduction of oxygen to water. Recently an antibody against the alternative oxidase has been prepared that reacts with a group of proteins in the 35–37 kDa range on Western blots[19]. Schonbaum et al.[20] showed that substituted hydroxamic acids such as SHAM (salicylhydroxamic acid) inhibited the activity of the cyanide-insensitive oxidase. Since then propyl

gallate and disulfiram have also been used. No inhibitor will be absolutely specific for a single enzyme, therefore care in the interpretation of results with inhibitors must be taken. For example, SHAM inhibition of lipoxygenase activity in germinating seeds was at one time mistaken for activity of the alternative oxidase.

Inhibition of the cytochrome pathway may divert the flux of electrons to the alternative oxidase, but inhibition of the alternative oxidase by SHAM does not lead to any significant diversion of electrons to the cytochrome pathway. Bahr and Bonner[21] proposed that ubiquinone would have to be almost completely reduced before there would be diversion of electrons to the alternative pathway. However, Dry et al.[22] varied the rate of succinate oxidation in mitochondria from soya bean cotyledons using the inhibitor malonate and used a Q electrode to determine the extent of ubiquinone reduction. They measure the ratio Qr : Qt, where Qr was reduced Q and Qt the total potential reduction of Q. There was a linear relationship between Q reduction and oxygen uptake during state 3 respiration (Figure 1.16). This respiration did not engage the alternative pathway. However, the relationship was non-linear in state 4. Where antimycin A was added to inhibit the cytochrome

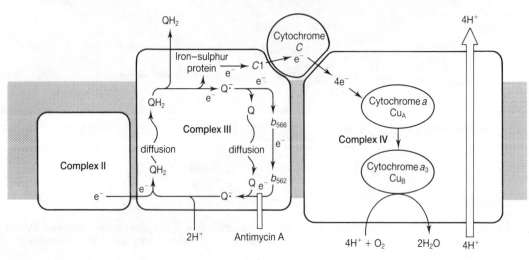

Figure 1.15. Electron transfer from complex II to oxygen via ubiquinone, the cytochrome b–c complex, cytochrome c and complex IV (cytochrome oxidase). The proton-motive Q cycle allows two protons to be transferred from the matrix to the inter-membrane space for each electron moving through the b–c complex. Q denotes fully oxidized ubiquinone, QH_2, fully reduced ubiquinone and Q^- the semiquinone intermediate which is reduced to QH_2 by an iron–sulphur centre of the dehydrogenase (complex II), resulting in the uptake of two protons from the matrix. The QH_2 thus formed 'diffuses' in the membrane and is then oxidized at the cytoplasmic side of the membrane by a reaction in which the Rieske protein (iron–sulphur protein) of complex III transfers one electron from QH_2 to cytochrome c_1 and a second electron from the Q^- thus formed to cytochrome b_{566}, forming Q and releasing two H^+ into the external medium. The Q diffuses in the membrane and is reduced by cytochrome b_{562} at the matrix side. The transfer of an electron from cytochrome b is sensitive to antimycin A. The reduction of molecular oxygen by cytochrome c oxidase requires four electrons and is the only irreversible step in oxidative phosphorylation. Modified with permission from R. Douce, 1985, *Mitochondria in Higher Plants: Structure, Function and Biogenesis*, Academic Press, London.

pathway, engagement of the alternative oxidase occurred at approximately 60% Q reduction— lower than predicted by Bahr and Bonner. However, in agreement with their predictions the alternative pathway was not engaged unless there was a high substrate flux to the mitochondria and rapid reduction of Q or restriction of the cytochrome pathway.

The role of the alternative oxidase in plants is unclear except in thermogenic tissues, where its activity results in the uncoupling of respiration from ATP synthesis so that energy is released as heat. In mitochondria from the sterile appendix of voodoo lilies (*Sauromatum guttatum*), it has been found that over the 5 days leading up to

thermogenesis the activity of the alternative oxidase increases dramatically, whereas cytochrome oxidase remains stable until the day before thermogenesis, when its activity drops by over 90%, effectively preventing significant flow via the cytochrome chain[23]. In thermogenic tissues there is a clear role for the alternative oxidase.

In certain other tissues Lambers[24] proposed an energy overflow hypothesis in which respiration via the alternative pathway was proposed to oxidize sugars in excess of those required for the production of carbon skeletons for growth, ATP for energy, osmoregulation, and storage of carbohydrate reserves. However, there is no good

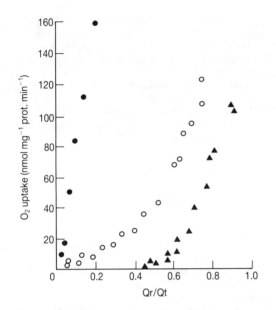

Figure 1.16. Dependence of total respiratory flux on quinone redox state in purified soya bean cotyledon mitochondria under various conditions. (●) State 3 conditions, (○) state 4 conditions, (▲) in the presence of 3.8 μM antimycin A. Reproduced from I.B. Dry *et al.*, 1989, *Arch. Biochem. Biophys.*, **273**, 148–157, by permission of Academic Press.

evidence that the flux of sucrose to tissue exceeds that which can be effectively utilized. The non-phosphorylating pathways of electron transport allow great flexibility in the proportion of NADH, ATP and carbon skeletons produced by respiration. However, this is quite distinct from the concept of an overflow pathway that responds simply to excess substrate. These pathways may be of particular importance in special situations, for example CAM and C_4 plants where rapid decarboxylation of malate is required to provide the substrate for photosynthesis. While a great deal of knowledge has been gained over the last 30 years, many questions as to how plants obtain and utilize their energy still remain unanswered.

REFERENCES

1. Nicholls, D.G. (1982) *Bioenergetics: An Introduction to the Chemiosmotic Theory*. Academic Press, London.

2. Walker, D. (1979) *Energy, Plants and Man*. Readers in Plant Productivity, Packard Publishing, Chichester.

3. Ort, D.R. and Good, N.E. (1988) Textbooks ignore photosystem II-dependent ATP formation: Is the Z scheme to blame? *Trends in Biochemical Science*, **13**, 467–469.

4. Ghanotakis, D.F. and Yocum C.F. (1990) Photosystem II and the oxygen-evolving complex. *Annual Review of Plant Physiology and Plant Molecular Biology*, **41**, 255–276.

5. Barber, J. (1987) Photosynthetic reaction centres: a common link. *Trends in Biochemical Science*, **12**, 321–326.

6. Anderson, J.M. and Anderson, B. (1988) The dynamic photosynthetic membrane and regulation of solar energy conversion. *Trends in Biochemical Science*, **13**, 351–355.

7. Newsholme, E.A. and Crabtree, B. (1973) Metabolic aspects of enzyme activity regulation. *Symposia of the Society for Experimental Biology*, **27**, 429–460.

8. Kacser, H. (1988) Control of Metabolism. In *The Biochemistry of Plants, Vol 11: Biochemistry of Metabolism* (ed. D.D. Davies), pp. 39–67. Academic Press, London and New York.

9. Kruger N.J. and Beevers, H. (1985) Synthesis and degradation of fructose 2, 6-bisphosphate in endosperm of castor bean seedlings. *Plant Physiology*, **77**, 358–364.

10. ap Rees, T. (1985) The organization of glycolysis and the oxidative pentose phosphate pathway in plants. In *Higher Plant Cell Respiration* (eds R. Douce and D.A. Day), pp. 391–417. Springer-Verlag, Berlin and New York.

11. Bryce, J.H. and ap Rees T. (1985) Rapid decarboxylation of the products of dark fixation of CO_2 in roots of *Pisum* and *Plantago*. *Phytochemistry*, **24**, 1635–1638.

12. Copeland L. and Turner J.F. (1988) The regulation of glycolysis and the pentose phosphate pathway. In *The Biochemistry of Plants, Vol 11: Biochemistry of Metabolism* (ed. D.D. Davies), pp. 107–128 Academic Press, London and New York.

13. Ball, K.L. and ap Rees, T. (1988) Fructose 2,6-bisphosphate and the climacteric in bananas. *European Journal of Biochemistry*, **177**, 637–641.

14. ap Rees, T., Green, J.H. and Wilson, P.M. (1985) Pyrophosphate:fructose 6-phosphate 1-phosphotransferase and glycolysis in non-photosynthetic tissues of higher plants. *Biochemical Journal*, **227**, 299–304.

15. ap Rees, T. (1987) Compartmentation of plant metabolism. In *The Biochemistry of Plants, Vol 12: Physiology of Metabolism* (ed. D.D. Davies),

pp. 87–115. Academic Press, London and New York.

16. Bryce, J.H. and Day, D.A. (1990) Tricarboxylic acid cycle activity in mitochondria from soybean nodules and cotyledons. *Journal of Experimental Botany*, **41**, 961–967.

17. Wiskich, J.T. (1980) Control of the Krebs cycle. In *The Biochemistry of Plants, Vol 2: Metabolism and Respiration* (ed. D.D. Davies), pp. 243–278. Academic Press, London and New York.

18. Douce, R. (1985) *Mitochondria in Higher Plants: Structure, Function and Biogenesis*. Academic Press, London and New York.

19. Elthon, T.E., Nickels, R.L. and McIntosh, L. (1989) Monoclonal antibodies to the alternative oxidase of higher plant mitochondria. *Plant Physiology*, **89**, 1311–1317.

20. Schonbaum, G.R., Bonner, W.D., Jr, Storey, B.T. and Bahr, J.T. (1971) Specific inhibition of the cyanide-insensitive respiratory pathway in plant mitochondria by hydroxamic acids. *Plant Physiology*, **47**, 124–128.

21. Bahr, J.T. and Bonner, W.D., Jr (1973) Cyanide-insensitive respiration II. Control of the alternate pathway. *Journal of Biological Chemistry*, **248**, 3446–3450.

22. Dry, I.B. Moore, A.L., Day, D.A. and Wiskich, J.T. (1989) Regulation of alternative pathway activity in plant mitochondria: nonlinear relationship between electron flux and the redox poise of the quinone pool. *Archives of Biochemistry and Biophysics*, **273**, 148–157.

23. Elthon, T.E. and McIntosh, L. (1986) Characterization and solubilisation of the alternative oxidase of *Sauromatum guttatum* mitochondria. *Plant Physiology*, **82**, 1–6.

24. Lambers, H. (1980) The physiological significance of cyanide-resistant respiration in higher plants. *Plant Cell Environment*, **3**, 293–302.

FURTHER READING

Davies, D.D. (ed.) (1988) *The Biochemistry of Plants, Vol 11: Biochemistry of Metabolism*. Academic Press, London and New York.

Douce, R. (1985) *Mitochondria in Higher Plants: Structure, Function and Biogenesis*. Academic Press, London and New York.

Douce, R. and Day, D.A. (eds) (1985) *Higher Plant Cell Respiration*. Springer-Verlag, Berlin and New York.

Gregory, R.P.F. (1989) *Biochemistry of Photosynthesis*. Wiley, Chichester.

Nicholls, D.G. (1982) *Bioenergetics: An Introduction to the Chemiosmotic Theory*. Academic Press, London.

2

THE CALVIN CYCLE AND PHOTORESPIRATION

R.C. Leegood

Department of Animal and Plant Sciences, University of Sheffield, UK

THE CALVIN CYCLE

In algae and in higher plants there is only one primary carboxylating mechanism which results in the net synthesis of carbon compounds. The Calvin cycle, or reductive pentose phosphate pathway, is common to all plants—C_3, C_4 and crassulacean acid metabolism (CAM)—although C_4 and CAM plants have auxiliary mechanisms of carbon fixation. Virtually all of the organic carbon in the biosphere has passed through the Calvin cycle. The Calvin cycle, which is located in the chloroplasts, comprises 13 reactions catalysed by 11 enzymes (Figure 2.1) and has four principal features:

1. Carboxylation of ribulose 1,5-bisphosphate (RuBP) to yield glycerate 3-phosphate, and

Plant Biochemistry and Molecular Biology. Edited by P.J. Lea and R.C. Leegood
© 1993 John Wiley & Sons Ltd

catalysed by ribulose 1,5-bisphosphate carboxylase–oxygenase (Rubisco).

2. Reduction of glycerate-3-phosphate to triose phosphate, at the expense of ATP and NADPH.

3. Regeneration of the primary acceptor, RuBP,

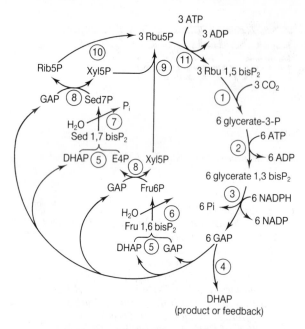

Figure 2.1. The Calvin cycle. The Calvin cycle can be divided into three phases. The first phase is carboxylation catalysed by Rubisco (1). The second phase is the reductive phase by which glycerate 3-phosphate is converted to triose phosphate, catalysed by glycerate 3-phosphate kinase (2) and NADP⁺-dependent glyceraldehyde phosphate dehydrogenase (3). The third phase is regeneration of the carbon dioxide acceptor in the sugar-phosphate shuffle, in which five C_3 units are rearranged to form three C_5 units (xylulose 5-phosphate (Xyl5P) and ribose 5-phosphate (Rib5P) by the actions of triose phosphate isomerase (4), aldolase (5), fructose bisphosphatase (6), sedoheptulose bisphosphatase (7) and transketolase (8). Xyl5P and Rib5P are then converted to ribulose 5-phosphate (Rbu5P) and ribulose 1,5-bisphosphate (Rbu 1,5bisP₂ or RuBP) by ribulose 5-phosphate 3-epimerase (9), ribulose 5-phosphate isomerase (10) and ribulose 5-phosphate kinase (11). Glyceraldehyde phosphate (GAP) dehydrogenase, the bisphosphatases and product synthesis recycle the Pi required for continued ATP synthesis. From Leegood (1990), by permission of Academic Press.

in which $5 \times C3$ molecules (triose phosphates) are rearranged to yield $3 \times C5$ molecules (pentose phosphates) in the 'sugar phosphate shuffle'. A further ATP molecule is required to convert pentose phosphate into RuBP, thus 3ATP and 2NADPH are required for each molecule of carbon dioxide fixed.

4. Autocatalysis: the cycle acts as an autocatalytic 'breeder' reaction and, for every three turns of the cycle, one molecule of triose phosphate is generated from three molecules of carbon dioxide (Figure 2.1). Triose phosphate may either be utilized in the synthesis of starch or sucrose or may re-enter the Calvin cycle to form more of the primary acceptor, RuBP. In the last case, the cycle generates more carbon dioxide acceptor than it consumes.

Formulation of the Calvin Cycle

In the mid-1940s, the radioactive isotope of carbon ^{14}C became available in quantities sufficient to investigate the path of carbon in photosynthesis. During the 1950s, Melvin Calvin (who was later awarded a Nobel Prize), Andrew Benson and their colleagues also employed newly developed techniques in two-dimensional paper chromatography to separate the labelled compounds and to study the sequence of labelling of carbon compounds in green algae such as *Chlorella* and *Scenedesmus*. Uniform suspensions of these organisms could be illuminated in the presence of $^{14}CO_2$ and samples collected by running the suspension into boiling ethanol. Such suspensions are much simpler to use for these experiments than are leaves. As the period of illumination was shortened fewer compounds became labelled. When the radioactivity in them was plotted as a fraction of the total ^{14}C fixed, one three-carbon compound, glycerate 3-phosphate, showed a negative slope which extrapolated to 100% of the total carbon fixed at zero time (Figure 2.2), indicating that this was the first product of carbon assimilation. A great deal of painstaking research led to the identification of other intermediates in the cycle and of the initial acceptor as a five-

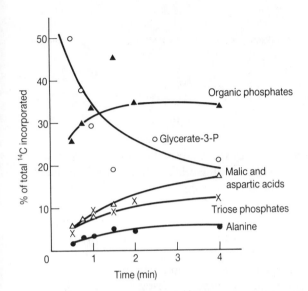

Figure 2.2. Incorporation of ^{14}C into products of photosynthesis in *Scenedesmus* after short periods of illumination. From Calvin *et al.* (eds) (1951), reproduced by permission of Cambridge University Press.

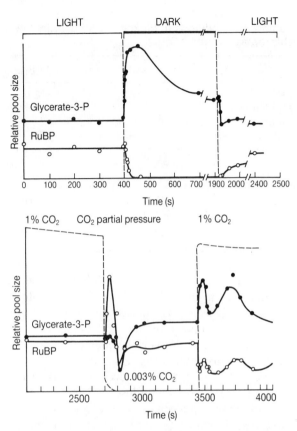

Figure 2.3. Changes in the pool sizes of RuBP and glycerate 3-phosphate in *Scenedesmus* as affected by light–dark changes and by changes in the partial pressure of carbon dioxide. (From Bassham and Calvin, 1957, by permission of Prentice Hall.)

carbon compound, RuBP. Transients confirmed that RuBP was the primary acceptor. For example, if the concentration of carbon dioxide was lowered suddenly, restricting the availability of one of the substrates of Rubisco, then glycerate 3-phosphate (the product) fell and RuBP (the other reactant) rose. Conversely, during a light-to-dark transition, lack of ATP and NADPH stopped glycerate 3-phosphate reduction and regeneration of RuBP. Therefore, the content of RuBP fell and that of glycerate 3-phosphate rose (Figure 2.3).

Control of the Calvin Cycle

The Calvin cycle lies at the interface between electron transport and product synthesis and it must therefore be very responsive to changes in the relationship between supply of, and demand for, the products of electron transport (e.g. during light transients) and during longer-term changes in environmental conditions (e.g. acclimation to sun and shade or to low temperature). The activities of Calvin cycle enzymes are there-

fore under several forms of control. Light-activation mechanisms and metabolite modulation allow rapid short-term modulation of enzyme activity, while protein synthesis (which, with the exception of the large subunit of Rubisco, is wholly under the control of the nuclear genome) allows long-term modulation of the amounts of Calvin cycle enzymes.

The conditions in the illuminated chloroplast stroma are quite different in the light and in the dark (Table 2.1). The pH of the stroma increases as protons are pumped into the intrathylakoid space, and the concentration of Mg^{2+} increases because it acts, with other ions, as a counterion to H^+ uptake. The availability of reductants such as

Table 2.1. Conditions in darkened and illuminated chloroplast stroma.

	Dark	Light
pH	7.3	8.0
Mg^{2+} (mM)	1–3	3–6
Thioredoxin (% reduced)	8–30	62–77

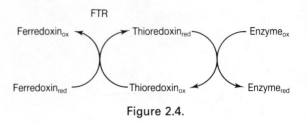

Figure 2.4.

ferredoxin and NADPH also increases as electron transport commences. All these factors are crucial in promoting the activities of enzymes such as Rubisco, fructose 1,6-bisphosphatase and ribulose 5-phosphate kinase in the light.

Of particular importance is the regulation of enzymes by reductive activation. Many enzymes function only in the reduced form, but in photosynthetic organisms a virtue has been made out of a necessity by linkage of the catalytic activity of the Calvin cycle and other enzymes to the availability of photosynthetically generated reductants. Four enzymes of the Calvin cycle (fructose 1,6-bisphosphatase, sedoheptulose 1,7-bisphosphatase, ribulose 5-phosphate kinase and NADP–GAP dehydrogenase) and several other enzymes involved in chloroplast metabolism, such as $NADP^+$-dependent malate dehydrogenase and phenylalanine ammonia lyase, undergo reductive activation (Table 2.2), while glucose 6-phosphate dehydrogenase is inactivated by reductant and is inactive in the illuminated stroma.

Electrons for reductive activation are provided by photosynthetic electron transport to ferredo-

xin. Ferredoxin reduces thioredoxin in a reaction catalysed by ferredoxin–thioredoxin reductase (FTR) and thioredoxin, in its turn, reduces the enzyme (Figure 2.4).

In vitro activation of enzymes can be mimicked by the addition of thiol-reducing agents such as dithiothreitol. FTR is an iron–sulphur protein comprising two dissimilar subunits and contains a reducible disulphide bridge. It catalyses the reduction of both types of chloroplast thioredoxin: thioredoxin *f*, which activates fructose bisphosphatase preferentially, and thioredoxins *mb* and *mc*, which activate NADP–malate dehydrogenase. All three thioredoxins have a molecular weight of 12 kDa and possess a single reducible disulphide bridge per monomer.

REACTIONS OF THE CALVIN CYCLE

Reactions Catalysed by Rubisco

As the name RuBP carboxylase–oxygenase implies, Rubisco is bifunctional. Besides catalysing the addition of carbon dioxide to RuBP, with the formation of two molecules of glycerate 3-phosphate, Rubisco catalyses the addition of oxygen to RuBP with the formation of one molecule of glycerate 3-phosphate and one of glycollate 2-phosphate (Figure 2.5). Carbon dioxide behaves as a competitive inhibitor of the oxygenase reaction and oxygen as a competitive inhibitor of the carboxylase reaction. The $K_m(CO_2)$ of Rubisco is accordingly higher in air (20 μM) than in nitrogen (10 μM) and the $K_m(CO_2)$ is comparable to the concentration of carbon dioxide at the site of carboxylation. The $K_m(O_2)$ is about 550 μM, which is roughly equiva-

Table 2.2. Light-dependent activation of Calvin cycle enzymes in pea leaves. Activities in μmol h^{-1} mg^{-1} chlorophyll.

	Dark	Light	Activation (fold)
Ribulose 5-phosphate kinase	29	1160	40
Sedoheptulose 1,7-bisphosphatase	2.8	29	12
Fructose 1,6-bisphosphatase	2.5	37	15

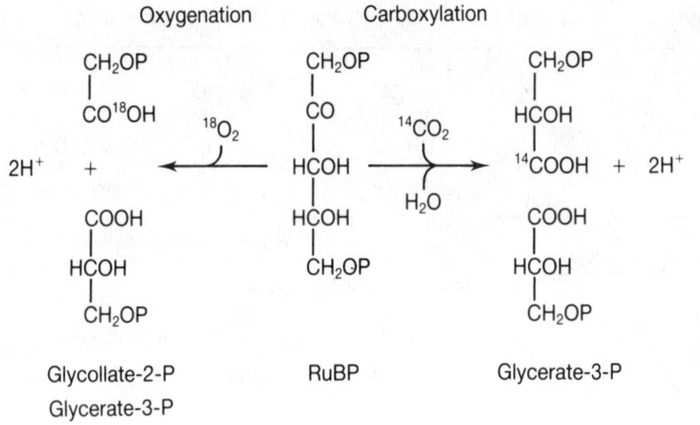

Figure 2.5.

lent to the oxygen concentration in solution in equilibrium with air. Oxygenase activity is nevertheless dependent upon the presence of carbon dioxide, since the enzyme is activated by carbon dioxide at an activator site which is distinct from the catalytic site (see below). The $K_m(CO_2)$ is somewhat higher in C_4 species (28–63 μM) than it is for C_3 and CAM species (8–26 μM).

Rubisco is the most abundant protein in the world; there are about 10 kg for every person on earth. This is because Rubisco is a very inefficient catalyst with a low specific activity (1 $\mu mol\ min^{-1}\ mg^{-1}$ protein). Therefore, large amounts of Rubisco are required to support high photosynthetic rates. Up to a quarter of the total protein, and half of the soluble protein, in a leaf is invested in Rubisco. This also has important implications for the nitrogen economy of plants.

In bacteria, the enzyme comprises two large subunits but in algae and higher plants eight large subunits (50–55 kDa), containing the catalytic site, are supplemented by eight small subunits (12–15 kDa), which are thought to play a regulatory role. The total molecular mass is about 550 kDa. The large subunit is synthesized in the chloroplast, whereas the small subunit is made on cytoplasmic 80S ribosomes and imported into the chloroplast.

Unlike many other carboxylases the reaction mechanism of Rubisco is uniquely based on an ene–diol mechanism, comprising five separate steps (Figure 2.6):

1. Deprotonation of carbon 3 (enolization) to yield a 2,3-enediol.
2. Carboxylation to yield a six-carbon β-ketoacid intermediate, 3-keto-2-carboxyarabinitol bisphosphate.
3. Hydration to a gem–diol form.
4. Deprotonation, resulting in C—C cleavage.
5. Protonation to give a second molecule of glycerate 3-phosphate.

The mechanism of oxygenation is less clear, but oxygen probably attacks the ene–diol, splitting it into two. Oxygenation is viewed as an inevitable consequence of the reaction mechanism since the enzyme isolated from any source—higher plants, algae or bacteria—possesses both carboxylase and oxygenase activities, although the relative activities vary. Rubisco originally evolved in an atmosphere rich in carbon dioxide and poor in oxygen. Subsequent reversal of these gases in the atmosphere has occurred because of photosynthesis and has led to the development of photorespiration (see below) as a means of retrieving glycollate formed by oxygenation. The ratio of the carboxylase to oxygenase activity is lowest in the bacteria and algae and highest in higher plant crop species. This difference in the relative rates of carboxylation and oxygenation has led to suggestions that the enzyme may yet be improved (by selectively diminishing oxygenase activity) through genetic manipulation.

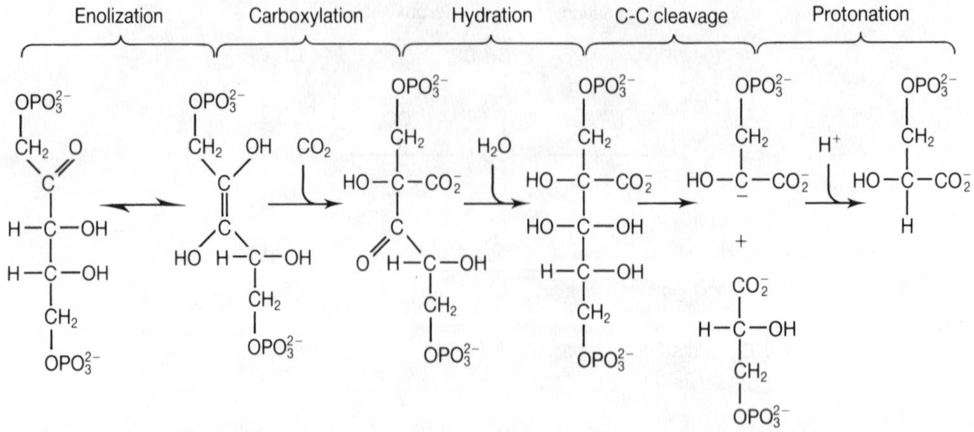

Figure 2.6. Intermediates in the reaction mechanism of Rubisco. From Andrews and Lorimer (1987), by permission of Academic Press.

Regulation of Rubisco

Under physiological conditions, the activity of Rubisco is regulated in response to changes in environmental conditions, such as light and temperature.

Research in the 1970s showed that, while the $K_m(CO_2)$ of carboxylation by a leaf was about 10 μM, *in vitro* the $K_m(CO_2)$ was about ten times higher. However, the activity of the enzyme *in vitro* could be increased, and its apparent $K_m(CO_2)$ decreased, by preincubation with carbon dioxide and Mg^{2+}. This was shown to involve binding of carbon dioxide to a lysine group on the enzyme. This forms a carbamate which can be stabilized by Mg^{2+} and certain phosphorylated intermediates. This lysine group (lysine[201]) reacts with a carbon dioxide molecule which is quite distinct from that involved in catalysis. The reaction is promoted by Mg^{2+} and, because it involves abstraction of a proton, by alkaline pH. These are similar conditions to those which occur in the illuminated stroma (Table 2.1; Figure 2.7).

An enigma was that the concentrations of carbon dioxide required for this preincubation (10

mM is often employed) were much higher than the concentrations of carbon dioxide (10 μM) which would normally be present in the chloroplast stroma. It was not clear how carbamylation could occur under physiological conditions.

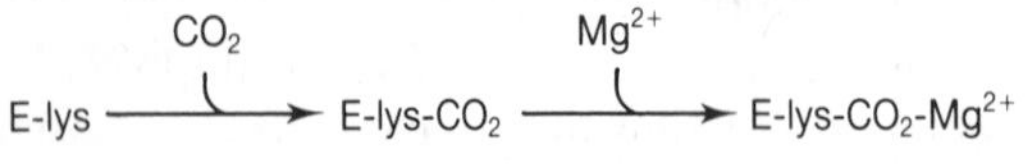

In 1983 a mutant of *Arabidopsis thaliana* was discovered which was unable to activate Rubisco. Its rate of photosynthesis was low in low carbon dioxide, it exhibited small light-induced changes in Rubisco activity and the activation state of Rubisco was low even when leaves were illuminated in high (1%) carbon dioxide, yet the kinetic parameters and *in vitro* activation of the enzyme were identical in extracts made from the mutant. These observations inferred that some activating factor was missing in the mutant which was unnecessary under *in vitro* conditions, but which was necessary under *in vivo* conditions. Two-dimensional gels of proteins extracted from chloroplasts showed that two polypeptides (47 and 50 kDa) were missing in the mutant. If Rubisco, ATP, RuBP, carbon dioxide and Mg^{2+} were incubated with this protein, called Rubisco activase, then activation occurred at low carbon dioxide concentrations, with a K_{act} of 4 μM. The ATP dependence of the activase may be important in linking the activation state of Rubisco to the PFD. Its precise role may be to remove bound RuBP from the Rubisco active site, so that it is in a state susceptible to carbamylation (Figure 2.8).

$$E\text{-lys} \xrightarrow{\quad CO_2 \quad} E\text{-lys-}CO_2 \xrightarrow{\quad Mg^{2+} \quad} E\text{-lys-}CO_2\text{-}Mg^{2+}$$

Figure 2.7.

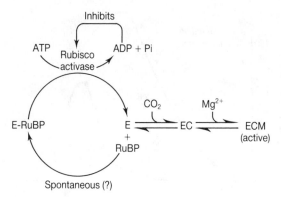

Figure 2.8. Possible role of Rubisco activase in the activation of Rubisco. The activase displaces RuBP (and possibly the tight-binding inhibitor, carboxy-arabinitol 1-phosphate) from the active site, permitting carbamylation of the enzyme and formation of the active complex with carbon dioxide and Mg^{2+}. From Portis, 1988, by permission of the American Society of Plant Physiologists.

Rubisco is also regulated by a quite different mechanism involving a natural tight-binding inhibitor. This was discovered when it was found that it was impossible to activate Rubisco from some plants with carbon dioxide and Mg^{2+} when extracts were made from darkened leaves, whereas the enzyme could be fully activated in extracts from plants which were brightly illuminated. This type of dark inactivation is common in C_3 plants, particularly in legumes such as bean (e.g. *Phaseolus vulgaris)* and soya bean, and also tobacco and potato, although it is not marked in spinach, wheat or barley or in C_4 plants. However, it is dramatic in CAM plants, such as pineapple. In these cases, light activation additionally involves the removal of an endogenous tight-binding inhibitor, which can remain bound to Rubisco, even through a number of purification steps.

The inhibitor is 2-carboxyarabinitol 1-phosphate (Figure 2.9), which has a K_d of 32 nM and occupies the active site of Rubisco, mimicking the six-carbon intermediate (Figure 2.6) and preventing activation of Rubisco by carbon dioxide and Mg^{2+} and turnover of the enzyme. In *P. vulgaris*, a specific phosphatase results in rapid

$$\begin{array}{c} CH_2OP \\ | \\ HO-C-CO_2^- \\ | \\ CHOH \\ | \\ CHOH \\ | \\ CH_2OH \end{array}$$

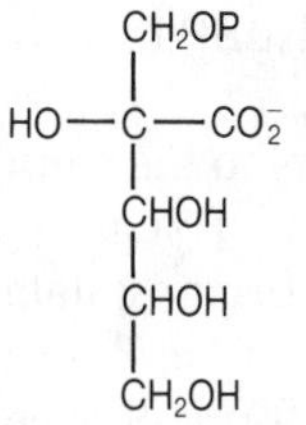

Figure 2.9. Carboxyarabinitol 1-phosphate.

degradation of carboxyarabinitol 1-phosphate to carboxyarabinitol, but the precise pathway of synthesis and degradation has not yet been fully formulated or its regulation identified. One possibility is that Rubisco activase is also involved in displacing this inhibitor from the active site.

Reduction of Glycerate 3-Phosphate to Triose Phosphate

Glycerate 3-phosphate is reduced to triose phosphate in two steps which constitute a freely reversible reaction. The first, catalysed by glycerate 3-phosphate kinase, results in the ATP-dependent formation of glycerate 1,3-bisphosphate. This does not accumulate, but is immediately reduced, in a reaction catalysed by glyceraldehyde 3-phosphate dehydrogenase. Glyceraldehyde 3-phosphate dehydrogenase is an enzyme which is regulated in a complex manner and which is light-activated by the thioredoxin system. High ATP:ADP and NADPH:NADP ratios are required to drive these freely reversible reactions to form triose phosphates:

glycerate 3-P + ATP → glycerate 1,3-bisP$_2$ +ADP

glycerate 1,3-bisP$_2$ + NADPH →
glyceraldehyde 3-P + NADP$^+$ + P$_i$

Triose phosphate isomerase then catalyses the interconversion of glyceraldehyde 3-phosphate and dihydroxyacetone phosphate, which are collectively termed triose phosphates. At equilibrium the ratio of dihydroxyacetone phosphate to glyceraldehyde 3-phosphate is about 22:1.

Regeneration of RuBP from Triose Phosphate

There are two parts to the shuffle which results in the rearrangement of sugar-phosphates. In the first part, fructose bisphosphatase, sedoheptulose bisphosphatase, transketolase and aldolase convert five C3 units (triose phosphate) into three C5 units (pentose phosphates). In the second part, pentose phosphates are converted to RuBP in reactions catalysed by phosphoribose isomerase and phosphoribulokinase, the latter utilizing ATP.

Fructose 1,6-bisphosphatase

$$Fru1,6BP + H_2O \rightarrow F6P + P_i$$

The chloroplastic fructose bisphosphatase differs from the cytosolic enzyme (and mammalian and yeast enzymes) in being insensitive to inhibition by AMP and in being regulated by the ferredoxin–thioredoxin system. The native oxidized enzyme is a tetramer (M_r 160 kDa) which contains two disulphide bridges that are cleaved upon activation. Reduction of the enzyme results in a decrease in the K_m for the $[FBP^{4-}.Mg^{2+}]^{2-}$ complex from 130 to 6 μM, although the maximal velocity of the enzyme is unaltered. In its oxidized form the enzyme has a pH optimum of pH 8.8, whereas the reduced form is optimally active at pH 7.5–8.5 (Figure 2.10). Hence the oxidized enzyme is completely inactive at the pH obtaining in the darkened stroma. This pH sensitivity does not result from changes in V_{max} but is due to the fact that the substrate for the enzyme is the $[FBP^{4-}.Mg^{2+}]^{2-}$ complex and that the concentration of this complex is pH dependent.

Sedoheptulose 1,7-bisphosphatase

$$Sed1,7BP_2 + H_2O \rightarrow Sed7P + P_i$$

The enzyme from spinach is a dimer (M_r 66 kDa) comprising two identical 35 kDa subunits. The enzyme is quite distinct from fructose bisphospha-

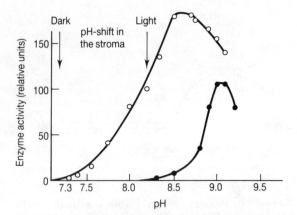

Figure 2.10. pH dependence of the activity of oxidized (●) and reduced (○) stromal fructose 1,6-bisphosphatase, showing how enzyme activation also depends upon the light-dependent pH shift in the stroma. From Baier and Latzko (1975), by permission of Elsevier Science Publishers BV.

tase and is immunologically unrelated, but like it, is light-activated by the thioredoxin system.

Ribulose 5-phosphate kinase

$$Rbu5P + ATP \rightarrow RuBP + ADP + H^+$$

The purified enzyme from wheat and spinach is a dimer of 83 kDa comprising two identical subunits. This enzyme is regulated by the ferredoxin–thioredoxin system. It has been shown that thiol regulation involves intra-subunit reduction of two cysteine residues (Cys-16 and Cys-55). The purified enzyme is very sensitive to pH and redox control, and the oxidized enzyme has an activity at pH 6.8 (conditions approximating to the darkened stroma), which is only 2% of the activity of the reduced enzyme at pH 7.8 (an approximation of the illuminated stroma). The enzyme shows complex control by metabolites, including inhibition by glycerate 3-P^{2-} (K_i 2 mM). Inhibition by glycerate 3-phosphate *in vivo* is enhanced by the decrease in stromal pH upon darkening, which favours the formation of glycerate 3-P^{2-}, rather than the glycerate 3-P^{3-} ion which predominates at the pH of the illuminated stroma.

TRANSPORT ACROSS THE CHLOROPLAST ENVELOPE

The chloroplast has a double envelope (Figure 9.7). The outer envelope contains porins and is freely permeable to quite large molecules (up to 10 kDa). The inner envelope provides a more selective barrier to the movement of compounds such as dicarboxylic and amino acids, sugars and adenylates. Proteins can also be imported from the cytosol. Regulation of the transport of Calvin cycle intermediates across the chloroplast envelope is crucially important to control of their metabolism, since the majority of metabolic intermediates of the Calvin cycle also occur as intermediates in sucrose synthesis, which is an exclusively cytosolic process, or in the oxidative pentose pathway. The chloroplast envelope provides an effective barrier to the movement of most of these compounds between the two cell compartments so that, for example, the metabolism of hexose phosphate in sucrose synthesis and in the Calvin cycle can be separately controlled. If isolated chloroplasts are allowed to fix carbon dioxide *in vitro*, it is apparent (a) that triose phosphate is the only significant product that is transported to the medium and (b) that in order to generate triose phosphate, chloroplasts require a supply of P_i. Chloroplasts, therefore, import P_i and export triose phosphate via a phosphate translocator (Figure 2.11).

Table 2.3. Kinetic constrants of the phosphate translocator in spinach chloroplasts (C_3), in maize mesophyll chloroplasts (C_4) and in pea-root plastids.

Compound	Chloroplasts		Pea-root plastids
	C_3	C_4	
K_m P_i	0.32	0.045	0.18
K_i dihydroxyacetone phosphate	0.20	0.084	0.11
K_i glycerate 3-phosphate	0.36	0.053	0.31
K_i glycerate 2-phosphate	3.8	0.073	–
K_i phosphoenol-pyruvate	4.1	0.086	0.20
K_i glucose 6-phosphate	13.0	–	0.33

Early evidence for the existence of the phosphate translocator came from studies by Walker with isolated chloroplasts. Carbon dioxide fixation is accompanied by a pronounced induction period or 'lag', which largely reflects the time taken for autocatalytic build-up of Calvin cycle intermediates. This lag can be shortened by those intermediates of the Calvin cycle which enter the chloroplast via the phosphate translocator (e.g. glycerate 3-phosphate and triose phosphates, but not hexose phosphates (Table 2.3)) and can be extended by super-optimal concentrations of P_i, which exerts its effect by enforcing the export of intermediates via the translocator, and delaying the process of autocatalysis (Figure 2.12).

Much of the characterization of the phosphate translocator has been done by Heldt and his colleagues. They were able to use techniques of silicone oil centrifugation for the study of transport across the chloroplast envelope that had been pioneered in the study of transport of metabolites into mitochondria. In this method a small plastic centrifuge tube contains a lower layer of perchloric acid, a middle layer of silicone oil (which is inert) and an upper layer containing a suspension of intact chloroplasts. After addition of radiolabelled metabolites to the chloroplasts in

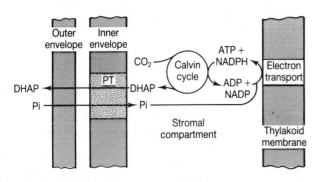

Figure 2.11. The role of the phosphate translocator (PT) in photosynthetic carbon assimilation.

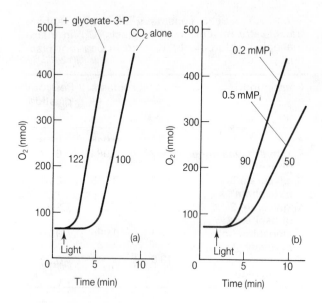

Figure 2.12. (a) Addition of 0.1 mM glycerate 3-phosphate causes a shortening of the induction lag in carbon dioxide-dependent oxygen evolution in isolated wheat chloroplasts. (b) Extension of the lag by super-optimal P_i. Numbers along the traces indicate rates of oxygen evolution in μmol h^{-1} mg^{-1} chlorophyll.

the upper layer, the tube can be centrifuged to remove the chloroplasts, through the silicone oil, to the lower layer, which stops further metabolism. Very short periods of exposure can be attained. The amount of metabolite taken up, and its internal concentration, can then be measured by removing the lower layer.

Properties of the Phosphate Translocator

1. The P_i translocator, with a molecular mass of 29 kDa, accounts for between 10% and 15% of the total protein of the inner envelope.
2. There is a strict counter-exchange of P_i for glycerate 3-P^{2-} or triose phosphate which is mediated by a translocator located in the inner membrane of the chloroplast envelope.
3. In chloroplasts from C_3 plants, such as spinach

and peas, the K_m of the translocator for glycerate 3-phosphate, triose phosphate and P_i is low, but is high for the transport of other sugar phosphates or phosphorylated metabolites (Table 2.3).

4. The translocator only recognizes glycerate 3-P^{2-}. The majority of glycerate 3-phosphate in the illuminated stroma is in the 3^- form, since the pH of the illuminated stroma is about pH 8.0 and the pK for glycerate 3-phosphate is 7.4. Hence only triose phosphate is exported to any appreciable degree under physiological conditions.
5. The properties of the phosphate translocator differ in different systems. In mesophyll chloroplasts of C_4 plants, the translocator also has a high affinity for phosphoenolpyruvate (PEP), a feature which is not observed in the C_3 phosphate translocator. On the other hand, in the plastids of non-photosynthetic cells, such as pea seeds and potato tubers, which store starch, the phosphate translocator readily transports glucose 6-phosphate (Table 2.3).
6. The ability to export triose phosphate and import glycerate 3-phosphate allows the chloroplast to export ATP and reducing equivalents to the cytosol during photosynthesis.

The phosphate translocator plays a crucial role in the control of carbon partitioning. The return of phosphate from the cytosol to the chloroplast is the message that tells the chloroplast how fast sucrose synthesis (or other processes) is using the triose phosphate exported from the chloroplast. If the rate of sucrose synthesis is slowed (by, for example, sucrose accumulation or by the restriction of sucrose synthesis which occurs at low temperatures), the rate at which phosphate enters the chloroplast is reduced and its concentration falls. This has two effects. First, it will tend to stimulate starch synthesis, partly because ADP–glucose pyrophosphorylase is inhibited by P_i. Second, it can limit the production of ATP (Figure 2.11), and hence the rate of photosynthesis, which will be regulated downwards to meet the decreased demand for photosynthate.

Translocation of Other Metabolites Across the Chloroplast Envelope

Besides carbon, the chloroplast is a major site of synthesis of reduced nitrogen and sulphur and the synthesis of amino acids. It is also involved in exchange of other carbon compounds through the process of photorespiration. The principal translocators involved are:

1. Dicarboxylate and amino acid transport. A number of dicarboxylates (malate, succinate, 2-oxoglutarate (2OG), aspartate and glutamate) are transported across the chloroplast envelope. In chloroplasts of C_3 plants uptake of 2-oxoglutarate and release of glutamate is required for NH_3 assimilation via the glutamine synthetase/glutamate synthase (GS/GOGAT) pathway. This occurs by uptake of 2-oxoglutarate in exchange for stromal malate on one translocator, and release of glutamate in exchange for cytosolic malate on another (i.e. exchange of 2-oxoglutarate and glutamate with no net exchange of malate) (Figure 2.14). Glutamine and oxaloacetate, and possibly glycine and serine, are transported on separate carriers, while other amino acids are thought to enter the chloroplast by simple diffusion.
2. Exchange of the photorespiratory intermediates, glycolate and glycerate, occurs on a single counter-exchange translocator (Figure 2.14).
3. In chloroplasts from pea leaves, ATP exchanges for ADP, P_i and PEP. The adenylate translocator is less active in spinach chloroplasts.
4. Although hexose phosphates are not transported at appreciable rates, except in pea-root plastids, slow transport of sugars such as glucose, fructose and the disaccharide maltose occurs into spinach chloroplasts on a specific translocator.

PHOTORESPIRATION

When leaves of C_3 plants are illuminated, they not only fix carbon dioxide via the Benson–Calvin cycle, but they also evolve carbon dioxide in an oxygen-dependent process. This photorespiration (i.e. apparent light-stimulated respiration) occurs at rates which are substantially greater than rates of respiration in the dark, but true (mitochondrial) respiration is not stimulated to any large degree by light. Indeed, the similarities between true respiration and photorespiration extend little further than the terminology. Unlike respiration, photorespiration is an energy-wasting process which results in a decrease in the efficiency of net carbon assimilation. The process leading to carbon dioxide release originates in the bifunctional nature of Rubisco and the generation of glycollate 2-phosphate in the oxygenase reaction.

Inhibition of Carbon Assimilation by Oxygen

In 1920, Otto Warburg observed that photosynthetic carbon dioxide assimilation by cells of *Chlorella* was inhibited by oxygen. The rate of carbon dioxide assimilation by leaves of C_3 plants is similarly inhibited by oxygen, since net carbon dioxide assimilation is stimulated by decreasing the oxygen concentration from air levels (21% oxygen) to 1–2% oxygen (Figure 2.13). The lower net rate of carbon dioxide uptake in air is due to the loss of carbon dioxide through photorespiration which occurs at the same time as photosynthetic carbon dioxide assimilation. Photorespiratory losses can be very high. At temperatures above 30 °C, net carbon assimilation may be depressed by as much as 60%. Similar

Table 2.4. Effect of oxygen and carbon dioxide concentration on growth of a C_3 plant, *Mimulus cardinalis*.

[CO$_2$] (p.p.m.)	Dry weight increase (mg per plant per 10 days)	
	21% O$_2$	2% O$_2$
110	10	150
320	565	1076
640	804	1144

large changes may be observed in plant growth
and productivity when photorespiration is suppressed (Table 2.4). Rates of photorespiration in
C_3 plants vary considerably with environmental
conditions, but range up to 5 μmol carbon dioxide
m^{-2} s^{-1} in saturating light at temperatures
between 20 and 30 °C. Such rates exceed rates
of respiration in the dark by between two- and
fourfold.

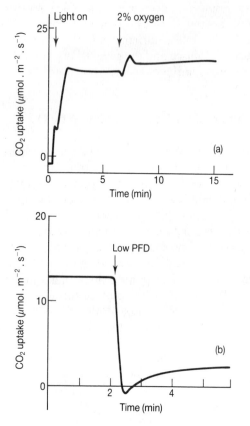

Figure 2.13. Characteristics of gas exchange in a
spinach (C_3) leaf. The vertical axis is the rate of
carbon dioxide uptake. (a) Upon illumination, there
is a rise to a steady rate of carbon dioxide uptake
(induction). If the gas phase is changed from 21%
to 2% oxygen (but carbon dioxide is kept constant
at 350 μl l^{-1}), there is an increase in the rate of
uptake. (b) If a leaf in steady-state photosynthesis is
darkened, or the photon flux density (PFD) is
decreased (here from 1300 to 130 μmol quanta m^{-2}
s^{-1}), a 'carbon dioxide burst' is observed, when the
rate of carbon dioxide evolution briefly exceeds the
subsequent rate of gas exchange.

The Origin of Glycollate

For many years the metabolic origin of glycollate
was unclear. In 1950 Benson and Calvin had
shown that, when *Chlamydomonas* was illuminated in the absence of carbon dioxide following a
period of photosynthesis in $^{14}CO_2$, both glycollate
and glycine became heavily labelled. In darkness
label disappeared from these compounds. In
algae such as *Chlamydomonas* an increase in the
amount of oxygen or a lowering of the concentration of carbon dioxide in the light enhances the
production of glycollate. Any explanation for the
origin of glycollate therefore has to accommodate
the influence of oxygen and carbon dioxide. The
sugar-phosphates of the Calvin cycle were identified as the precursors of glycollate. For example,
when cells of *Chlamydomonas* or isolated chloroplasts are illuminated in the presence of limiting
amounts of carbon dioxide, they convert a large
proportion of their endogenous sugar-phosphates
to glycollate, which appears in the suspending
medium. When $^{14}CO_2$ is supplied the labelling of
sugar-phosphates always precedes that of glycollate.

Wilson and Calvin proposed that the origin of
glycollate lay in the transketolase reaction (Figure
2.1). In this reaction, a two-carbon moiety
(CHO—CHOH) transferred between sugar-
phosphates is attached to thiamine pyrophosphate. In this intermediate state the two-carbon
moiety is susceptible to oxidation to glycollate *in
vitro* by oxidants such as hydrogen peroxide,
although oxygen is not incorporated directly.
However, subsequent evidence, notably the discovery of the oxygenase activity of Rubisco by
Bowes, Hageman and Ogren, has shown convinc-
ingly that the transketolase reaction is not the *in
vivo* source of glycollate.

1. If leaves or isolated chloroplasts are illuminated in the presence of $^{18}O_2$, ^{18}O is incorporated into the carboxyl group of glycollate.
   The enrichment of ^{18}O in glycollate
   approaches that of the oxygen supplied within
   a few seconds. This evidence is consistent with
   the operation of Rubisco as an oxygenase and
   the formation of glycollate as an immediate

product. The direct incorporation of oxygen into glycollate is not consistent with an origin in the transketolase reaction.

2. If triose phosphate is supplied to isolated chloroplasts, then light-driven ATP synthesis is necessary for glycollate formation from it. These data suggest that ribulose 5-phosphate or RuBP must be the precursors of glycollate, rather than other sugar-phosphates of the Benson–Calvin cycle which may be generated from triose phosphate in the absence of ATP.

3. Chloroplasts contain a specific phosphatase for glycollate 2-phosphate. Mutants of *Arabidopsis thaliana* lacking this enzyme accumulate glycollate 2-phosphate in air, but do not form any glycollate, showing that glycollate 2-phosphate is the only significant precursor of glycollate *in vivo*.

All of the glycollate generated in photosynthesis is therefore believed to originate in the oxygenation of RuBP. Once generated by the oxygenase reaction, glycollate 2-phosphate cannot be utilized in the reactions of the Benson–Calvin cycle or within the chloroplast. It is salvaged in a sequence of reactions which involves cooperation between several subcellular compartments: the chloroplasts, the mitochondria and the peroxisomes (Figure 2.14). The important feature of this mechanism for salvaging glycollate 2-phosphate is that it is inefficient. Not only is 25% of the carbon in glycollate 2-phosphate lost as carbon dioxide, but NH_3 is also released when glycine is decarboxylated to serine.

The Recovery of Glycollate

Glycollate 2-phosphate generated by Rubisco is rapidly hydrolysed within the chloroplast stroma by a specific phosphatase. Glycollate is trans-

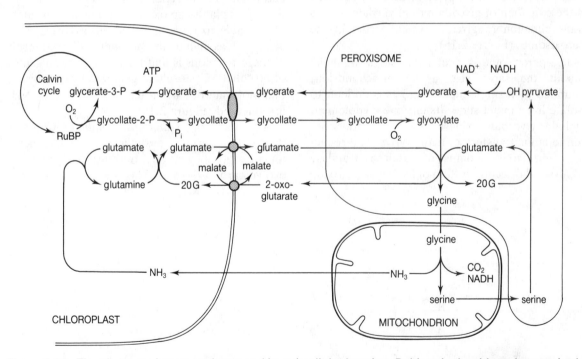

Figure 2.14. The photorespiratory pathway and its subcellular location. Rubisco in the chloroplast can both oxygenate and carboxylate RuBP. Oxygenation leads to the generation of glycollate 2-phosphate from which glycerate 3-phosphate is regenerated in the photorespiratory pathway. During this process, one quarter of the carbon in glycollate is lost as carbon dioxide, together with half the nitrogen, as NH_3, in glycine.

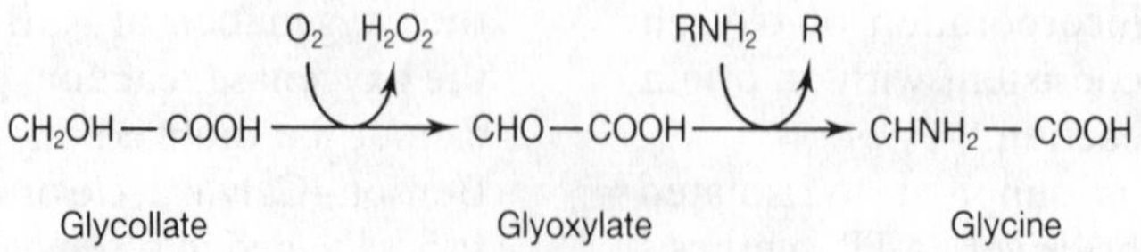

$$O_2 \quad H_2O_2 \qquad\qquad RNH_2 \quad R$$

$$CH_2OH—COOH \longrightarrow CHO—COOH \longrightarrow CHNH_2—COOH$$

Glycollate   Glyoxylate   Glycine

Figure 2.15.

ported out through the chloroplast envelope on a specific carrier in exchange for glycerate. The amounts of glycollate 2-phosphate and glycollate present in leaves are usually rather low, although some algae and bacteria excrete relatively massive amounts of glycollate when transferred to conditions in which carbon dioxide is strongly limiting, or if the concentration of oxygen is suddenly increased. This excretion of glycollate presumably occurs because the algae lack the capacity to metabolize glycollate when grown under conditions in which photorespiration is low or absent.

The next stage in the photorespiratory pathway is the oxidation of glycollate to glyoxylate and its transamination to glycine, which occurs in the peroxisomes (Figure 2.15).

The peroxisome is a relatively small organelle (about the same size as a mitochondrion) bounded by a single membrane. It is difficult to isolate in an intact state. Its enzyme complement includes glycollate oxidase, glyoxylate reductase, serine aminotransferase, hydroxypyruvate reductase and large amounts of catalase. Catalase catalyses the decomposition of hydrogen peroxide (which is toxic and is generated by the oxidation of glycollate) to water:

$$2H_2O_2 \xrightarrow{\text{catalase}} 2H_2O + O_2$$

The next stage in the photorespiratory pathway is the movement of glycine from the peroxisomes to the mitochondria, where glycine is decarboxylated to serine (Figure 2.16).

The decarboxylation of glycine is an oxidative process which involves transfer of a methylene (—CH_2—) group to tetrahydrofolate. In isolated mitochondria it is coupled to electron transport and results in the generation of ATP. Mitochondria of C_3 plants can oxidize glycine at high rates, comparable to rates of oxidation of Krebs cycle acids such as malate and succinate. The net result of these reactions is that for every two molecules of glycollate 2-phosphate generated by Rubisco, one molecule of carbon dioxide and one of NH_3 are released (Figure 2.16).

A mutant of *Arabidopsis thaliana* which is deficient in serine transhydroxymethylase accumulates glycine in air, but also continues to evolve carbon dioxide. This may be due to non-

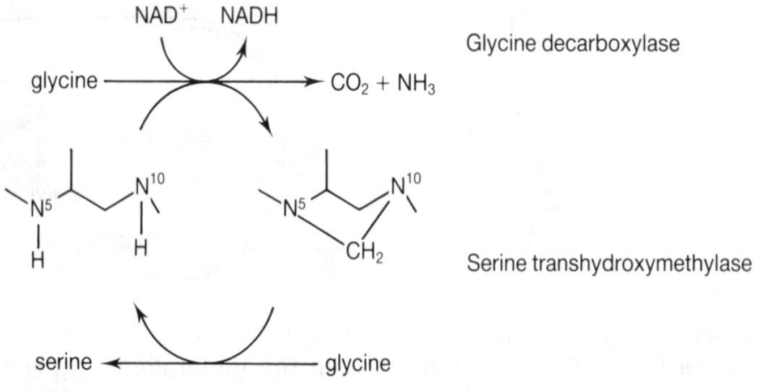

$$NAD^+ \quad NADH$$

glycine $\longrightarrow$ $CO_2 + NH_3$

Glycine decarboxylase

Serine transhydroxymethylase

serine $\longleftarrow$ $\longleftarrow$ glycine

Figure 2.16.

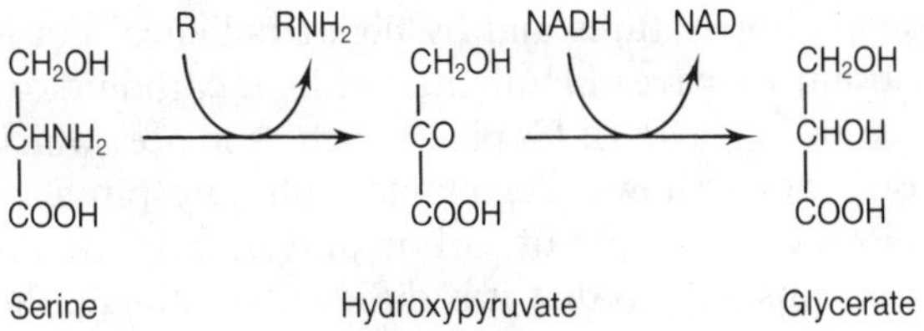

Figure 2.17.

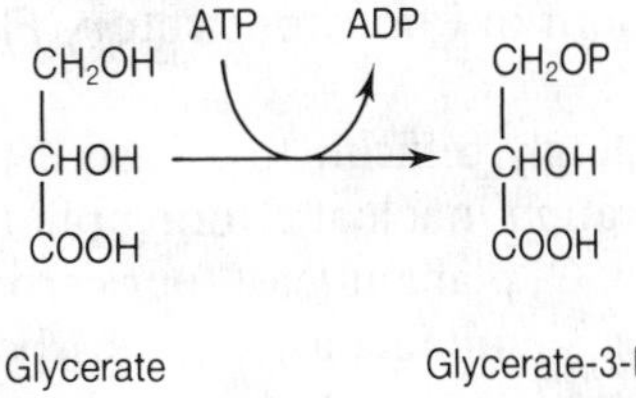

Figure 2.19.

enzymic decarboxylation of glyoxylate to formate (e.g. by hydrogen peroxide). However, this carbon dioxide evolution in the mutant ceases if the leaves are supplied with ammonia, indicating that alternative mechanisms for the release of carbon dioxide occur only when there is a shortage of amino-group donors. The last stage of the photorespiratory pathway is conversion of serine to glycerate 3-phosphate (Figure 2.17). These reactions can be seen if leaves or isolated chloroplasts are illuminated in the presence of $^{18}O_2$ and the $^{18}O_2$ content is measured in other intermediates, it is found that glycollate is labelled first and is

followed by labelling of glycine and then serine. Glycine and serine rapidly attain the same ^{18}O enrichment as glycollate (Figure 2.18). Like glycollate, they are labelled only in the carboxyl groups. This evidence suggests a precursor–product relationship between these intermediates. Glycerate and glycerate 3-phosphate are labelled more slowly still, but never reach the same specific activity as glycollate, which is fully consistent with the dual origin of glycerate 3-phosphate both in the oxygenation and carboxylation of RuBP.

The last reaction in the photorespiratory pathway occurs following uptake of glycerate from the cytosol across the chloroplast envelope and into the chloroplast stroma, in exchange for glycollate. This is the phosphorylation of glycerate, catalysed by glycerate kinase (Figure 2. 19). Three-quarters of the carbon which was originally incorporated into phosphoglycollate has therefore been salvaged in this pathway, and the glycerate 3-phosphate formed can re-enter the Calvin cycle.

The Reassimilation of NH_3

Apart from the loss of carbon dioxide, NH_3 is also released in photorespiration at rates equal to the photorespiratory loss of carbon dioxide (Figure 2.14). Since NH_3 is both a precious resource and is also toxic to plant tissues, the leaf must have an efficient system for its recovery. This is done by the GS/GOGAT system (chapter 7). The rates of reassimilation of NH_3 required are an order of magnitude higher than the rates of primary net assimilation of NH_3 (e.g. after reduction of nitrate). The photorespiratory nitrogen cycle can therefore be regarded as a major pathway of nitrogen metabolism in the leaves of C_3 plants.

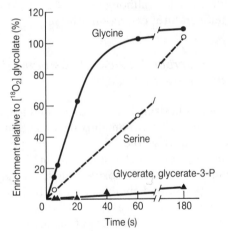

Figure 2.18. Kinetics of labelling of photorespiratory intermediates following a pulse of $^{18}O_2$ to a spinach leaf. The enrichment in each compound is expressed rëative to the enrichment of glycollate by $^{18}O_2$. Note that glycine rapidly attains the same enrichment as glycollate, and is followed by serine. Glycerate 3-phosphate does not reach the same degree of enrichment, because it is also generated in the carboxylation reaction by Rubisco. From Berry et al, 1978, by permission of the American Society of Plant Physiologists.

Measurement of Photorespiratory Fluxes

It is exceedingly difficult to measure the rate of photorespiration in a leaf. During photosynthesis in a leaf of a C_3 plant at least four carbon dioxide fluxes occur simultaneously: (a) carbon dioxide fixation in the Benson–Calvin cycle; (b) carbon dioxide release by respiration (i.e. glycolytic and mitochondrial activity); (c) carbon dioxide release in photorespiration through glycine decarboxylation; and (d) carbon dioxide fixation by PEP carboxylase. Although the magnitude of respiration in the light has been, and remains, a controversial subject, there is little doubt that respiration in the light does occur and that it *may* occur at rates equal to those which occur in darkness. While the chloroplast may export both ATP and reducing equivalents to the cytosol via the triose phosphate/glycerate 3-phosphate shuttle, activity of the Krebs (tricarboxylic) acid cycle in the mitochondria is necessary for the provision of carbon skeletons for biosynthesis. Several methods can give a qualitative idea of the magnitude of photorespiration, but a precise measurement of the flux through the photorespiratory pathway is much more difficult. Most methods are open to objection simply because they involve perturbation of steady-state photosynthesis in the leaf. Thus photorespiratory intermediates will not be metabolized in the same manner and at the same rates under the altered conditions, e.g. in low light, in darkness or in 2% oxygen.

Post-illumination Carbon Dioxide Burst

When photosynthesizing leaves are darkened (or the photon flux density is decreased), they display an initially rapid evolution of carbon dioxide (the 'carbon dioxide burst') which decreases after a few minutes to a steady dark rate of respiratory carbon dioxide uptake (Figure 2.13). Carbon dioxide evolution upon darkening is probably due to continued turnover of endogenous pools of glycine and glycollate which have been built up during photorespiration in the light. The carbon dioxide burst is increased in leaves of C_3 plants by high temperatures and by high irradiance, but is absent in leaves illuminated in high carbon dioxide or in leaves of C_4 plants such as maize, which do not show detectable photorespiration. However, the rate of carbon dioxide evolution is not constant, so that it is difficult to estimate the maximum rate. The method also presupposes that the pool sizes of glycine, glycollate, etc. bear a linear relation to the flux through the photorespiratory pathway in the light.

Stimulation of Photosynthetic Carbon Assimilation by Decreased Oxygen

A change in the oxygen concentration from 21% to 2% increases the rate of carbon dioxide assimilation in leaves of C_3 plants (Figure 2.13), but usually has little influence on the rate of carbon dioxide uptake by leaves of C_4 plants. Photorespiration is almost entirely suppressed in 2% oxygen and the transition between air and 2% oxygen can give a measure of the previous rate of photorespiration, although other factors, notably low temperature, can modify the response.

Efflux of Carbon Dioxide into Carbon Dioxide-free Air

If photosynthesizing leaves are exposed to carbon dioxide-free air, carbon dioxide evolution continues, reflecting the dissipation of pools of photorespiratory metabolites as in the post-illumination carbon dioxide burst. A slightly more sophisticated modification of this method requires exposure of leaves to $^{14}CO_2$, sufficient to achieve steady-state labelling of the intermediates of the photosynthetic and photorespiratory cycles, followed by measurement of the release of $^{14}CO_2$ into carbon dioxide-free air in the light or in darkness.

Uptake of $^{14}CO_2$

If, under steady-state conditions, a leaf photosynthesizing in $^{12}CO_2$ is suddenly exposed to

$^{14}CO_2$, the initial rate of $^{14}CO_2$ uptake is greater than the net rate of $^{12}CO_2$ uptake, the difference being due to the loss of carbon dioxide in photorespiration. Thus the specific activity of $^{14}CO_2$ supplied will gradually fall as it is diluted by $^{12}CO_2$ released during photorespiration. The method relies upon the presence of a substantial lag between carbon dioxide uptake and its release in the photorespiratory pathway. The times commonly employed are 15–30 s but even these short times are not sufficient to ensure that none of the $^{14}CO_2$ incorporated during such a time has been re-released in the photorespiratory pathway. Moreover, the shorter the time of exposure to $^{14}CO_2$ the greater the errors involved in determining the specific activity of the $^{14}CO_2$ supplied in the leaf, because of the time taken for uniform mixing with $^{12}CO_2$ already present inside and outside the leaf. This method does have the advantage that no disturbance of steady-state photosynthesis occurs.

Uptake of $^{18}O_2$

Oxygen uptake by leaves can be distinguished from oxygen evolution by the use of two isotopes of oxygen, $^{16}O_2$ and $^{18}O_2$, and measurement in a mass spectrometer. However, uptake of $^{18}O_2$ by leaves of C_3 plants is only partially inhibited by raising the concentration of carbon dioxide, because oxygen is also reduced directly by the electron transport chain in the Mehler reaction, which may occur at rates comparable to the rate of photorespiration.

Photorespiratory Mutants

Since photorespiration can be entirely inhibited in leaves by provision of high carbon dioxide or low oxygen, mutants with defects in the photorespiratory pathway can easily be identified since they grow normally in carbon dioxide-enriched air but die in air. Seeds treated with a mutagen, such as azide, are grown in a carbon dioxide-enriched atmosphere (2% CO_2). Those with a lesion in the photorespiratory pathway can be recognized by symptoms which are apparent upon transfer to air. Such plants are immediately transferred back to carbon dioxide-enriched air for diagnosis of the mutation. Several mutants of *Arabidopsis thaliana* and of barley have been isolated (Table 2.5).

Besides providing further evidence for the pathway of photorespiration, these mutants have also served to emphasize several points. The first is that inhibition of the photorespiratory pathway is lethal. Once generated in the oxygenase reaction, glycollate must be retrieved. This stresses the point that chemical intervention in the photorespiratory pathway will also be lethal. Another is the lack of mutants which have Rubisco oxygenase activity lowered or absent. This is consistent with the view that oxygenase activity is entirely unavoidable because it is an inevitable consequence of the reaction mechanism of carboxylation.

Table 2.5. Some photorespiratory mutants found in *Arabidopsis thaliana* and in barley. All the lesions are lethal when the plants are grown in air. Note that several of the lesions involve deficiencies in the recovery of ammonia released in photorespiration.

Lesion	Symptoms
Phosphoglycollate phosphatase	Glycollate 2-P accumulates in the chloroplast and depletes stromal P_i
Catalase	Hydrogen peroxide generated in the peroxisomes accumulates and destroys cell
Serine trans-hydroxymethylase	Accumulates glycine in air
Dicarboxylic acid translocator	Inability to transfer 2-oxoglutarate between cytosol and chloroplast; NH_3 accumulates
Glutamate synthase	Accumulation of glutamine
Glutamine synthetase	Accumulation of NH_3

Factors Affecting the Rate of Photorespiration

The rate of photorespiration is promoted by increases in temperature and irradiance and by an increase in the ratio of oxygen : carbon dioxide. As we have seen, the change in the rate of photorespiration with changing ratios of oxygen : carbon dioxide is due to competition between these substrates at the active site of Rubisco. Hence photorespiration is largely suppressed if the oxygen concentration is reduced relative to the concentration of carbon dioxide or if the carbon dioxide concentration is increased relative to the concentration of oxygen. These changes in the rate of photorespiration are evident in increased rates of net carbon dioxide uptake as well as enhanced growth (Figure 2.13, Table 2.4).

The photorespiratory rate increases with light intensity because the underlying rate of photosynthesis is higher. The rate of photorespiration also increases as the temperature is raised. This is partly due to higher rates of photosynthesis at higher temperatures, but at higher temperatures photorespiration increases markedly as a proportion of total photosynthesis. The influence of temperature on the rate of photorespiration can be explained in terms of changes in the relative concentrations of oxygen and carbon dioxide. Carbon dioxide is not only less soluble at higher temperatures but oxygen is relatively more soluble than carbon dioxide as the temperature is raised, and oxygenation is therefore favoured at higher temperatures (Table 2.6). Temperature-dependent changes in the kinetic constants of Rubisco may also underlie the increase in the rate of photorespiration at higher temperatures.

Table 2.6. Solubilities of oxygen and carbon dioxide in water in equilibrium with air at different temperatures.

Temperature	Solubility (μM)		Ratio
	21% O_2	0.035% CO_2	
10	348	17	20.5
20	299	13	23.0
30	230	9	25.5
40	224	8	28.0

The Role of Photorespiration

At first sight, C_3 plants do not appear to need photorespiration. They grow quite normally in carbon dioxide-enriched air; indeed, carbon dioxide enrichment (to approx. 1000 μl l^{-1} carbon dioxide in air) is commonly employed to improve the yields of glasshouse crops such as tomatoes, in which yield increases may be as much as 50%. Similarly, mutants of *Arabidopsis thaliana* and of barley which are deficient in an enzyme of the photorespiratory pathway grow normally in carbon dioxide-enriched air. It can therefore be inferred that photorespiration is not a necessary pathway for the biosynthesis of intermediates such as glycine and serine. It is apparent that the cost of net fixation of carbon dioxide rises enormously at high rates of photorespiration. In the absence of oxygenation, fixation of each molecule of carbon dioxide into triose phosphate via the Calvin cycle requires three ATP and two NADPH molecules (Figure 2.1). As the compensation point is approached, when carbon dioxide fixation is equalled by the rate of carbon dioxide released, the cost of net fixation of carbon dioxide (i.e. the cost of manufacturing carbohydrate) tends towards infinity, but at the compensation point ten ATP and six NADPH molecules are still required to fix carbon dioxide into RuBP and to recycle the photorespired carbon dioxide. The fact that photorespiration can work at the compensation point, dissipating energy, but merely recycling carbon dioxide, may be of benefit to the leaf, preventing a phenomenon known as photoinhibition. When leaves are brightly illuminated in the absence of carbon dioxide they become photoinhibited because light energy which is absorbed cannot be utilized efficiently. In the absence of carbon fixation, the energy is dissipated by reduction of oxygen, with the generation of toxic oxygen radicals and, when all reaction centres are reduced, by photodestruction of pigments. In the short term such processes are irreversible and lead to a loss of photosynthetic capacity. The value of recycling carbon dioxide through the photorespiratory pathway may lie in harmless dissipation of light energy when the intercellular carbon dioxide concentration is low, as might occur, for example, when stomata close during water stress. In C_4 plants, which lack the capacity for high rates of photorespiration, the intercellular concentration of carbon dioxide must fall to much lower concentrations than in C_3 plants before the rate of carbon dioxide assimilation is reduced, so that photorespiratory recycling of carbon dioxide may not be necessary.

In summary, photorespiration can potentially cause a large loss of carbon in C_3 species. Up to 60% of newly fixed carbon may be re-released as carbon dioxide in environments which favour photorespiration, for example, in high light and at high temperatures obtaining in tropical and subtropical climates, or at the low concentrations of carbon dioxide which may occur in some aquatic environments. Because atmospheric concentrations of CO_2 have, in the past, been even later than they are now, there has therefore been considerable selection pressure for mechanisms which can reduce photorespiration. Since carbon dioxide and oxygen compete for the same active site on Rubisco, one obvious strategy for reducing photorespiration is the development of a means of concentrating carbon dioxide relative to oxygen at the site of carboxylation. Several mechanisms have evolved independently and in unrelated families. These include C_4 photosynthesis, crassulacean acid metabolism and algal carbon dioxide-concentrating mechanisms (chapter 3).

REFERENCES

1. Baier, D. and Latzko, E. (1975) Properties and regulation of C-1-fructose-1, 6-diphosphatase from spinach chloroplasts. *Biochim. Biophys. Acta,* **396**, 141–148.
2. Berry, J.A., Osmond, C.B. and Lorimer, G.H. (1978) Fixation of $^{18}O_2$ during photorespiration. *Plant Physiol.,* **62**, 954–967.
3. Calvin, M., Bassham, J.A., Benson, A.A., Lynch, V.H., Ouellet, C., Schou, L., Stepka, W. and Tolbert, N.E. (1951). Carbon dioxide fixation and photosynthesis. SEB Symposium **5**, 284–305.
4. Leegood, R.C. (1990) Enzymes of the Calvin cycle. In *Methods in Plant Biochemistry* (ed. P.J. Lea), Vol 3, pp. 15–37. Academic Press, London.
5. Portis, A.R. (1988) Purification and assay of Rubisco activase from leaves. *Plant Physiol,* **88**, 1008–1014.

FURTHER READING

Andrews, T.J. and Lorimer, G.H. (1987) Rubisco: Structure, mechanisms, and prospects for improvement. In *The Biochemistry of Plants,* Vol. 10 (eds M.D. Hatch and N.K. Boardman), pp. 132–218. Academic Press, New York.

Bassham, J.A. and Calvin, M. (1957) *The Path of Carbon in Photosynthesis.* Prentice-Hall, New Jersey.

Burris, R.H. and Black, C.C. (eds) (1975) *CO_2 Metabolism and Plant Productivity.* University Park Press, Baltimore.

Edwards, G.E. and Walker, D.A. (1983) C_3, C_4: Mechanisms, and Environmental Regulation of, Photosynthesis. Blackwell Scientific, Oxford.

Flügge, U.-I. and Heldt, H.W. (1991) Metabolite translocators of the chloroplast envelope. *Annual Review of Plant Physiology and Molecular Biology,* **42**, 129–144.

Heber, U. and Heldt, H.W. (1981) The chloroplast envelope: Structure, function and role in leaf metabolism. *Annual Review of Plant Physiology,* **32**, 139–168.

Lea, P.J., Robinson, S.A. and Stewart, G.R. (1990) The enzymology and metabolism of glutamine, glutamate and asparagine. In *The Biochemistry of Plants,* Vol 16 (eds B.J. Miflin and P.J. Lea), pp. 121–159. Academic Press, San Diego.

Leegood, R.C., Walker, D.A. and Foyer, C.H. (1985) Regulation of the Benson–Calvin cycle. In *Photosynthetic Mechanisms and the Environment* (eds J. Barber and N.R. Baker), pp. 189–258. Elsevier, Amsterdam.

Lorimer, G.H. and Andrews, T.J. (1980) The C_2 chemo- and photorespiratory carbon oxidation cycle. In *Biochemistry of Plants* (eds M. D. Hatch and N.K. Boardman), pp. 329–374. Academic Press, New York.

Ogren, W.L. (1984) Photorespiration: pathways, regulation and modification. *Annual Review of Plant Physiology,* **35**, 415–442 (1984).

Portis, A.R. (1992) Regulation of ribulase 1,5-bisphosphate carboxylase/oxygenase activity. *Annual Review of Plant Physiology,* **43**, 415—437.

Salvucci, M.E. (1989) Regulation of Rubisco activity *in vivo. Physiologia Plantarum,* **77**, 164–171.

Woodrow, I.E. and Berry, J.A. (1988) Enzymatic regulation of photosynthetic CO_2 fixation. *Annual Review of Plant Physiology,* **39**, 533–594.

3

CARBON DIOXIDE-CONCENTRATING MECHANISMS

R.C. Leegood

Department of Animal and Plant Sciences, University of Sheffield, UK

C_4 PLANTS: THE ELIMINATION OF PHOTORESPIRATION

C_4 plants possess a carbon dioxide pump which generates an elevated concentration of carbon dioxide in the vicinity of Rubisco. The carbon dioxide pump is based upon a cycle of carboxylation and decarboxylation involving the generation of C_4 acids, and is split between two compartments: the mesophyll cells and the bundle-sheath cells. The bundle-sheath cells form a relatively gas-tight compartment in which carbon dioxide is

Plant Biochemistry and Molecular Biology. Edited by P.J. Lea and R.C. Leegood
© 1993 John Wiley & Sons Ltd

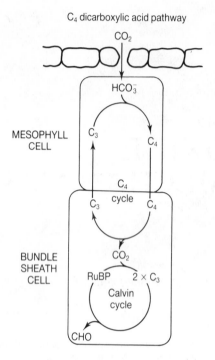

C_4 dicarboxylic acid pathway

CO_2

HCO$_3^-$

MESOPHYLL CELL

C_3

C_4

C_4 cycle

C_3

C_4

BUNDLE SHEATH CELL

CO_2

RuBP   2 × C_3

Calvin cycle

CHO

Figure 3.1. Simplified version of the pathway of carbon dioxide fixation in leaves of C_4 plants, showing the spatial separation of fixation of carbon dioxide by PEP carboxylase prior to fixation in the Calvin cycle. From Leegood and Osmond, 1990[1], by permission of Longman.

concentrated for assimilation via Rubisco and the Calvin cycle (Figure 3.1). The biochemistry of the C_4 pathway is, therefore, tightly integrated with anatomical adaptations.

C_4 species are predominantly tropical and sub-tropical and occur in 17 families of higher plants. They include a large number of important crop species such as maize, millet, sorghum and sugar-cane, as well as eight out of the ten worst weeds of the world.

The Elucidation of the C_4 Pathway

Investigations Using $^{14}CO_2$

In the mid-1960s investigations of the pathway of carbon dioxide fixation in sugar-cane and in other tropical grasses were made by Hatch and Slack in Australia. These followed the findings by Kortschak and his colleagues in Hawaii and Karpilov and his colleagues in Russia that short-term photosynthetic fixation of $^{14}CO_2$ in maize and sugar-cane resulted in extensive formation of [^{14}C]malate. In the work of Hatch and Slack, $^{14}CO_2$ was fed to leaves which were undergoing steady-state photosynthesis, similar to Calvin's experiments with algae (chapter 2). When leaves were pulsed with $^{14}CO_2$ it was found that four-carbon acids such as malate and aspartate were the first stable products of carbon dioxide fixation—hence the term 'C_4' photosynthesis. The curve for percentage incorporation of label for C_4 acids (malate and aspartate) extrapolated back to 100% of the $^{14}CO_2$ incorporated at the beginning of the pulse (Figure 3.2). The kinetics of labelling of glycerate 3-phosphate (the three-carbon product of Rubisco) indicated that no ^{14}C was incorporated into it at the beginning of the pulse, in contrast to the kinetics of labelling observed in C_3 plants, in which glycerate 3-phosphate is the first product of carbon fixation (chapter 2).

In other experiments, leaves were pulsed with $^{14}CO_2$ for about 30 s and then transferred back into $^{12}CO_2$. By this means the incorporation into products of the $^{14}CO_2$ incorporated in the pulse could be followed during the 'cold ($^{12}CO_2$)' chase. At the onset of the chase label was rapidly lost from the C_4 acids such as malate, while incorporation of label into glycerate 3-phosphate and hexose phosphates continued for a short time and then declined, consistent with these compounds being intermediates in the pathway of carbon assimilation and accepting ^{14}C from the C_4 acids. ^{14}C continued to accumulate in starch and sucrose during the chase, as would be expected of compounds which are the end-products of carbon assimilation in the leaf (Figure 3.2).

Detailed analysis of the labelling of the individual carbon atoms of malate and glycerate 3-phosphate during the pulse with $^{14}CO_2$ showed that malate was labelled first in the C4 position (Table 3.1). Label spread to the other carbons in

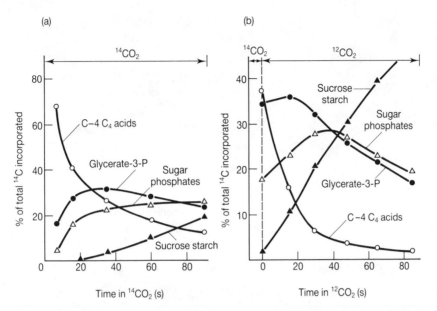

Figure 3.2. Kinetic analysis of the pathway of incorporation of $^{14}CO_2$ in leaves of maize (*Zea mays*). (a) Time-course of $^{14}CO_2$ incorporation; (b) pulse-chase in $^{12}CO_2$ following 35 s in $^{14}CO_2$. Both are plotted as a percentage of total ^{14}C incorporated. From Hatch and Osmond, 1976, by permission of Springer-Verlag.

malate more slowly than to carbon 1 of glycerate 3-phosphate implying that glycerate 3-phosphate was labelled by transfer from the C4 carboxyl group of malate.

Although most plant tissues, photosynthetic and non-photosynthetic alike, show labelling of C_4 acids by $^{14}CO_2$ in the dark, labelling of C_4

Table 3.1. Percentage distribution of ^{14}C in the individual carbon atoms of glycerate 3-phosphate and malate during a pulse of $^{14}CO_2$ given to an illuminated maize leaf.

Time in $^{14}CO_2$ (s)	Malate			Glycerate 3-phosphate	
	CI	C2 + C3	C4	CI	C2 + C3
3	10	0	90	98	2
10	16	0	84	96	4
150	33	25	42	45	55

acids is small, variable and not light dependent in the leaves of C_3 plants. In contrast, the labelling of C_4 acids in the leaves of C_4 plants is massive and is largely light dependent.

The rapid labelling of carbon atom 4 of malate by $^{14}CO_2$ is consistent with carboxylation catalysed by phosphoenolpyruvate (PEP) carboxylase, and subsequent reduction of the oxaloacetate formed to malate or transamination to form aspartate (Figure 3.3). (Oxaloacetate is not shown in Figure 3.2 because leaves usually contain a very small pool of oxaloacetate which turns over extremely rapidly. Oxaloacetate is also unstable. However, if the killing and extraction procedures are modified so as to preserve oxaloacetate, as its 2,4-dinitrophenyl hydrazone derivative, its behaviour is consistent with its being the first product of carbon dioxide fixation in C_4 plants.)

PEP carboxylase probably occurs universally in plant cells, which accounts for the labelling of C_4 acids even in darkened plant tissues. In the leaves of C_4 plants, however, the activity of PEP car-

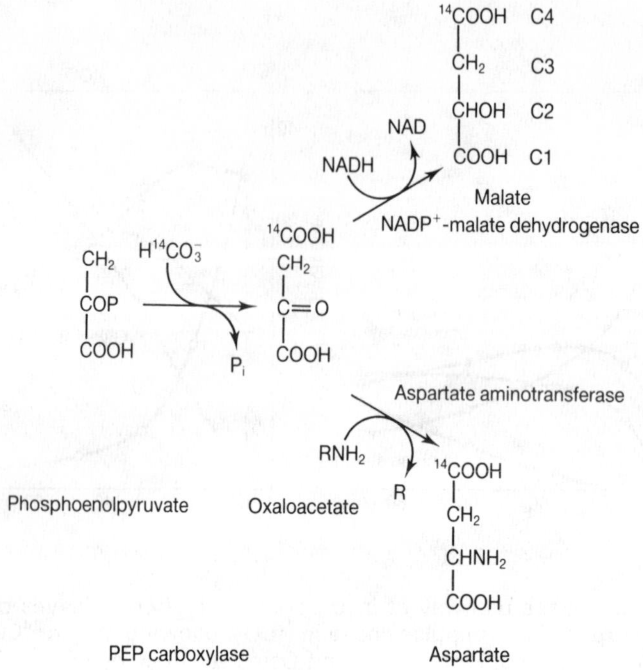

$^{14}COOH$ C4

CH_2 C3

$CHOH$ C2

$COOH$ C1

Malate

NAD

NADH

$NADP^+$-malate dehydrogenase

$^{14}COOH$

CH_2

$C=O$

$COOH$

$H^{14}CO_3^-$

P_i

CH_2

COP

$COOH$

Aspartate aminotransferase

RNH_2

R

$^{14}COOH$

CH_2

$CHNH_2$

$COOH$

Phosphoenolpyruvate   Oxaloacetate

PEP carboxylase   Aspartate

Figure 3.3.

boxylase can be as much as a hundredfold greater than in the leaves of C_3 plants (Table 3.2) and is more than adequate to account for observed rates of carbon dioxide fixation by the leaves of C_4 plants. Unlike Rubisco, PEP carboxylase utilizes HCO_3^- rather the carbon dioxide as substrate. The affinity for HCO_3^-) is high ($K_m(HCO_3^-)$ is $10 \, \mu M$ (equivalent to a carbon dioxide concentration of $2 \, \mu M$ at pH 7.0)) and, like Rubisco, the equilibrium position of the reaction is favourable

Table 3.2. Distribution of enzymes between the mesophyll and bundle-sheath of the leaves of C_4 plants.

Enzyme	Activity ($\mu mol \, h^{-1} \, mg^{-1}$ chlorophyll)		
	Mesophyll cells	Bundle-sheath cells	C_3 species
Enzymes of the C_4 cycle			
PEP carboxylase	2000	25	25–100
pyruvate P_i dikinase	230	7	n.d.
NADP–malate dehydrogenase	600	n.d.	30–300
NADP–malic enzyme	2–40	320–1000	50
Enzymes of the Calvin cycle			
Rubisco	5–24	260–560	200–500
ribulose 5–phosphate kinase	50	1400–3800	1000
ribose 5–phosphate isomerase	50	970	1000
fructose 1,6–bisphosphatase	11	100	300

n.d., not detectable.

$(\Delta G = -29 \text{ kJ mol}^{-1})$. However, the crucial feature of PEP carboxylase, when compared with Rubisco, is that it does not act as an oxygenase.

Leaf Anatomy in C_4 Plants

In the majority of C_4 plants the photosynthetic cells are organized in two concentric cylinders. Although the vascular bundle of C_3 grasses is surrounded by two concentric rings of bundle-sheath cells, the inner thick-walled mesotome contains no chloroplasts and the outer paren-chyma sheath contains few chloroplasts, which play no special role in carbon assimilation. C_4 grasses and dicots have a single or double bundle-sheath, which is distinguished from the C_3 bundle-sheath by the abundance of large chloroplasts. Thin-walled mesophyll cells which also contain chloroplasts radiate out from the bundle-sheath (Figure 3.4). The cell types are arranged so that a mesophyll cell is never more than one cell removed from a bundle-sheath cell. The bundle-sheath cell walls are heavily thickened, are often suberized and are rich in plasmodesmatal con-nections with the mesophyll cells. Only the meso-phyll cells are in contact with the intercellu-

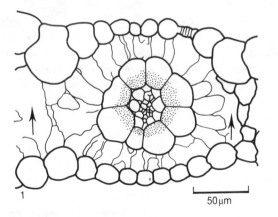

Figure 3.4. Transverse section of the leaf of an NAD^+–malic enzyme-type C_4 grass (*Panicum miliaceum*). Note the thick-walled bundle-sheath and the thin-walled, radially arranged mesophyll cells. The chloroplasts are arranged in a centripetal position in the bundle-sheath. Arrows indicate intercellular air spaces.

lar air spaces. This arrangement of the chloren-chyma is termed Kranz anatomy (Kranz means 'wreath' in German).

This distinctive leaf anatomy has two particu-larly important consequences. The first is that carbon dioxide may be concentrated within the bundle-sheath, whose thickened walls have an extremely low permeability to gases and sub-strates. The second is that, because the majority of mesophyll cells are immediately adjacent to bundle-sheath cells, division of labour and cooperation between the two cell types is possi-ble.

Enzyme Distribution Between the Cell Types

Measurement of enzyme activity in the leaf tis-sues of C_4 plants reveals a profound difference in the enzyme complement between the mesophyll and the bundle-sheath cells. Since the bundle-sheath is very resistant to mechanical disruption, it is relatively easy to extract the leaf sequentially—the thin-walled mesophyll cells being broken first, followed after harsher grinding by the release of the contents of the bundle-sheath. In the early 1970s, enzymic digestion pro-cedures were developed which allowed the selec-tive digestion of the thin mesophyll cell walls but not of the bundle-sheath by the use of cellulase and pectinase. Pure fractions of each cell type, mesophyll protoplasts and bundle-sheath strands, were isolated and the enzymic distribution deter-mined. It is clear that the mesophyll cells contain all of the PEP carboxylase in the leaf and that the bundle-sheath cells contain all of the Rubisco (Table 3.2) and most of the enzymes of the Calvin cycle.

The carboxylation of PEP thus occurs in the mesophyll cells, with the formation of OAA and malate (or aspartate); malate is transferred to the bundle-sheath cells where it is decarboxylated to form pyruvate, a reaction which in maize is catalysed by $NADP^+$–malic enzyme in the chlor-oplast:

$$malate + NADP^+ \rightarrow pyruvate + NADPH + CO_2$$

Subsequently, carbon dioxide is fixed in the Calvin cycle. The pyruvate is then returned to the mesophyll chloroplast, where it is phosphorylated to PEP, thus completing the cycle of carboxylation and decarboxylation. The phosphorylation of pyruvate is catalysed by the chloroplastic enzyme, pyruvate phosphate dikinase, an enzyme which had previously been known only in bacteria, but is now known to occur in all C_4 and in some C_3 and CAM plants:

$$\text{pyruvate} + \text{ATP} + P_i \xrightarrow[\text{pyruvate } P_i \text{ dikinase}]{} \text{PEP}$$
$$+ \text{AMP} + PP_i$$

$$PP_i \xrightarrow[\text{pyrophosphatase}]{} 2P_i$$

$$\text{AMP} + \text{ATP} \xrightarrow[\text{adenylate kinase}]{} 2\text{ADP}$$

The reaction catalysed by pyruvate P_i dikinase has a ΔG close to zero. However, removal of the products, pyrophosphate and AMP, means that the synthesis of PEP is favoured. It should also be noted that, since AMP, not ADP, is the product of the reaction, the effective cost of the conversion of each molecule of pyruvate to PEP is two molecules of ATP.

The Operation of the C_4 System

The C_4 system comprises two cycles: the Calvin cycle and the C_4 cycle (Figure 3.1). Regeneration of PEP occurs in the C_4 cycle, but there is no net fixation of carbon dioxide, as regeneration involves the re-release of carbon dioxide. The C_4 cycle is not, therefore, autocatalytic in the way that the Benson–Calvin cycle is; it cannot regenerate more substrate than it consumes. This can be demonstrated if the overall reactions in each cell are added together. The net effect of the C_4 cycle can be seen to be the transfer of carbon dioxide from one compartment to another at the expense of two molecules of ATP per molecule of carbon dioxide transferred. The C_4 cycle is therefore an ATP-driven carbon dioxide pump.

mesophyll cell (outside):

$$\text{pyruvate} + \text{NADPH} + 2\text{ATP} + CO_3 \rightarrow$$
$$2\text{ADP} + \text{malate} + \text{NADP}^+ + 2P_i$$

bundle-sheath cell (inside):

$$\text{malate} + \text{NADP}^+ \rightarrow \text{pyruvate} + CO_2 + \text{NADPH}$$

sum:

$$2\text{ATP} + CO_{2 \text{ (outside)}} \rightarrow 2\text{ADP} + CO_{2 \text{ (inside)}} + 2P_i$$

Differences in the C_4 mechanism

The C_4 pathway is thought to have arisen between 7 and 30 million years ago in response to a decline in atmospheric carbon dioxide to levels of about $200\,\mu l\ 1^{-1}$. Such conditions would have greatly favoured photorespiration. In general, none of the reactions, regulations or transport processes in C_4 or crassulacean acid metabolism (CAM) plants is unique. C_4 photosynthesis is distributed among a wide range of unrelated plants and it is believed that the mechanism evolved independently many times. This has resulted in differences in the mechanism of C_4 photosynthesis in different groups of C_4 plants.

C_4 plants have been classified into three subgroups based on the different enzymes which decarboxylate C_4 acids in the bundle-sheath (Figure 3.5). NADP^+–malic enzyme is responsible for malate decarboxylation in plants such as maize, sorghum, sugar-cane and *Digitaria sanguinalis* (digit grass). In *Amaranthus* spp. and *Panicum miliaceum* (millet), NAD^+–malic enzyme is the predominant decarboxylase. In *Spartina* spp. and *Panicum maximum* (guinea grass), PEP carboxykinase decarboxylates oxaloacetate deriving from malate. NAD^+–malic enzyme is located in the mitochondria, while PEP carboxykinase is a cytosolic enzyme.

In each case the net result is the same in that a C_4 acid is decarboxylated to yield pyruvate or PEP, but the energy cost of the C_4 cycle may differ. The energetic cost in C_4 NAD^+ and NADP^+–malic enzyme species is 5ATP and

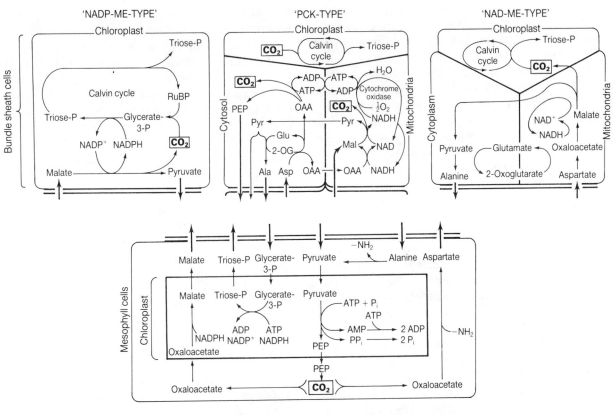

Figure 3.5. Pathways of carbon dioxide assimilation in different subgroups of C_4 plants. From Hatch and Osmond, 1976, by permission of Springer-Verlag.

2NADPH per carbon dioxide molecule fixed (compare with C_3 plants where the cost is 3ATP plus 2NADPH per carbon dioxide molecule fixed; in the C_4 plant the extra 2ATP is used to drive the C_4 cycle).

All of the decarboxylases catalyse reversible reactions:

$$\text{malate} + NAD^+ \underset{NAD^+\text{--malic enzyme}}{\longleftrightarrow} \text{pyruvate} + CO_2 + NADH$$

$$\text{malate} + NADP^+ \underset{NADP^+\text{--malic enzyme}}{\longleftrightarrow} \text{pyruvate} + CO_2 + NADPH$$

$$OAA + ATP \underset{\text{PEP carboxykinase}}{\longleftrightarrow} PEP + ADP + CO_2$$

There are other differences between the three subgroups of C_4 plants. The relative proportion of malate and aspartate formed from oxaloacetate in the mesophyll cells differs. $NADP^+$–malic enzyme-type species tend to form malate, whereas PEP carboxykinase and NAD^+–malic enzyme-type species tend to form aspartate rather than malate, although this also depends upon factors such as the nitrogen nutrition of the plant. The formation of either malate or aspartate has important consequences for the energetics of the C_4 pathway because, unlike aspartate, malate can carry reducing equivalents from the mesophyll to the bundle-sheath. Aspartate formers must therefore have photosystem II in the bundle-sheath to generate the reductant required for carbon assimilation. In NAD^+–malic enzyme-type plants the carbon fluxes through the mitochondria are equivalent to the rate of photosynthesis, i.e. several-fold greater than those during photorespiration in C_3 plants and 10–20-fold greater than respiratory carbon fluxes in leaves.

In PEP carboxykinase-type C_4 plants, a much more complicated interaction between malate decarboxylation in mitochondria and OAA decarboxylation in the cytoplasm occurs. It has been suggested that the ATP required for PEP carboxykinase in the cytoplasm may arise from the mitochondrial oxidation of malate and oxidative phosphorylation (Figure 3.5).

There are also distinct differences in subcellular morphology between the different types. In the majority of NAD^+–malic enzyme species and in $NADP^+$–malic enzyme-type dicots, the chloroplasts are centripetally located, but in PEP-carboxykinase species and in $NADP^+$–malic enzyme-type grasses the chloroplasts are centrifugally located. $NADP^+$–malic enzyme-type species have a single mesotome sheath, whereas other C_4 types have a double bundle-sheath. The functional significance of these and other anatomical differences remains unclear. However, a further distinguishing feature of $NADP^+$–malic enzyme species is the presence of agranal chloroplasts, which lack thylakoid stacking, in the bundle-sheath. These chloroplasts are deficient in photosystem II. Bundle-sheath chloroplasts have high chlorophyll a/b ratios (up to 10) compared to mesophyll chloroplasts (chlorophyll a/b ratio about 2.5). The thylakoids of these agranal chloroplasts catalyse photosystem I-mediated cyclic electron transfer, and therefore generate ATP, but do not generate appreciable amounts of reductant (NADPH). Malate from the mesophyll therefore provides the reductant for carbon assimilation (Figure 3.5). The reason for this deficiency in photosytem II in the bundle-sheath is probably that the lack of oxygen evolution in the gas-tight bundle-sheath is of benefit in maintaining a high ratio of carbon dioxide to oxygen, further reducing the possibility of photorespiration occurring in the bundle-sheath.

The fixation of carbon dioxide by Rubisco in the bundle-sheath results in the generation of two molecules of glycerate 3-phosphate. While half of the NADPH requirement for glycerate 3-phosphate reduction can be met by malate decarboxylation, which derives from reductant generated in the mesophyll, the remaining half of the glycerate 3-phosphate is exported to the mesophyll chloroplasts for reduction, and triose phosphate returns to the bundle-sheath. Thus the Calvin cycle is also split between the two cell types. It should be noted that although other C_4 subgroups forming aspartate maintain a high capacity for photosynthetic oxygen evolution in the bundle-sheath, they also export glycerate 3-phosphate to the mesophyll, where they have the enzymes for glycerate 3-phosphate reduction (Figure 3.5). This process is also likely to result in diminished NADPH consumption, and hence oxygen evolution, in the bundle-sheath.

Intercellular Transport in C_4 Plants

Transport between the mesophyll and bundle-sheath cells is of crucial importance in C_4 plants. Although the bundle-sheath cell wall is essentially impermeable to substrates, transport is facilitated by the presence of numerous plasmodesmata. In some C_4 plants, the cell walls between the bundle-sheath and mesophyll contain suberin, but in all it seems likely that the thickened cell wall is itself a major barrier to the diffusion of solutes and gases. On the wall between mesophyll and bundle-sheath cells there are extensive pit-fields with plasmodesmata which provide symplastic connections between cells. The plasmodesmata exclude large molecules, such as the cytoplasmic enzymes, because they have an exclusion limit of about 800 Da. By restricting the movement of gas into and out of the compartment, a high carbon dioxide concentration can be maintained in the bundle-sheath cells, thus eliminating photorespiration. The carbon dioxide concentration in the bundle-sheath has been estimated at 60 μM at pH 7.5 or 20 μM at pH 8 (equivalent to 2000 μl l^{-1} carbon dioxide in air) compared to 7-8 μM in C_3 leaves of C_3 plants. The leakage of CO_2/HCO_3^- from the bundle-sheath has been estimated at only 10% of total transfer. Furthermore, any carbon dioxide which does leak back to the mesophyll is refixed by PEP carboxylase.

The operation of C_4 photosynthesis requires that compounds such as malate, aspartate and pyruvate should move between the bundle-sheath and mesophyll cells at rates equal to rates of carbon dioxide assimilation. This is also the case for two metabolites of the Calvin cycle: glycerate 3-phosphate and triose phosphate. Transport could be active (e.g. ATP driven) or passive (e.g. diffusion driven). Current evidence suggests that both mechanisms are responsible for the intercellular transport of metabolites in maize. When the leaf is illuminated, large intercellular gradients of malate, glycerate 3-phosphate and triose phosphates are built up. The existence of these gradients has been inferred from the fact that the leaf accumulates very large amounts of glycerate 3-phosphate and triose phosphates during steady-state photosynthesis (concentrations of triose phosphate are typically 500–1000 nmol mg^{-1} chlorophyll in leaves of C_4 plants and 50 nmol mg^{-1} chlorophyll in leaves of C_3 plants). Direct measurements of metabolites have also been made in leaves which have been fractionated into mesophyll and bundle-sheath fractions, either by rolling the leaf, which squeezes out mesophyll sap, or by fractionation of the leaf by sequential grinding in liquid nitrogen, which first releases a fraction enriched in mesophyll cell contents. The difference in concentration between the two cell

types is of the order of 10 mM—enough to drive metabolite transport between the two cell types at rates sufficient to support observed rates of photosynthesis (Table 3.3). Even though an intercellular gradient is not evident from measurements of total cellular pyruvate (Table 3.3), light-dependent transport of pyruvate occurs on a specific carrier in mesophyll chloroplasts of C_4 plants, so that pyruvate is taken up from the mesophyll cytosol.

Comparatively little is known about transport across the envelope of the bundle-sheath chloroplast, but the mesophyll chloroplasts of C_4 plants have different transport properties compared to their C_3 counterparts. One feature which distinguishes chloroplastic transport of phosphorylated intermediates in C_4 plants from that in C_3 plants is the direction of transport across the chloroplast envelope. During photosynthesis in C_4 plants, the mesophyll chloroplasts *import* glycerate 3-phosphate and *export* triose phosphate, and the bundle-sheath chloroplasts *export* glycerate 3-phosphate and *import* triose phosphate which has been reduced by the mesophyll chloroplasts (Figure 3.5). In addition to the exchange of glycerate 3-phosphate, triose phosphate and P_i, the mesophyll chloroplasts also catalyse the export of PEP, formed in the chloroplast by the action of pyruvate P_i dikinase, in exchange for P_i to sustain PEP carboxylase in the cytoplasm. The evidence indicates that exchange of PEP/P_i and glycerate 3-phosphate/triose phosphate occurs on a common

Table 3.3. Concentrations of metabolites in the mesophyll and bundle-sheath cells of maize leaves during steady-state photosynthesis. Note that active uptake of pyruvate occurs into the mesophyll chloroplasts, so that the gradient is concealed when whole cell measurements are made.

Metabolite	Concentration (mM)		Gradient (mM)
	Mesophyll	Bundle-sheath	
Glycerate 3-phosphate	6.6	16	9.4
Triose phosphate	14.8	7.0	7.8
Malate	20.0	5.0	15.0
Aspartate	5.2	6.6	1.4
Pyruvate	10.3	8.8	1.5
PEP	3.0	0.6	2.4

translocator in the chloroplast envelope. In contrast, the phosphate translocator in spinach chloroplasts appears both to recognize, and transport, PEP poorly (chapter 2).

There is a dicarboxylate translocator which transports malate, but, as in spinach chloroplasts, a separate translocator for oxaloacetate which is not strongly inhibited by malate (K_m oxaloacetate 60 μM, K_i malate 8 mM). This therefore allows oxaloacetate uptake by chloroplasts from a cytosol which has high concentrations of malate and low concentrations of oxaloacetate. The inner envelope of mesophyll chloroplasts is also specialized, and different from C_3 plants, because it comprises a series of anastomosing tubules called the peripheral reticulum. The function of this membrane elaboration has never been clear, but it may simply be a means of increasing the surface area of the chloroplast envelope for transport.

Regulation of the C_4 Pathway

C_4 photosynthesis requires the coordinated operation of two interconnecting metabolic cycles. This complex process not only involves cytoplasmic and chloroplastic locations in two cell types, but the pathways are also interconnected, because interconversion of PEP and glycerate 3-phosphate can occur in the mesophyll.

As in C_3 plants, certain enzymes of the C_4 pathway are regulated by light. $NADP^+$–malate dehydrogenase is subject to redox control by the thioredoxin system (chapter 2), and is virtually inactive in darkened leaves. $NADP^+$–malic enzyme is sensitive to malate, pH and Mg^{2+} in a manner which indicates that it may be largely inactive in the darkened stromal environment of the bundle-sheath chloroplast. PEP carboxykinase is inhibited by fructose 1,6-bisphosphate (F1,6BP), glycerate 3-phosphate and triose phosphate. However, by far the most complex and well-studied regulation is that of pyruvate P_i dikinase and PEP carboxylase.

PEP carboxylase from C_4 plants exists as a tetramer of four identical subunits, each 110 kDa. It has sigmoidal kinetics and is activated by hex-

ose phosphate and triose phosphate and is strongly inhibited by malate (Figure 3.6). This means that when carbon utilization by the bundle-sheath falls, malate will accumulate and lower the activity of the enzyme, and hence the carbon dioxide pump. Conversely, when carbon output from the bundle-sheath rises, triose phosphate and hexose phosphate (destined for the synthesis of sucrose) will rise in the mesophyll and promote PEP carboxylase activity. It should be noted that modulation of PEP carboxylase by these metabolites provides a channel of communication between the Calvin cycle and the C_4 cycle.

There is an additional level of control of PEP carboxylase because its sensitivity to metabolites, and the K_m(PEP) are themselves subject to modification by the phosphorylation state of the

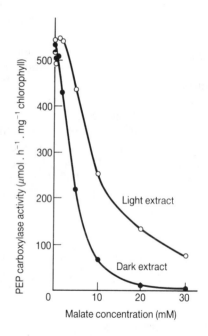

Figure 3.6. The sensitivity of PEP carboxylase activity to malate in extracts of illuminated and darkened maize leaves. Inhibition of the enzyme extracted from illuminated leaves occurs at higher concentrations of malate, because light activation, by phosphorylation, of PEP carboxylase decreases the sensitivity of the enzyme to malate. From Doncaster and Leegood, 1987[2], by permission of the American Society of Plant Physiologists.

enzyme. PEP carboxylase extracted from the leaves of illuminated leaves is less sensitive to inhibition by malate than is the enzyme isolated from darkened leaves (Figure 3.6), in which the enzyme would be less active *in vivo*. This change in malate sensitivity has been studied in maize and sorghum, and occurs when an N-terminal serine residue on the enzyme is phosphorylated by a protein kinase. Dephosphorylation, by a protein phosphatase (of the mammalian type 2A, which is sensitive to the inhibitor, okadaic acid), results in an increase in malate sensitivity. In addition the protein kinase itself undergoes diurnal changes in activity, possibly as a result of protein turnover. PEP carboxylase is thus regulated in a very complex way, in a manner which is analogous to that occurring in CAM plants.

Pyruvate P_i dikinase is another enzyme of the C_4 pathway which is regulated by phosphoryla-tion. It undergoes reversible phosphorylation of the histidine residue at the catalytic site and of a threonine residue two residues away from the histidine residue. Phosphorylation and dephos-phorylation of the threonine residue is catalysed by a single regulatory protein, a rare example of a protein which is bifunctional—it has two active sites. Phosphorylation and dephosphorylation of the histidine residue can be accomplished revers-ibly by ATP plus P_i (phosphorylation) and by AMP plus PP_i (dephosphorylation). Phosphoryla-tion of the histidine residue leads to a less active form of the enzyme. It is also significant, and very unusual, that the catalytic reaction features in the regulation of the activity of this enzyme. This has implications for *in vivo* regulation, since when the substrate accumulates the enzyme will be acti-vated or maintained in an active state. The com-plexity of this regulatory system nevertheless results in a response to light *in vivo* very much like the response of enzymes modulated by the thioredoxin system (Figures 3.7 and 3.8).

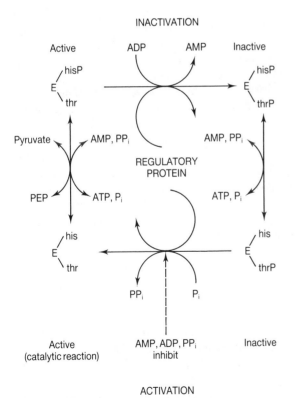

Figure 3.7. Regulation by pyruvate P_i dikinase (E) in C_4 plants.

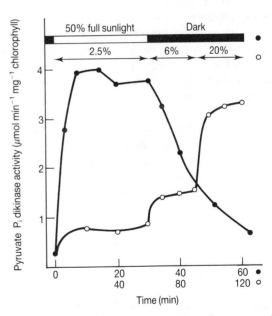

Figure 3.8. Changes in the activity of pyruvate P_i dikinase in maize leaves during dark–light transi-tions and at different photon flux densities. Photon flux densities are shown as percentages of full sunlight. From Hatch (1981)[3], by permission of Balaban International Scientific Services.

Sucrose and Starch Synthesis in C$_4$ Plants

Further division of labour between the bundle-sheath cells and the mesophyll cells is apparent in the sites of product synthesis. In leaves of maize, starch is usually exclusively located in the bundle-sheath chloroplasts. If the plants are illuminated continuously, however, starch may also be synthesized in the mesophyll chloroplasts, together with induction of the enzymes necessary for its synthesis. On the other hand, *Digitaria* spp., which are also NADP$^+$–malic enzyme plants, synthesize both sucrose and starch in the mesophyll compartment. Sucrose synthesis appears to occur predominantly in the mesophyll in maize. During $^{14}CO_2$ fixation, labelled sucrose appears first in the mesophyll cells. The major portions of sucrose phosphate synthetase, fructose 6-phosphate-2-kinase, fructose 2,6-bisphosphate and F2,6BP are present in the mesophyll cells of maize leaves. Since sucrose is made in the mesophyll this also means that sucrose must be transported through the bundle-sheath cell back to the phloem.

Sucrose originates from triose phosphate which is generated, in turn, from glycerate 3-phosphate produced in the bundle-sheath. It has already been noted that the triose phosphate content of mesophyll cells is remarkably high because of its role in diffusion-driven transport between the mesophyll and bundle-sheath. Sucrose synthesis and transport from the mesophyll to the bundle-sheath thus compete for triose phosphate. For this reason, sucrose synthesis proceeds at its maximum rate in maize only when triose phosphate is present at much higher concentrations than it would be in C$_3$ plants. The cytosolic fructose bisphosphatase (chapter 4) in maize has a lower affinity for F1,6BP than its C$_3$ counterpart. For example, the K_m(F1,6BP) is 3 μM in spinach and 20 μM in maize. The enzyme is also more sensitive to the concentration of F2,6BP, with a K_m (F1,6BP) of 250 μM in spinach and 3500 μM in maize in the presence of 10 μM F2,6BP.

The regulation of starch synthesis has also been modified in C$_4$ plants. ADP glucose pyrophosphorylase from spinach chloroplasts is activated by glycerate 3-phosphate and inhibited by P$_i$, with ratios of glycerate 3-phosphate/P$_i$ for half-maximal activation typically being less than 1.5. ADP glucose pyrophosphorylase from maize leaves requires a ratio of glycerate 3-phosphate to P$_i$ of between 7 and 10 for half-maximal activation in the bundle-sheath and even higher ratios in the mesophyll. The metabolite gradients which develop during photosynthesis lead to lower glycerate 3-phosphate/P$_i$ ratios in the mesophyll than in the bundle-sheath which, together with the relatively low activities of enzymes of starch synthesis, would appear to limit synthesis of starch in the mesophyll relative to the bundle-sheath.

CRASSULACEAN ACID METABOLISM

CAM is an adaptation which, in the majority of species in which it occurs, allows the survival of plants in habitats with a periodic, inconstant supply of water. The paradigm CAM plants are the stem-succulent Cactaceae or Euphorbiaceae from semi-arid regions, but the adaptation is known to be distributed among a range of plants from widely different families, for example, the Cactaceae (probably all species are CAM), Crassulaceae (*Bryophyllum, Kalanchoe, Sedum*), Euphorbiaceae, Aizoaceae (*Mesembryanthemum, Lithops*), Liliaceae (*Aloe, Yucca*), Agavaceae (*Agave*) from arid regions to the epiphytic Bromeliaceae (e.g. *Tillandsia*) or Orchidaceae of tropical rainforests, as well as in two species of epiphytic fern. Some 50% of known CAM plants are epiphytes. This wide taxonomic and ecological distribution of CAM suggests that, like C$_4$ photosynthesis, the mechanism has arisen independently many times during the course of evolution.

CAM, as the name implies, was first recognized in the Crassulaceae. CAM plants are characterized by a massive nocturnal fixation of carbon dioxide into an organic acid such as malate. The malic acid is decarboxylated during the day to generate an internal reservoir of carbon dioxide (Figure 3.9). This mechanism allows the plant to

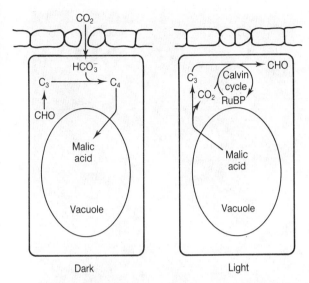

Figure 3.9. The mechanism of crassulacean acid metabolism (CAM) photosynthesis, showing the temporal separation of fixation of carbon dioxide by PEP carboxylase prior to fixation in the Calvin cycle. Nocturnal carbon dioxide fixation leads to malate storage in the vacuole. During the day, malate is decarboxylated to provide carbon dioxide for fixation in the Calvin cycle (compare with Figure 3.1). From Leegood and Osmond, 1990, by permission of Longman.

accumulate carbon dioxide from the air when stomata are open at night, and the temperature, and hence water loss, are at their lowest, and allows photosynthetic carbon dioxide fixation (via the Calvin cycle) to proceed behind closed stomata during the day. The process therefore results in minimal water loss. The presence of high concentrations of carbon dioxide (1%) during fixation in the light also means that the extent of photorespiration is greatly reduced in CAM plants. Like C_4 photosynthesis, CAM does not constitute an alternative to carbon dioxide fixation by the Calvin cycle because the CAM mechanism cannot catalyse a net fixation of carbon dioxide. Carbon dioxide released by the decarboxylation of malate is therefore refixed in the Calvin cycle to generate triose phosphate, which can be converted to carbohydrate.

Succulence is extremely common among CAM plants, but this does not mean that all succulent

plants are CAM plants. Succulence is a means of water storage and is recognized by the presence of voluminous water-storing parenchymatous tissues and an increase in volume relative to surface area. In many CAM plants, leaves, stems or both may be succulent. Although some CAM plants are not outwardly succulent (e.g. 'air-plants' such as *Tillandsia*), they nevertheless exhibit what has been termed cellular succulence, a feature of all CAM plants, in that the cells contain large vacuoles which may account for 90% or more of the total cell volume (Figure 3.10).

In CAM plants, both PEP carboxylase and Rubisco are present in all chloroplast-containing cells. The activities of these enzymes are thus regulated in time, rather than in space, with PEP carboxylase active in the dark, but inactive for much of the light period, and Rubisco active only in the light.

The Evidence for Crassulacean Acid Metabolism

The nocturnal accumulation of acid in the Crassulaceae was first recognized in the seventeenth century. In 1815 Heyne remarked that the leaves of *Cotyledon calycina* tasted remarkably bitter: "in the morning as acid as sorrel, if not more so; as the day advances, they lose their acidity and are tasteless about noon and become bitterish towards evening". Leaves of CAM plants show remarkably large changes in titratable acidity during diurnal cycles (Figure 3.11), most of which can be accounted for by the synthesis of malic acid. Thus when leaves are supplied with $^{14}CO_2$ in the dark, as much as 90% of the label incorporated may be confined to malate.

The incorporation of carbon dioxide into malate is catalysed by PEP carboxylase, as in C_4 plants. The leaves of CAM plants contain very high activities of PEP carboxylase, comparable to those observed in C_4 plants. The substrate for the reaction, PEP, derives from carbohydrate breakdown. During the diurnal cycle there is a stoichiometric interconversion of starch and malate.

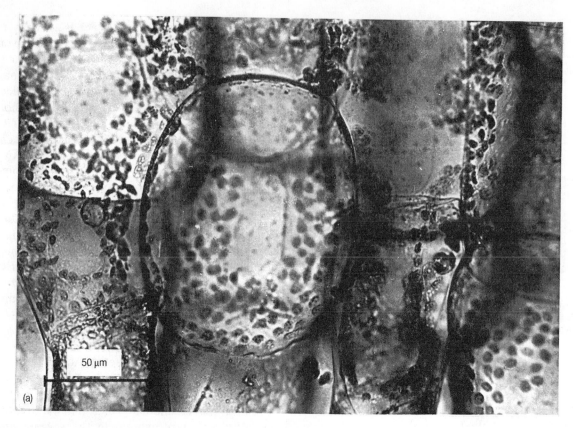

Figure 3.10. Mesophyll cells from a cross-section through the leaf of *Kalanchoe daigremontiana*. An example of a water-storing, photosynthetic mesophyll, typical of CAM tissues. From Kluge and Ting, 1978[4], by permission of Springer-Verlag.

The Nocturnal Conversion of Carbohydrate to Malate

Many CAM plants store starch as the carbohydrate reserve, but some, such as pineapple, *Ananas comosus*, store soluble sugars in the leaves. Carbohydrate is rapidly degraded at the onset of darkness (Figure 3.11). It is not yet clear whether starch breakdown is predominantly amylolytic or phosphorolytic in CAM chloroplasts. It is likely that conversion of glycerate 3-phosphate to PEP occurs in the the cytoplasm, after export from the chloroplast via the phosphate translocator. In the cytoplasm, the enzymes phosphoglycerate mutase and enolase convert glycerate 3-phosphate to PEP. PEP carboxylase is also present in the cytoplasm. Reduction of the oxaloacetate formed by

carboxylation would then be accomplished by the cytoplasmic NADH-dependent malate dehydrogenase.

The amount of starch in the leaves bears an inverse relationship to the amount of malate (Figure 3.11), indicating that interconversion occurs. If starch is pre-labelled with $^{14}CO_2$ then this label is largely recovered in malate. If plants are placed in carbon dioxide-free air in the dark, in which conditions PEP cannot be utilized in carbon dioxide fixation, starch degradation is completely inhibited.

The amounts of malate formed during dark fixation, up to 100 μmol g^{-1} fresh weight, are extremely large by comparison with the content of organic acids in most plant tissues (cf. 1 μmol malate g^{-1} fresh weight in a maize leaf). In CAM

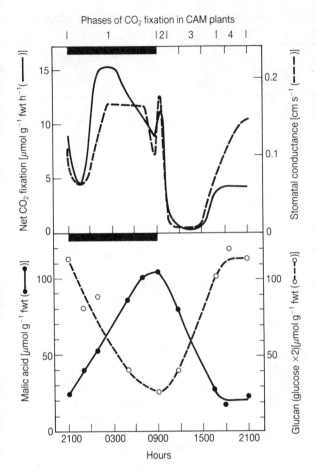

Figure 3.11. Phases of carbon dioxide fixation in a CAM plant (*Kalanchoe daigremontiana*) showing stomatal conductance and diurnal changes in malic acid and glucan content. From Osmond and Holtum, 1981, by permission of Academic Press.

plants the malic acid is stored principally in the cell vacuoles. Sequestration of malate in the vacuole prevents further metabolism of the malate and prevents excessive acidification of the cytoplasm. The vast amounts of malic acid which accumulate in CAM plants also indicate that it must be transported to the cell vacuole. Studies of the efflux kinetics of malate labelled after $^{14}CO_2$ incorporation and studies of isolated intact vacuoles bear out the view that most of the malate is present in the vacuole. However, the accumulation of malate in the vacuole against a concentra-

tion gradient must be driven by an active transport mechanism. Both ATP-dependent, and PP_i-dependent, H^+ transport across the tonoplast membrane of CAM plants has been shown. The proton transport system involved in malic acid accumulation in CAM plants is probably a $2H^+/$ATPase.

The Decarboxylation of Malate

Organic acids, such as malate, are decarboxylated during the day to provide carbon dioxide for fixation in the Calvin cycle. Prior to decarboxylation, malate must first be released from the vacuole to the cytoplasm. This is thought to be a passive process. During nocturnal malate accumulation in the vacuole, the undissociated acid accumulates and diffuses freely across the membrane of the tonoplast. It is likely that the malic acid storage capacity of the vacuole is set by the equilibrium between active influx and passive efflux. In theory, all of the malate could enter the Krebs (tricarboxylic acid) cycle, be converted to carbon dioxide and the carbon dioxide could then be refixed in the Calvin cycle. This does not appear to occur to any appreciable extent. Instead, malate is decarboxylated in specific decarboxylations yielding carbon dioxide and a C_3 compound (pyruvate or PEP) which is then converted back to starch.

CAM plants can be divided into two groups: those which decarboxylate malate via the reaction catalysed by NAD(P)-dependent malic enzyme, and those in which PEP carboxykinase is present (Table 3.4). The latter enzyme yields PEP, which can be immediately converted to glycerate 3-phosphate and then to starch in the chloroplasts. Decarboxylation by NAD(P)–malic enzyme yields pyruvate, which must first be converted to PEP. As in C_4 plants, this reaction is catalysed by pyruvate P_i dikinase, which is present in the chloroplasts. Pyruvate P_i dikinase is absent from CAM plants which have PEP carboxykinase as the major decarboxylating enzyme. However, unlike C_4 plants, the malic enzyme-type (ME–CAM) plants have a cytoplasmic $NADP^+$–ME

Table 3.4. Activities of some key enzymes in NAD(P)$^+$–malic enzyme (ME) and PEP carboxykinase (PEPCK)-type CAM plants.

Enzyme	Activity (μmol min^{-1} mg^{-1} chlorophyll)	
	NAD(P)ME–CAM	PEPCK–CAM
Malic acid synthesis		
PEP carboxylase	7.2	–
NAD$^+$–malate dehydrogenase	138.0	–
Maximum rate of acidification	0.54	–
Malic acid decarboxylation		
NADP$^+$–malic enzyme	1.0	1.6
NAD$^+$–malic enzyme	0.5–2.4	0.1–0.4
PEP carboxykinase	n.d.	4.7
Pyruvate P$_i$ dikinase	0.83	n.d.
Maximum rate of deacidification	0.74	1.8

n.d., not detectable.

which, with mitochondrial NAD$^+$–ME, participates in decarboxylation. Unlike C$_4$ plants, PEP carboxykinase-type CAM plants have very low malic enzyme activities and no pyruvate P$_i$ dikinase, but high activities of cytoplasmic PEP carboxykinase.

Triose phosphate is not apportioned simultaneously into starch and sucrose in CAM plants. The first task the plant performs is to restore its reserves of starch. Starch synthesis would be favoured by the large amount of glycerate 3-phosphate generated, which would activate ADP–glucose pyrophosphorylase. The ratio of glycerate 3-phosphate to P$_i$ is likely to very high during deacidification and hence to favour glucan synthesis via ADP–glucose pyrophosphorylase, partly because glycerate 3-phosphate greatly lowers the $S_{0.5}$(glucose 1-phosphate), and partly because P$_i$ does not completely inhibit the enzyme. Once starch synthesis is complete, net carbon dioxide fixation begins and sucrose is synthesized.

Gas Exchange in Crassulacean Acid Metabolism Plants

The pattern of gas exchange and metabolism in CAM plants has been described as comprising four phases (Figure 3.11).

Phase 1: Malic Acid Synthesis in the Dark

During this period PEP carboxylase is active and carbohydrate is converted to PEP. Fixation of carbon dioxide is insensitive to oxygen since PEP carboxylase is the enzyme active in carbon dioxide fixation. Malate accumulates in the vacuole. Towards the end of the dark period the rate of fixation declines. This may be due to inhibition of PEP carboxylase by accumulated malate, or to a decrease in the cytoplasmic pH.

Phase 2: The Initial Period in the Light

The stomata close gradually and fixation of external carbon dioxide ceases. There is an initial burst of carbon dioxide fixation when both carboxylases are operative and there is a lag in malic acid decarboxylation. The magnitude of the carbon dioxide burst is influenced by a number of environmental factors such as daylength, temperature and light intensity.

Phase 3: Malic Acid Decarboxylation

Decarboxylation occurs concurrent with a low rate of carbon dioxide fixation. During this period the rate of fixation of exogenous $^{14}CO_2$ is extremely slow, but oxygen inhibition of photosynthesis can be demonstrated. The internal carbon dioxide concentration rises to extremely high levels (e.g. 0.8% carbon dioxide has been measured in *Agave desertii*). Although the concentration of oxygen also rises due to photosynthetic oxygen evolution behind closed stomata, the ratio of oxygen to carbon dioxide is reduced well below that of air, so that photorespiration is suppressed. Under these conditions of high carbon dioxide, Rubisco operates at 80% of its maximum capacity.

Phase 4: Malate Decarboxylation Ceases, as does Glucan Accumulation

The internal concentration of carbon dioxide declines as consumption of carbon dioxide in photosynthesis exceeds production, net carbon dioxide fixation and sucrose formation. Fixation of exogenous carbon dioxide also occurs via Rubisco, since the plants show a high compensation point (50 μl l^{-1}) and oxygen inhibition of carbon assimilation occurs. PEP carboxylase slowly becomes active, possibly because the concentration of malate in the cytoplasm is lowered.

Regulation of Crassulacean Acid Metabolism

Regulation of any photosynthetic system is necessarily complex, but two major regulatory problems confront the CAM plant. The first is regulation of the fate of PEP and the avoidance of futile cycles of fixation and release of carbon dioxide resulting in the hydrolysis of ATP. The second is the apportioning of triose phosphate production by the choroplast in the light between starch and sucrose synthesis.

As in C_4 plants, pyruvate P_i dikinase is probably inactive in the dark. Little is known concerning the regulation of PEP carboxykinase in CAM plants but regulation is not achieved by compartmentation, since both PEP carboxylase and PEP

carboxykinase are present in the cytoplasm. In NAD(P)–malic enzyme-type plants, cycling is avoided by strict regulation of the activity of pyruvate phosphate dikinase. In all CAM plants, PEP carboxylase must be regulated in the light so that carbon dioxide released by malate decarboxylation is not immediately refixed, which would again constitute a futile cycle. Similarly, pyruvate kinase must be regulated in the dark so that it does not compete with PEP carboxylase for the available PEP.

PEP carboxylase from CAM plants is distinguished from the enzyme present in the leaves of C_3 and C_4 plants by its high V_{max} (like C_4 plants) and relatively low K_m(PEP) (like C_3 plants). Like PEP carboxylase from all other sources, the enzyme is strongly inhibited by malate, and glucose 6-phosphate is a positive effector, increasing the affinity for PEP by as much as tenfold. There are marked diurnal changes in the properties of the enzyme. One of the most important of these is a change in the sensitivity to malate inhibition (Table 3.5, Figure 3.12).

In a manner analogous to the regulation of PEP carboxylase in the leaves of C_4 plants, but reversed in its response to light, PEP carboxylase extracted from the leaves of darkened plants is less sensitive to inhibition by malate than is the enzyme isolated from illuminated leaves. Again, this change in malate sensitivity occurs when the enzyme is phosphorylated by a protein kinase in darkness and dephosphorylated by a type 2A protein phosphatase in the light. Phosphorylation and dephosphorylation of PEP carboxylase in CAM plants also occurs independently of light under the control of diurnal rhythms.

Table 3.5. Changes in the properties of PEP carboxylase (measured at pH 7.5) extracted from illuminated and darkened leaves of *Mesembryanthemum crystallinum*.

Treatment	V_{max}	K_m(PEP) (μM)	K_i(malate) (μM)
Light	23.8	900–1250	4–5
Dark	25.6	90–220	60–86

Data from Winter (1981)

Crassulacean Acid Metabolism as a Survival Mechanism

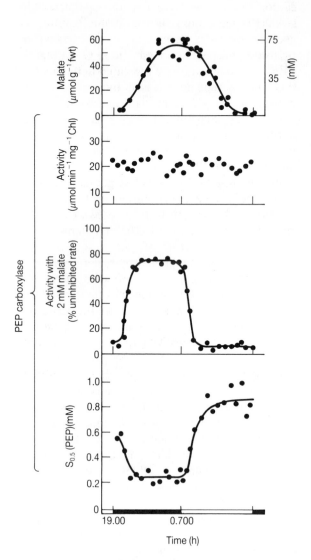

Figure 3.12. Changes in the properties of PEP carboxylase extracted from leaves of *Mesembryanthemum crystallinum* at different times of the diel cycle of CAM. The concentration of malic acid in the vacuole was calculated assuming 85% water content. Although the total activity of PEP carboxylase remains unchanged, the decreased affinity for PEP and the increased sensitivity to malate ensure that the enzyme is unhibited during deacidification in the light. Redrawn from Winter, 1982[5], by permission of the American Society of Plant Physiologists.

CAM is an adaptation which in most cases allows survival of plants in arid areas which receive only periodic supplies of water. These can be semi-deserts, shallow soils or the epiphytic environment in tropical forests. In the Namib Desert and similar coastal deserts more than 80% of the plant species are CAM. Succulence is a means of water storage which permits survival of many CAM plants through a dry period. On the other hand, if leaf-succulent plants experience profound water stress, leaves may simply be shed. Some CAM plants exhibit seasonal changes in morphology between stem succulence and leaf succulence.

Water use efficiency in CAM plants is dramatically improved because night-time fixation of carbon dioxide occurs at temperatures which are much lower than daytime temperatures. *Agave desertii* typically experiences a 25 °C day and a 5 °C night. Transpiration through open stomata is therefore dramatically reduced at lower night-time temperatures, because of the lower vapour pressure deficit of the air compared to the leaf. For example, in the case of *Agave desertii* the water vapour concentration difference (and therefore the flux of water vapour for a given stomatal aperture) between the leaf and the air is about seven times greater at 25 °C than at 5 °C. The stomatal frequency is also much lower in CAM plants than it is in C_3 or C_4 plants (Table 3.6).

One of the outstanding features of CAM is its extreme metabolic flexibility. If sufficient water is available, then significant daytime uptake of carbon dioxide may occur in leaf-succulent plants, and may contribute significantly to the net carbon gain by the plant. Many plants show facultative CAM. Pineapple is largely C_3 (and remarkably productive) when irrigated, particularly when night temperatures are high (which induces stomatal closure). *Mesembryanthemum crystallinum* and other members of the Aizoaceae can be induced to shift from C_3 to CAM photosynthesis by irrigating with saline water or by withholding water. Seeds of *M. crystallinum*, which inhabits arid cliffs by the Dead Sea, germinate when water is plentiful during January and February. The plant develops as a C_3 seedling, during which period $\delta^{13}C$ values (see below) are $-25‰$. The onset of the dry season leads to the induction of CAM. This can be recognized by dramatic changes such as the development of nocturnal malic acid accumulation, induction of key enzymes such as pyruvate P_i dikinase, PEP carboxylase and $NADP^+$–malic enzyme and a fall in $\delta^{13}C$ values to $-16‰$.

In some CAM plants (e.g. stem-succulent plants such as *Opuntia*), a prolonged lack of water leads to complete sealing of the stomata and shedding of the roots. Although there is no gas exchange, the content of organic acids continues to fluctuate. This reflects internal recycling of carbon dioxide (generated by respiration and photorespiration). This CAM idling allows survival under conditions of extreme drought, it prevents photoinhibition (which might occur if the intercellular concentration of carbon dioxide were extremely low) and also allows the plant to respond immediately to water availability. On the other hand, some CAM plants show fluctuations in organic acid content with little or no nocturnal carbon dioxide fixation. This strategy, known as CAM cycling, may keep the plant poised for the

Table 3.6. Some characteristics of C_3, C_4 and CAM plants.

	C_3	C_4	CAM
Typical species of economic importance	Wheat, barley, rice, potatoes	Maize, millet, sugar-cane, sorghum	Pineapple
% world flora (species number)	89	<1	10
Typical habitat	Widely distributed (dominant in, e.g., forests)	Warm to hot, open sites (grassland)	Xeric sites (includes epiphytes)
First product of CO_2 fixation	Glycerate 3-phosphate	Malate (two fixation processes separated in space)	Malate (two fixation processes separated in time)
Anatomy	If present, bundle-sheath not green	bundle-sheath with chloroplasts (Kranz anatomy)	Cellular or tissue succulence
Photorespiration	Up to 40% of photosynthesis	Not detectable	Not detectable
CO_2 compensation point	$40–100\ \mu l\ l^{-1}$	$0–10\ \mu l\ l^{-1}$	$0–10\ \mu l\ l^{-1}$ (in dark)
Intracellular $[CO_2]$ in light $(\mu l\ l^{-1})$	200	100	10 000
Stomatal frequency	2000–31 000	10 000–16 000	100–800
Water-use efficiency (g CO_2 fixed/kg H_2O transpired)	1–3	2–5	10–40
Maximum growth rate (g $m^{-2}\ d^{-1}$)	5–20	40–50	0.2
Maximum productivity (t $ha^{-1}\ y^{-1}$)	10–30	60–80	Generally less than 10

sudden onset of drought. Epiphytes especially may be subject to frequent periodic droughts, since they may be fully hydrated and desiccated within the course of a single day.

C₃–C₄ INTERMEDIATES

Over twenty species of plants exhibit photosynthetic characteristics which are intermediate between C_3 and C_4 plants in that they show reduced rates of photorespiration and carbon dioxide compensation points in the range 7–15 μl l^{-1} compared with typical values in C_3 plants of 50 μl l^{-1} and in C_4 plants of less than 5 μl l^{-1}. Although all show a degree of 'Kranz' anatomy, both mesophyll and bundle-sheath cells contain Rubisco. The one enzyme which does show a clear compartmentation between the mesophyll and bundle-sheath is glycine decarboxylase, the step which results in photorespiratory carbon dioxide release (chapter 2). As in C_4 plants, the mesophyll cell mitochondria of *Moricandia arvensis* and many other 'intermediate' plants have low activities of glycine decarboxylase. It has been proposed that shuttling of glycine from the mesophyll to the bundle-sheath and the return of serine to the mesophyll occurs (Figure 3.13), although the operation of such a shuttle has yet to be demonstrated. Efficient refixation of photorespiratory carbon dioxide in the bundle-sheath cells would then be responsible for reducing the carbon dioxide compensation point.

Two main types of C_3–C_4 intermediate have been distinguished. In one type (e.g. *Panicum milioides* and *Moricandia arvensis*), there is no evidence for a functional C_4 acid cycle which donates carbon dioxide from C_4 acids to the Calvin cycle. However, in others (such as *Flaveria anomala* and *Neurachne minor*) appreciable activities of the C_4 pathway enzymes PEP carboxylase, pyruvate P_i dikinase and NADP$^+$–malic enzyme are found. These plants show varying capacities to fix ^{14}C into C_4 acids during short-term exposure to $^{14}CO_2$, and to transfer this to products of the Calvin cycle, suggesting a limited capacity for a C_4 pathway which would occur in addition to the glycine shuttle. The mechanisms obtaining in these plants have prompted speculation that they represent steps in the evolution of true C_4 photosynthesis.

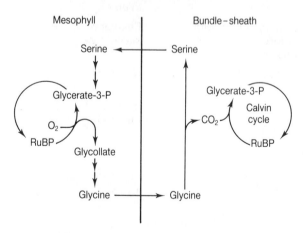

Figure 3.13. Possible scheme for photosynthesis in C_3–C_4 intermediates which lack C_4 metabolism. Glycine is shuttled to the bundle-sheath, where decarboxylation occurs, trapping photorespired carbon dioxide within the bundle-sheath.

THE AVAILABILITY OF CARBON DIOXIDE IN THE AQUATIC ENVIRONMENT

In nature, carbon dioxide-limiting conditions are commonly encountered in aquatic environments. This can occur for various reasons. (a) The diffusive resistance to movement of carbon dioxide in water is 10^4 greater than in air. The boundary layer resistance to gaseous diffusion is accordingly very much higher in water. (b) Inorganic carbon is available as carbon dioxide, HCO_3^- and CO_3^{2-}, but the species present is strongly pH dependent. Alkaline waters will be rich in HCO_3^- and CO_3^{2-}, but poor in free carbon dioxide. In poorly buffered, low-alkalinity waters, the carbon dioxide concentration may also be reduced by phytoplankton to below the concentration which would normally be in equilibrium with air. (c) Besides a limitation on carbon dioxide availability, the oxygen concentration may increase substantially as a result of photosynthesis by phytoplankton, increasing to as much as 30%. Consequently aquatic plants may be simultaneously confronted with low carbon dioxide and high oxygen, which will lead to increased rates of photorespiration.

One strategy which overcomes these problems is active HCO_3^- uptake. This can increase internal carbon dioxide and is probably important in all waters, except those of naturally low alkalinity. This carbon dioxide-concentrating mechanism is induced on transfer from carbon dioxide-rich to carbon dioxide-limiting conditions. The total pool of internal inorganic carbon is higher in *Chlamydomonas* grown in low carbon dioxide than in the alga grown in high carbon dioxide for a given external concentration of carbon dioxide (Figure 3.14). Adapted cells are able to achieve carbon dioxide concentrations ten times those of the external solution. The carbon dioxide-concentrating mechanism has two components: (a) a transmembrane carrier for inorganic carbon (HCO_3^-); and (b) high activities of carbonic anhydrase. Mutants of *Chlamydomonas* deficient in both of these components have been isolated. Both mutants have rates of photosynthesis at limiting concentrations of carbon dioxide which are much lower than the wild-type. The mutant deficient in carbonic anhydrase behaves similarly to the wild-type treated with ethoxyzolamide, an inhibitor of carbonic anhydrase.

Some land plants that have become adapted to the aquatic habit have also developed, to greater or lesser degrees, the ability to take up HCO_3^- and have developed gas-filled lacunae in their leaves in which high concentrations of carbon dioxide may develop.

Few aquatic plants bear the structural features of the C_4 pathway. However, it does appear that a number of aquatic macrophytes and algae possess particularly high levels of PEP carboxylase activity. Plants such as *Isoetes howellii* (which grows in densely populated shallow pools that are prone to carbon dioxide limitation) have been shown to accumulate malic acid as a result of carbon dioxide fixation at night. Decarboxylation occurs the following day in a manner which resembles CAM. The conventional view of CAM is that it is a strategy to improve water-use efficiency. This is clearly not the case in these aquatic plants. Acid metabolism in these circumstances alleviates the problems involved in obtaining carbon dioxide from the aqueous environment. For this reason it has sometimes been termed aquatic CAM.

C_3, C_4 AND CRASSULACEAN ACID METABOLISM COMPARED

Response to Carbon Dioxide and the Carbon Dioxide Compensation Point

The response of photosynthesis to carbon dioxide concentration differs markedly in C_3 and C_4 plants. Carbon dioxide assimilation in leaves of C_4 plants is saturated in air (Figure 3.15) but the rate increases in C_3 plants with increases in the concentration of carbon dioxide to well above ambient levels. This is because the carbon dioxide-concentrating mechanism of C_4 photosynthesis allows saturation of the rate of carbon dioxide assimilation at relatively low intercellular concentrations of carbon dioxide. During photo-

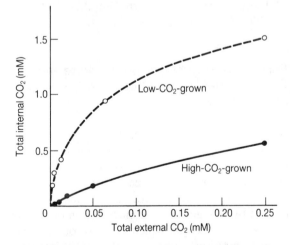

Figure 3.14. The response of internal carbon dioxide to external carbon dioxide for high and low carbon dioxide grown cells in *Chlamydomonas reinhardtii*. From Badger *et al.* (1977)[6].

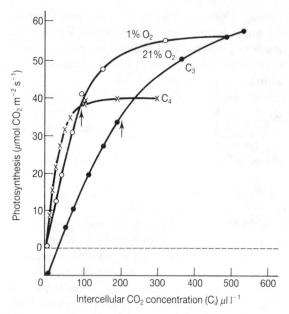

Figure 3.15. Response of photosynthesis to inter-cellular carbon dioxide in *Amaranthus retroflexus* (C₄,X) and in *Chenopodium album* (C₃, ●,○). From Lemon, 1984[7], by permission of Westview Press, Colorado.

synthesis in air the stomatal conductance is adjusted so as to achieve rather constant intercel-lular concentrations of carbon dioxide in C_3 and C_4 species (Figure 3.15). C_4 species operate at an intercellular carbon dioxide concentration of about 100 $\mu l\ l^{-1}$ carbon dioxide, which is just sufficient to achieve saturation of carbon dioxide assimilation, compared to 250 $\mu l\ l^{-1}$ carbon diox-ide in C_3 plants.

If a plant is placed in a closed container and is left to photosynthesize, it continues to reduce the carbon dioxide concentration until a stable con-centration is reached. At the carbon dioxide com-pensation point net photosynthesis is zero, i.e. photosynthetic carbon dioxide uptake is balanced by the processes of photorespiration plus respira-tion. Since the carbon dioxide compensation point (Γ) is the balance point between carbon dioxide uptake and carbon dioxide evolution, it will be strongly influenced by factors which alter the photorespiratory flux and hence the rate of carbon dioxide evolution. In C_3 plants, Γ nor-mally occurs at external carbon dioxide concen-

trations of between 40 and 100 $\mu l\ l^{-1}$ (Figure 3.15). By contrast, the compensation point in C_4 plants occurs at less than 5 $\mu l\ l^{-1}$. In CAM plants the compensation point varies between extremely low values (near zero) during dark fixation of carbon dioxide (phase I) to higher values during phase IV (50 $\mu l\ l^{-1}$).

If photorespiration is inhibited in C_3 plants, by reducing the oxygen concentration in air to 2%, then the compensation point falls to low values similar to those measured in C_4 plants (Figure 3.15) and the rate of carbon assimilation accord-ingly saturates at lower concentrations of carbon dioxide. The compensation point therefore changes markedly with changes in temperature, since higher temperatures encourage higher rates of photorespiration. At 15 °C, the compensation point in *Pelargonium zonale* is 40 $\mu l\ l^{-1}$ carbon dioxide, which rises to 70 $\mu l\ l^{-1}$ at 25 °C. Com-pensation point measurement under different conditions can therefore give some idea of the magnitude of photorespiration in C_3 plants.

Although carbon assimilation in C_4 plants is not affected by oxygen and the compensation point of C_4 plants is low, this does not mean that C_4 plants lack the apparatus of the photorespiratory cycle in the bundle-sheath. Both the bundle-sheath and mesophyll cells contain abundant peroxisomes, while mitochondria from the bundle-sheath are capable of glycine decarboxylation.

Responses to Light and Temperature and Quantum Yields for Carbon Dioxide Fixation in C₃ and C₄ Plants

The responses to temperature are different in C_3 and C_4 plants. Carbon dioxide assimilation shows an optimum of 20–30 °C in typical C_3 leaves, and 30–40 °C in typical C_4 leaves (Figure 3.16). Again this temperature dependency in C_3 leaves may be altered by changes in the concentration of carbon dioxide or of oxygen which increase or suppress photorespiration, just as the light-intensity dependence is altered. Thus at high concentra-tions of carbon dioxide the temperature response of a C_3 leaf becomes similar to that of a C_4 leaf

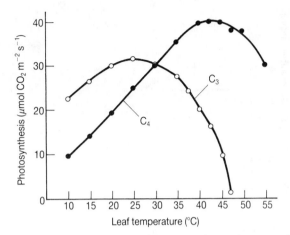

Figure 3.16. The temperature dependence of photosynthesis in air in the prairie grasses *Agropyron smithii* (C_3) and *Bouteloua gracilis* (C_4). From Edwards *et al*, 1985[8], by permission of Elsevier Science Publishers BV.

and the decline in carbon dioxide uptake in a C_3 leaf can be attributed to accelerated rates of photorespiratory loss.

It has already been seen that photorespiration results in a large increase in the energy required for net carbon dioxide fixation in C_3 species (chapter 2). With the proportion of oxygenation occurring in air at 25–30 °C the theoretical cost of net fixation of carbon dioxide in a C_3 plant is approximately equal to the theoretical cost of fixing carbon dioxide in C_4 photosynthesis (4.5 ATP and 3 NADPH in air in C_3 plants and 5 ATP and 2 NADPH in C_4 plants). If C_3 plants have rates of photorespiration below such levels (e.g. at lower temperatures or at a higher intercellular concentration of carbon dioxide) then this will be, energetically speaking, to their advantage compared to C_4 plants. On the other hand, if photorespiration exceeds such rates (e.g. at higher temperatures or at lower intercellular concentrations), C_3 plants will be at a disadvantage in terms of photosynthetic efficiency compared with C_4 plants. Such differences may be seen in the quantum yields measured for photosynthesis in C_3 and C_4 plants.

At rate-limiting photon flux densities, the amount of carbon dioxide taken up can be related to the amount of light absorbed by the quantum yield (moles of carbon dioxide fixed per mole of photons absorbed). The quantum yield is thus the initial slope of the curve relating carbon dioxide assimilation rate to light intensity. The inverse of the quantum yield is the quantum requirement for carbon dioxide assimilation, i.e the number of quanta required to fix a molecule of carbon dioxide. The quantum yield thus provides a direct measure of the energy required to fix carbon dioxide. Measurements of quantum yields in C_3 and C_4 plants have shown that in air at between 25 and 30 °C the quantum yield for both C_3 and C_4 species is comparable (about 0.053 mol carbon dioxide per mole of photons, equivalent to a quantum requirement of 19, but some variation in values is observed). In 2% oxygen the quantum yield rises to 0.07–0.08 in C_3 species as photorespiratory losses are suppressed, but remains constant in C_4 species (Figure 3.17). Thus in the absence of photorespiration C_3 plants are more efficient than C_4 plants in low light. The quantum yield of leaves of C_3 plants also changes with temperature, falling as the temperature is raised, but it remains constant in leaves of C_4 plants. These changes in quantum yield are entirely consistent with the known characteristics of photorespiration, which is abolished by low oxygen and which increases with temperature. As oxygen inhibition of photosynthesis increases, the quantum yield falls exponentially (Figure 3.17). In contrast, the cost of fixing carbon dioxide in a C_4 plant remains roughly constant in relation to changes in temperature and carbon dioxide concentration.

In CAM plants the quantum yield has been studied in *Sedum praealtum*, a malic enzyme-type species. During deacidification (phase II), the theoretical energetic requirement is 6 ATP and 2 NADPH per carbon dioxide molecule fixed (i.e. only slightly higher than C_3 and C_4 plants). The quantum yield is accordingly similar to that observed in C_3 and C_4 plants in air, being 0.062 mol carbon dioxide per mole of photons. However, during the late afternoon (phase IV) the quantum yield falls to 0.024 mol carbon dioxide per mole of photons (a quantum requirement of 41). This extremely high energetic requirement

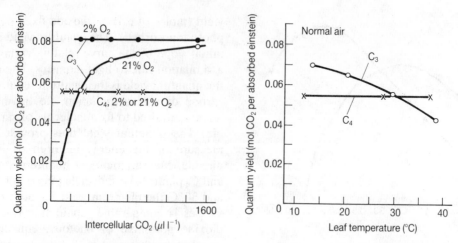

Figure 3.17. The quantum yield for net carbon dioxide uptake of C_3 (*Encelia californica*) and C_4 (*Atriplex rosea*) as a function of both carbon dioxide and oxygen concentration (left panel) and temperature in air (right panel). From Osmond *et al*, 1980[9], by permission of Springer-Verlag.

has been attributed to futile cycling through PEP carboxylase and NAD(P)–malic enzyme.

Water-use Efficiency

The C_4 mechanism brings advantages in water-use efficiency to C_4 plants. The fact that photosynthesis can operate at low intercellular concentrations of carbon dioxide, and hence lower stomatal conductance, means that C_4 plants can restrict water loss to a minimum. Thus water-use efficiency in C_4 plants is roughly double that of C_3 plants (Table 3.6). However, the C_4 mechanism does not confer any tolerance of water stress so that, in general, C_4 plants are no more tolerant of water stress than are C_3 plants.

Carbon Isotope Discrimination

[13]C is a naturally occurring stable isotope of carbon which is present in the atmosphere at concentrations of about 1% of total carbon dioxide. $^{13}CO_2$ diffuses more slowly than $^{12}CO_2$ in air and both PEP carboxylase and Rubisco discriminate against $^{13}CO_2$. However, the degree of discrimination by PEP carboxylase is slight compared to that by Rubisco. The extent of discrimination is

therefore an indication of the mechanism of carboxylation operating in photosynthesis.

The ratio of ^{13}C to ^{12}C in a sample can be measured by mass spectrometry. Discrimination against ^{13}C is expressed in units per thousand compared with a standard of belemnite limestone from South Carolina:

$$\delta^{13}C = \frac{^{13}C/^{12}C \text{ sample}}{^{13}C/^{12}C \text{ standard}} - 1 \times 10^3$$

Compared with the standard, the present atmosphere over open oceans shows a $\delta^{13}C$ value of $-7‰$. The fractionation value is the difference in $\delta^{13}C$ value between the source and the product. The fractionation value for diffusion in air is $+4‰$; for PEP carboxylase, $+2‰$; for Rubisco $+34‰$ and for the solution of carbon dioxide in water ($CO_2 + H_2O \leftrightarrow HCO_3^- + H^+$, catalysed by carbonic anhydrase) it is $-8‰$. $\delta^{13}C$ values in the products of photosynthesis are affected by fractionation at all these stages and are not simply attributable to the enzyme catalysing carboxylation. Thus although PEP carboxylase discriminates little, in C_4 and CAM plants $\delta^{13}C$ values are $-10‰$ to $-17‰$ (the value expected on the basis of carboxylation alone would be equal to $(-7 + (-2))$ $-9‰$). In C_3 plants values are typically in the region of $-29‰$. Although C_4 plants ultimately fix carbon dioxide via Rubisco, as in C_3 plants, the carbon dioxide-concentrating mechanism of C_4 plants means that Rubisco is effectively prevented from discriminating against $^{13}CO_2$. PEP carboxylase therefore determines the value of $\delta^{13}C$. In CAM plants the $\delta^{13}C$ values are variable. If the plant fixes carbon dioxide exclusively in the CAM mode, with no daytime fixation, then $\delta^{13}C$ values are like those of C_4 plants. On the other hand, if CAM is facultative or if appreciable daytime fixation occurs, then the $\delta^{13}C$ values are closer to, or equal to, those of C_3 plants (see above). $\delta^{13}C$ values are a function of the ratio of intercellular carbon dioxide to the external concentration of carbon dioxide and can therefore be used as a measure of the integrated water-use efficiency of, for example, crop plants. Xylem parasites such as mistletoes show lower $\delta^{13}C$ values than their hosts because they have a lower water-use efficiency than the host. $\delta^{13}C$ values of collagen in bone have been used by archaeologists to date the introduction of maize into the diet of North American Indians at AD 1000.

REFERENCES

1. Leegood, R.C. and Osmond, C.B. (1990) The flux of metabolites in C_4 and CAM plants. In *Plant Physiology, Biochemistry and Molecular Biology* (eds. D.T. Dennis and D.H. Turpin), pp. 274–298. Longman, Harlow.
2. Doncaster, H.D. and Leegood, R.C. (1987) Regulation of phosphoenolpyruvate carboxylase in maize leaves. *Plant Physiol.*, **84**, 82–87.
3. Hatch, M.D. (1981) Regulation of C_4 photosynthesis and the mechanism of light/dark modulation of pyruvate, phosphate dikinase activity. In *Photosynthesis IV*, pp. 227–236. Balaban International Science Services, Philadelphia.
4. Kluge, M. and Ting, I.P. (1978) *Crassulacean acid metabolism. Analysis of an ecological adaptation.* Springer-Verlag, Berlin.
5. Winter, K. (1982) Regulation of PEP carboxylase in CAM plants. In *Crassulacean Acid Metabolism* (eds I.P. Ting and M. Gibbs), pp. 153–169. American Society of Plant Physiologists, Rockville.
6. Badger, R.M., Kaplan, A. and Berry, J.A. (1977)
The internal CO_2 pool of *Chlamydomonas reinhardtii*: response to external CO_2. Carnegie Institution of Washington Yearbook 76, 362–366.
7. Lemon, E. (ed.) (1984) *CO_2 and Plants: the Response of Plants to Rising Levels of Atmospheric Carbon Dioxide.* Westview Press, Boulder, Colorado.
8. Edwards, G.E., Ku, M.S.B. and Monson, R.K. (1985) C_4 Photosynthesis and Its Regulation. In *Photosynthetic Mechanisms and the Environment* (eds J. Barber and N.R. Baker), pp. 287–327. Elsevier, Amsterdam.
9. Osmond, C.B., Björkman, O. and Anderson (1980a) *Physiological processes in plant ecology. Towards a synthesis with Atriplex.* Springer-Verlag, Berlin.

FURTHER READING

Edwards, G.E. and Ku, M.S.B. (1987) Biochemistry of C_3–C_4 intermediates. In *The Biochemistry of Plants*, Vol 10, (eds M.D. Hatch and N.K. Boardman), pp 275–325. Academic Press, New York.

Edwards, G.E. and Walker, D.A. (1983) *C_3, C_4: Mechanisms, and Environmental Regulation of, Photosynthesis.* Blackwell, Oxford.

Edwards, G.E., Nakomoto, H., Burnell, J.N. and Hatch, M.D. (1985) Pyruvate, Pi dikinase and NADP–malate dehydrogenase in C_4 photosynthesis: properties and mechanism of light/dark regulation. *Annual Review of Plant Physiology*, **36**, 255–286.

Jiao, J.-A. and Chollet, R. (1991) Posttranslational regulation of phosphoenolpyruvate carboxylase in C_4 and CAM plants. *Plant Physiology*, **95**, 981–985.

Hatch, M.D. (1988) C_4 photosynthesis: a unique blend of modified biochemistry, anatomy and ultrastructure. *Biochimica et Biophysica Acta*, **895**, 81–106.

Hatch, M.D. and Osmond, C.B. (1976) Compartmentation and transport in C_4 photosynthesis. *Encyclopedia of Plant Physiology*, Vol. 3 (eds C.R. Stocking and U. Heber), pp. 144–184. Springer-Verlag, Berlin.

Lüttge, U. (ed.) (1989) *Vascular Plants as Epiphytes*. Ecological Studies No. 76, Springer-Verlag, Berlin.

Nobel, P.S. (1988) *The Environmental Biology of Agaves and Cacti*. Cambridge University Press, New York.

Osmond, C.B. (1978) Crassulacean acid metabolism: a curiosity in context. *Annual Review of Plant Physiology*, **29**, 379–414.

Osmond, C.B. and Holtum, J.A.M. (1981) Crassulacean acid metabolism. In *Biochemistry of Plants*, Vol. 8 (eds M.D. Hatch and N.K. Boardman), pp, 283–328. Academic Press, New York.

Osmond, C.B., Winter, K. and Ziegler, H. (1980) Functional significance of different pathways of CO_2 fixation of photosynthesis. *Encyclopedia of Plant Physiology*, pp. 480–547.

Ting, I.P. (1985) Crassulacean acid metabolism. *Annual Review of Plant Physiology*, **36**, 595–622.

Ting, I.P. and Gibbs, M. (eds) (1982) *Crassulacean Acid Metabolism*. American Society of Plant Physiologists. Rockville, Maryland, USA.

Winter, K. (1985) Crassulacean acid metabolism. In *Photosynthetic Mechanisms and the Environment, Topics in Photosynthesis, Vol. 6* (eds J. Barber and N.R. Baker), pp. 329–387. Elsevier, Amsterdam.

4

CARBOHYDRATE CHEMISTRY

C.J. Smith

School of Biological Sciences, University of Wales, Swansea, UK

Plant Biochemistry and Molecular Biology. Edited by P.J. Lea and R.C. Leegood
© 1993 John Wiley & Sons Ltd

INTRODUCTION

In one form or another carbohydrates make up the bulk of the organic components of the living world. In plants they fulfil the same fundamental roles as they do in animal and microbial systems, and many of the pathways by which carbohydrates are metabolized are closely similar throughout the biological world. It is not surprising therefore to find that in plants oxidation of monosaccharides via glycolysis and related pathways provides the chemical energy of ATP as it does in mammalian tissues such as muscle or brain. Similarly the skeletons of carbohydrates are used in plants as the source of carbon for the synthesis of a wide range of other components (ultimately of course virtually all the carbon of biological molecules is derived from the carbohydrate product of photosynthesis), just as many microorganisms can synthesize the bulk of their constituents from simple sugars. Polymeric carbohydrates, in the form of starch and fructans, represent storage forms of carbon and energy that accumulate in seeds and tubers in a manner similar to the accumulation of glycogen in mammals and microorganisms. Some polysaccharides, such as cellulose, serve a structural function, providing support to the plant in the same way that chitin does in the exoskeletons of insects. In addition to these more obvious roles, carbohydrates are also important structural components of lipids (glycolipids) and a variety of proteins (glycoproteins) where their presence is often essential to the stability and activity of the protein.

Of course, as well as having features in common with the rest of the biological world there are some unique aspects of carbohydrate biochemistry in plants. Perhaps the most obvious of these is the presence of a photosynthetic system capable of fixing carbon dioxide and converting it to monosaccharides (chapter 2). While the independence from an external supply of carbohydrate that such a system gives to plants is a considerable advantage, it does not mean they are entirely free from the sorts of problem encountered by heterotrophs in their carbohydrate metabolism. Before being able to use carbohydrates, heterotrophs must transport them from the source (gut or extracellular space) and across the cell membrane. In the same way the carbohydrate products of photosynthesis must be transported out of the chloroplast and across the membrane surrounding it before they can be brought into the metabolic pathways of the cytoplasm. The system responsible for transport of carbohydrate out of the chloroplast is described in chapter 2. Here we will consider the mechanisms responsible for regulating production of carbohydrate within the chloroplast with its use outside in the cytoplasm.

Transport across cell membranes is only one of the problems that must be solved, and once the carbohydrate has been moved outside the chloroplast the plant is then faced with transporting it around the system before it can be used in growth and maintenance of all those tissues dependent upon it. The disaccharide sucrose, one of the major early products of photosynthesis, plays a fundamental role in transport of carbohydrate within plants. Its metabolism is central to the distribution of photosynthate and will be considered in the light of that role.

The polysaccharide starch is another major product of photosynthesis. We shall see that its synthesis occurs in two distinctly different situations. In the leaf it is synthesized within the chloroplast at times when the capacity of the chloroplast to synthesize carbohydrate exceeds its capacity to export it to the cytoplasm. Under these circumstances storage is short term, serving to buffer the synthetic capacity of the chloroplast. In contrast, starch synthesis in storage tissues such as seeds and tubers occurs over a longer period. In this case starch synthesis serves to build a supply of carbon and energy to be used at a later date, for instance during germination. The metabolism of starch under both circumstances is described in this chapter, in addition to the role and metabolism of a second class of storage polysaccharide, the fructans.

A further major difference between the carbohydrate metabolism of plants and animals is synthesis of a cell wall surrounding the plant cell. Composed of a mixture of complex polysaccharides, the cell wall represents a considerable investment to the plant in terms of carbohydrate metabolism, at times consuming a large proportion of the photosynthetic output. Considerable progress has been made in defining the molecular architecture of the wall and this will be considered in the appropriate section.

CARBOHYDRATE CHEMISTRY

Before considering the metabolic transformations of carbohydrates it is necessary to describe something of their structure and chemistry. What follows is not intended to be a comprehensive treatment of the topic, and a more extensive discussion may be found in Pigman and Horton[1].

Monosaccharide Structure

The number of monosaccharides that have been identified throughout the plant kingdom is considerable. Fortunately, however, those that occur in significant quantity and which will be dealt with here are relatively few. The structures of some of them, in the form of Fischer projection formulae, are presented in Figure 4.1.

Monosaccharides are polyhydroxy compounds possessing a carbonyl group in the form either of an aldehyde or a ketone. Those monosaccharides containing an aldehyde group are aldoses (e.g. glucose); those containing a ketone group are ketoses (e.g. fructose). Monosaccharides may be further classified as pentoses, hexoses, etc. according to the number of carbon atoms present. Glucose is therefore an example of an aldohexose, fructose a ketohexose, while arabinose, a common constituent of the polysaccharides of cell walls, is an aldopentose.

The carbonyl group is an important functional group of monosaccharides and in the Fischer projection formulae of Figures 4.1 and 4.2 the structures are drawn in accordance with the convention that places the carbonyl or potential carbonyl group at the top of the molecule. This top carbon is designated 1 (C1). One of the reactions the aldehyde group may undergo is oxidation to a carboxylic acid group, and when this occurs the molecule formed is termed an aldonic acid. Glucose, for example, gives rise to gluconic acid (Figure 4.2). Also shown in Figure 4.1 is glucuronic acid, a sugar that occurs as a component of pectic polysaccharides of cell walls. Uronic acids arise from the oxidation of the hydroxyl group on C6 of the sugar.

In contrast, reduction of the aldehyde or ketone group leads, in both cases, to the formation of a polyhydroxy alcohol or polyol (Figure 4.2). Polyols are a group of naturally occurring carbohydrate derivates that, in addition to other roles, function in the interconversions of some monosaccharides. Reduction of the aldehyde of glucose is straightforward and produces the polyol glucitol (Figure 4.2), but a complication arises when the ketone group of fructose is reduced because the ketone group is at C2 and two products can result in equal proportions— glucitol and mannitol (Figure 4.2). This situation arises because the final position of the hydroxyl group that is produced by reduction of the ketone may be projecting either to the right of the carbon atom, in which case the product is glucitol, or to the left, mannitol.

Stereoisomerism

The example of fructose reduction illustrates another aspect of carbohydrate chemistry— stereoisomerism. What has happened during reduction of the ketone at C2 of fructose is that an asymmetric carbon has been introduced into the molecule. An asymmetric carbon atom is one which has four different substituents attached to it. (In this case following reduction of the ketone the substituents are a hydrogen atom, a hydroxyl group, a carbon (C1) bearing an aldehyde group, and a second carbon (C3) bearing the remainder of the molecule.) The substituents on such an asymmetric or chiral carbon can be arranged around the carbon atom in either of two ways, such that one arrangement is a mirror image of the other and neither can be superimposed on the other. The two different forms of the molecule

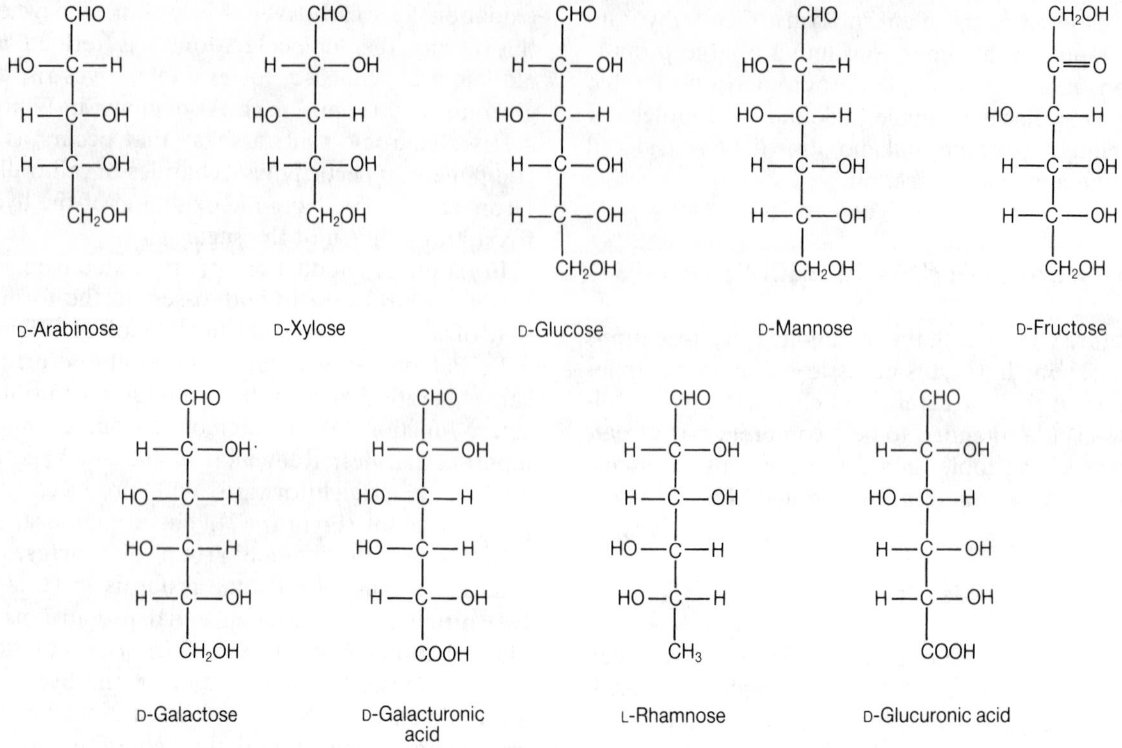

CHO · HO—C—H · H—C—OH · H—C—OH · CH₂OH

D-Arabinose

CHO · H—C—OH · HO—C—H · H—C—OH · CH₂OH

D-Xylose

CHO · H—C—OH · HO—C—H · H—C—OH · H—C—OH · CH₂OH

D-Glucose

CHO · HO—C—H · HO—C—H · H—C—OH · H—C—OH · CH₂OH

D-Mannose

CH₂OH · C═O · HO—C—H · H—C—OH · H—C—OH · CH₂OH

D-Fructose

CHO · H—C—OH· · HO—C—H · HO—C—H · H—C—OH · CH₂OH

D-Galactose

CHO · H—C—OH · HO—C—H · HO—C—H · H—C—OH · COOH

D-Galacturonic acid

CHO · H—C—OH · H—C—OH · HO—C—H · HO—C—H · CH₃

L-Rhamnose

CHO · H—C—OH · HO—C—H · H—C—OH · H—C—OH · COOH

D-Glucuronic acid

Figure 4.1. Fischer projection formulae for some of the naturally occurring monosaccharides of plants. They are all aldoses, with the exception of D-fructose, which is a ketose. The structures are shown with the carbonyl or potential carbonyl group at the top end of the molecule; this top carbon is C1.

are termed stereoisomers. Aldohexoses contain four such chiral carbon atoms, each of which can be varied independently of the other. In the case of glucitol and mannitol mirror images exist only at C2; at all the other carbon atoms the arrangement of the substituents is the same in both sugars. Sugar stereoisomers which differ in arrangement at only a single asymmetric centre are known as epimers (in the specific case of glucose and mannose, C2 epimers (Figure 4.1)).

The absolute arrangement or configuration of the four substituents at a chiral carbon is designated by the symbols D and L. Whether a particular chiral centre is D or L is determined by a convention that relates its configuration to D- or L-glyceraldehyde (Figure 4.3). Isomers stereochemically related to D-glyceraldehyde are designated D, those related to L-glyceraldehyde, L. Monosac-

charides with more than three carbon atoms contain more than one chiral centre, and where this occurs it is the configuration round the chiral carbon most remote from the carbonyl group that is used to assign a sugar to the D- or L- series. In carbohydrates, it is the penultimate carbon atom which is the chiral carbon most remote from the carbonyl group. Thus, in the case of D-glucose it is the configuration about C5 that is related to the configuration of D-glyceraldehyde.

Hemiacetal Formation and Ring Structure

Aldehydes and ketones can add hydroxyl compounds to the carbonyl group (Figure 4.4). If the hydroxyl-carrying compound is an alcohol the product of the addition is termed a hemiacetal.

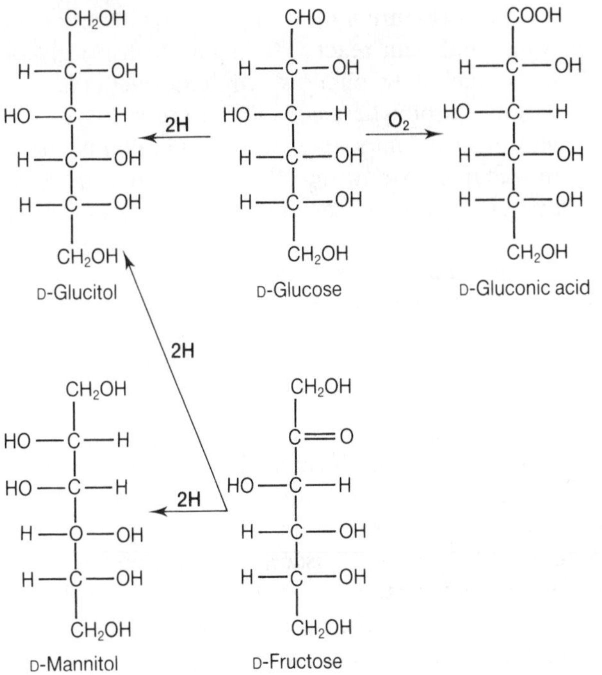

CH₂OH / D-Glucitol — CHO / D-Glucose — COOH / D-Gluconic acid

$$CH_2OH\ (D\text{-Glucitol}) \xleftarrow{2H} CHO\ (D\text{-Glucose}) \xrightarrow{O_2} COOH\ (D\text{-Gluconic acid})$$

D-Glucitol

2H

D-Mannitol — CH₂OH

$$\xleftarrow{2H}$$

D-Fructose — CH₂OH / C=O

$$\xleftarrow{2H}$$

Figure 4.2. Oxidation and reduction of the carbonyl group of monosaccharides. Note that reduction of the aldehyde of glucose gives rise only to the polyol glucitol. Reduction of the ketone group at C2 of fructose introduces a new asymmetric centre and so gives rise to two products: glucitol and mannitol.

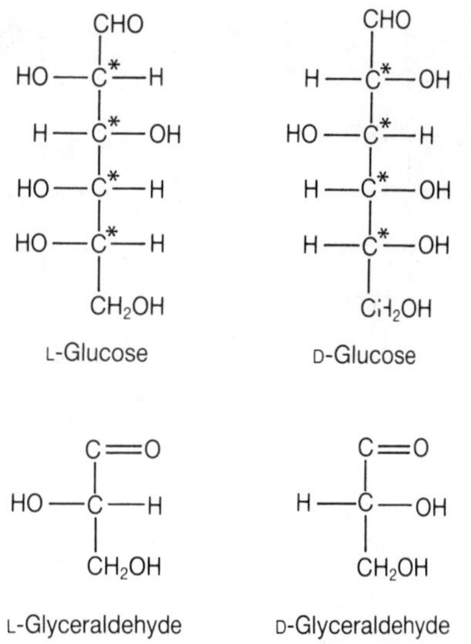

L-Glucose — CHO

D-Glucose — CHO

L-Glyceraldehyde — C=O

D-Glyceraldehyde — C=O

Figure 4.3. Stereoisomerism. The projection formulae of D- and L-glyceraldehyde, showing the arrangement of four different substituents around the chiral carbon atom such that D- and L-glyceraldehyde are mirror images of each other. The convention used is that horizontal substituents project forwards and vertical substituents backwards. In D- and L-glucose the chiral carbon atoms are marked with an asterisk. Assignment of a monosaccharide to the D- or L-series is based on the relation of the chiral carbon most remote from the carbonyl group with the configuration of D- or L-glyceraldehyde.

Addition to the hemiacetal of a further alcohol results in formation of an acetal (see 'Glycoside Formation'). Monosaccharides readily form intra-molecular hemiacetals in which both the carbonyl and hydroxyl group taking part in the addition belong to the same sugar molecule. The result is a ring structure (Figure 4.5). Aldohexoses can form five- or six-membered rings which are called furanoses or pyranoses based on their relationship with the five-membered ring structure of furan or the six-membered ring of pyran, respectively. For aldohexoses the pyranose form of the molecule predominates over the furanose or the open-chain structure.

A consequence of ring formation is the introduction of a new asymmetric centre at the carbon formerly bearing the carbonyl group, and two possible forms or anomers of the molecule exist. In the example of D-glucose this new centre is at C1 and in the Fischer projection formula ring closure may fix the position of the new hydroxyl group on the same side as the ring oxygen, in which case it is known as the α anomer, or on the opposite side, the β anomer (Figure 4.5).

The different ring forms of a sugar such as glucose, the furanose and pyranose forms, α and

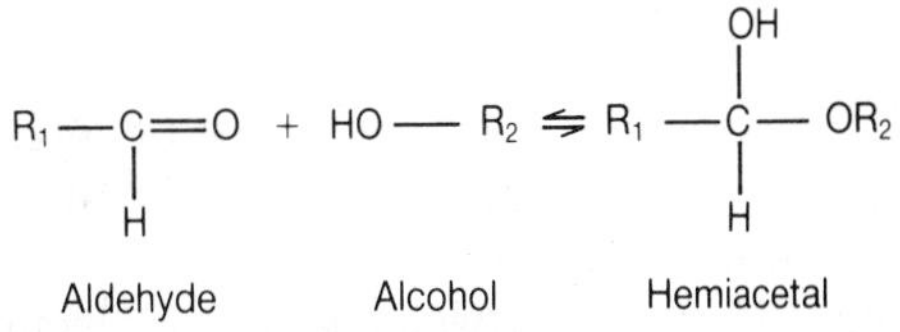

$$R_1-C(H)=O + HO-R_2 \rightleftharpoons R_1-C(H)(OH)-OR_2$$

Aldehyde   Alcohol   Hemiacetal

Figure 4.4.

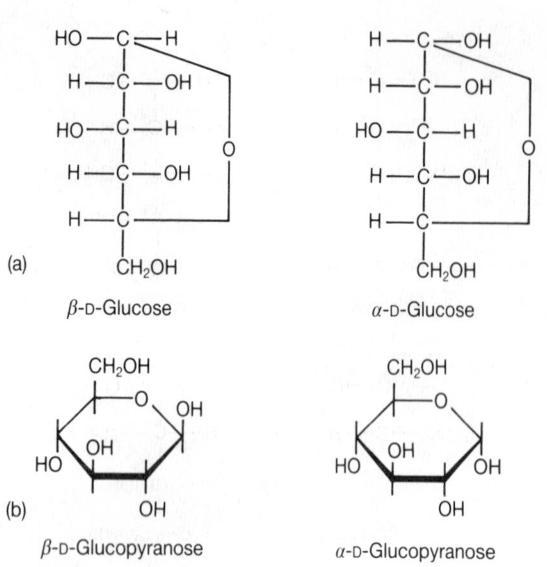

Figure 4.5. Cyclic or ring forms of D-glucose represented by projection (a) and Haworth (b) formulae.

β, make an equilibrium mixture in solution because they can be interconverted via the free aldehyde or ketone form. Thus an aqueous solution of glucose contains β-D-glucopyranose (c. 66%), α-D-glucopyranose (c. 33%) with small quantities of both furanose forms, α- and β-, and the free aldehyde. Once the anomeric hydroxyl is used in glycoside formation, however (see 'Glycoside Formation'), there is no free aldehyde and so interconversion is not possible and the ring form of the sugar is fixed.

A more complete representation of the ring structure of sugars is given by the Haworth projection formulae (Figure 4.5). In these representations the position of the anomeric hydroxyl is shown as being below the plane of the ring (α anomer) or above it (β).

Glycoside Formation

When the anomeric hydroxyl of a monosaccharide reacts with a second hydroxyl-containing molecule with the elimination of water, an acetal is formed. The bond between the two molecules is called a glycosidic linkage and the product is a glycoside (Figure 4.6). Many hydroxyl-containing compounds can react with sugars to form glycosides, including phenols, aliphatic alcohols and various carboxylic acids. There are many examples of such compounds occurring in plants but in this chapter we are mostly concerned with compounds where the second component is another sugar. When two sugars are joined via a glycosidic linkage the resulting glycoside is called a disaccharide. Disaccharide formation is a special case of an acetal (Figure 4.7). Formation of a glycosidic linkage always involves the anomeric hydroxyl of one of the sugars, but there are several hydroxyls on the second sugar that could be involved in the linkage. This means therefore that a number of different disaccharides could be produced from two monosaccharides and the precise nature of the disaccharide formed will be determined by the activity of the enzyme catalysing formation of the linkage as well as the nature of the sugars themselves.

Note that once the anomeric carbon has been substituted (in this case by linkage to another sugar) the anomeric form of the molecule is fixed, giving rise to α- or β-linkages depending upon the form of the sugar involved.

A sugar that is linked to another molecule by way of its carbonyl group is termed a glycosyl residue and is no longer a reducing sugar. Maltose, a disaccharide that features as a major intermediate during the mobilization of starch reserves, contains two glucose residues and so may be named specifically D-glucosyl-D-glucose (Figure 4.6), the convention being to name the non-reducing residue, the glycosyl unit, first. (The ending 'osyl' indicates the non-reducing residue, the ending 'ose' indicates that the sugar is still in possession of a reducing group.)

A more complete nomenclature involves specifying the carbon atoms bearing the carbonyl and the hydroxyl group involved in the linkage, in addition to the ring structure and anomeric form of the glycosyl residue. (In a disaccharide the ring form of the sugar bearing the carbonyl involved in linkage is fixed, while the ring forms of the other sugar may interconvert through the unsubstituted carbonyl group.) Hence a complete name for the

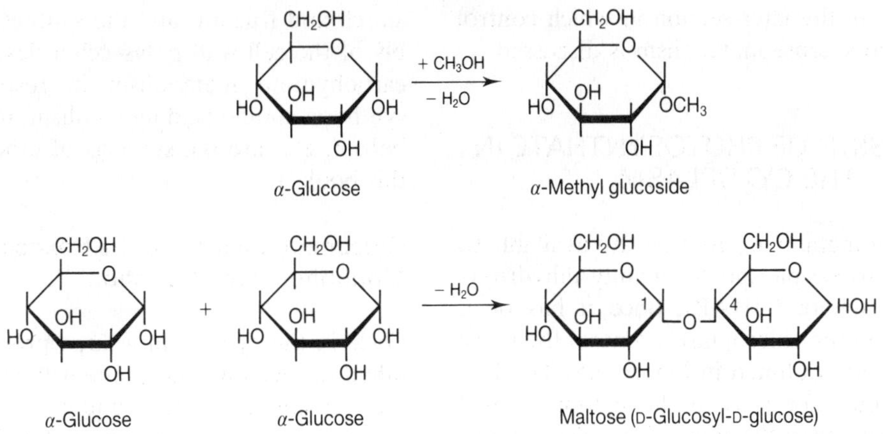

Figure 4.6. Formation of a glycosidic linkage between two monosaccharides. The glycoside product is called a disaccharide. A complete name for maltose involves describing the anomer (α), the isomer (D), the ring form (pyranosyl) of the unsubstituted sugar and the linkage (1→4) between the two monosaccharides.

disaccharide maltose is α-D-glucopyranosyl-(1→4)-D-glucose. The arrow points from the glycosyl residue towards the reducing terminus.

When more than two sugar residues are joined the product may be described by the general term oligosaccharide, though an oligosaccharide may be named more specifically according to the number of residues present as a tri-, tetra-, penta-saccharide, etc. When the number of residues is greater than about 20 the term polysaccharide is applied and 'an' is used as a suffix in the name. Cellulose for example is a glucan, while a polymer of xylose is a xylan.

PHOTOSYNTHESIS AS THE SOURCE OF CARBOHYDRATE

Fixation and reduction of carbon dioxide by the reactions of the Calvin cycle (chapter 2) lead to the production of phosphorylated sugars located in the stroma of the chloroplast. They represent the source of the carbon that will be used to synthesize the vast array of components required for growth and maintenance of the plant. There is a problem, however, in that the majority of the metabolic pathways that will make use of this source of carbon are located outside the chloroplast, and so the carbon in some form or other must be transported across the chloroplast membrane. Details of the mechanism by which carbohydrate is transported from the stroma across the barrier of the inner chloroplast membrane and into the cytoplasm will be found in chapter 2. At this point, however, it should be emphasized that the translocator responsible is a key element in maintaining the metabolic connection between the reactions of the stroma that produce the carbohydrate and those of the cytoplasm that make use of it. The significance of that connection will

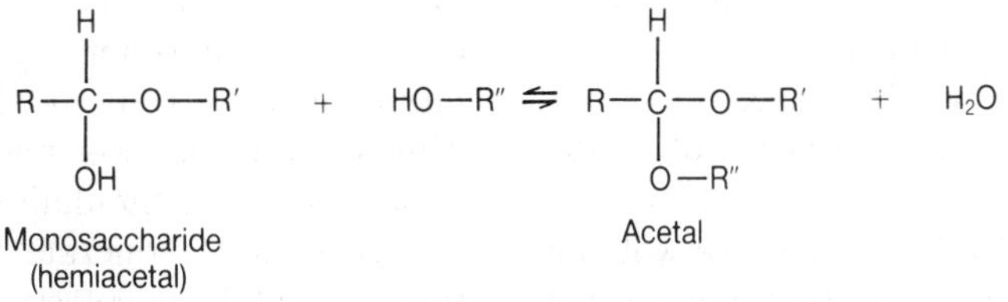

Figure 4.7.

be apparent in the later section in which control of starch and sucrose metabolism is discussed.

PROCESSING OF PHOTOSYNTHATE IN THE CYTOPLASM

A number of metabolic pathways are available to process the triose phosphate, mainly dihydroxy-acetone phosphate (DHAP), once it has been exported from the chloroplast. Some of the more central ones are outlined in Figure 4.8. The fate of the carbon once it is in the cytoplasm will depend in part upon the stage of development of the particular leaf tissue. In immature leaves much of the photosynthate will be retained within the leaf and will be used to synthesize new cellular components. For example, measurements of photosynthate distribution in leaves of white lupin show that during its first 10 days, in addition to retaining all the photosynthate it produces, the developing leaf imports carbon from other sources. In this situation the carbon fluxes will be primarily along those pathways leading to synthesis of lipid, the pentose phosphate pathway (nucleic acids), glycolysis (ATP and pyruvate), the tricarboxylic acid (TCA) cycle, amino acid synthesis and the sugar nucleotide oxidation pathway (cell wall components); very little of the fixed carbon will be diverted to the synthesis of sucrose for transport out of the leaf, or to synthesis of starch or fructans for storage. In contrast, much of the photosynthate of mature leaves is exported, in the form of sucrose, to other parts of the plant, particularly the non-photosynthetic parts such as root, shoot apex and the developing storage organs (fruits, seeds, etc.). Transition from an importer to a net exporter of photosynthate occurs in the white lupin at a point 10 days after leaf initiation, with a peak of export occurring some 10 days later. Thus the major flux of carbon in mature leaves will be directed towards the synthesis of sucrose for transport out of the leaf.

In the remainder of this chapter we will be mainly considering the metabolism of sucrose, the formation and role of the storage polysaccharides starch and fructan, and the synthesis and assembly of the cell wall polysaccharides. The roles of carbohydrate metabolism in respiration, lipid synthesis, amino acid metabolism, nitrogen metabolism, etc. are the subjects of other chapters in this book.

Gluconeogenesis and the Hexose Monophosphate 'Junction'

Once in the cytoplasm a large proportion of the dihydroxyacetone phosphate will be converted by the gluconeogenetic sequence of reactions (Figure 4.8) to hexose phosphates. The sequence involves isomerization of some of the DHAP to glyceraldehyde 3-phosphate and combination of these two trioses in an aldol condensation to form fructose 1,6-bisphosphate. Through the activity of fructose 1,6-bisphosphatase this is converted to fructose 6-phosphate, one of the three hexose monophosphates which together form an important metabolic junction (Figure 4.9) and which are generally regarded to be close to equilibrium with each other. The two other members of the junction are glucose 6-phosphate and glucose 1-phosphate formed from fructose 6-phosphate through the successive activities of phosphohexo-isomerase and phosphoglucomutase, respectively. Measurement of the concentration of these three metabolites *in vivo* permits calculation of the mass-action ratios which, when compared to the equilibrium constants for the two reactions, indicate that the reactions are poised close to their equilibrium positions. The junction represents a point of interaction between many of the reactions of carbohydrate metabolism and the metabolism of other cellular components. Carbon from any source that can be converted to one of the three hexose monophosphates can pass through this junction and can be used in glycolysis or any of the other metabolic sequences that interconnect with the pool (Figures 4.8 and 4.9). Thus gluconeogenesis, photosynthesis, hydrolysis of sucrose and degradation of starch, fructans or cell walls can all generate fluxes of carbon into the junction, while glycolysis, the pentose phosphate pathway and the synthetic sequences leading to

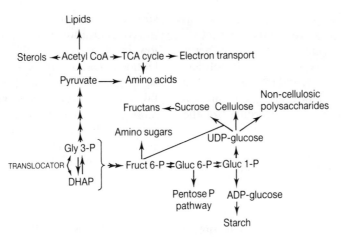

Figure 4.8. Metabolic pathways of the cytosol by which triose phosphate exported from the chloroplast may be processed. The flux generated through each one will depend upon the developmental stage of the leaf as well as the rates of photosynthesis achieved.

sucrose, starch, fructans and cell wall polysaccharides, in addition to such pathways as lipid synthesis, all behave as sinks taking carbon away from the junction. The direction of flux of the carbon will depend upon the metabolic demands and conditions prevailing in a particular tissue. When sucrose is present in excess, as is the situation when it is imported into developing grain or storage tissues such as tubers, for example, it will be broken down either by the activity of sucrose synthase or invertase to form UDP–glucose and

fructose in the case of the former, or glucose and fructose when cleavage is catalysed by invertase (see the later section on 'Transport of Sucrose: Unloading'). While the UDP–glucose may be used directly as a precursor of cell wall synthesis or cleaved in a reversal of the UDP–glucose pyrophosphorylase-catalysed reaction to yield glucose 1-phosphate, the fructose produced may enter the metabolic junction through the activity of hexokinase which phosphorylates the fructose to form fructose 6-phosphate. The carbon that enters the junction may then leave to be used in glycolysis or the pentose phosphate pathway, or be converted to glucose 1-phosphate and used in the formation of ADP–glucose for starch synthesis. If the demand for wall synthesis is high, however, and the impact of sucrose restricted, as may happen in the early stages of leaf development, a major flux out of the junction will be towards UDP–glucose formation.

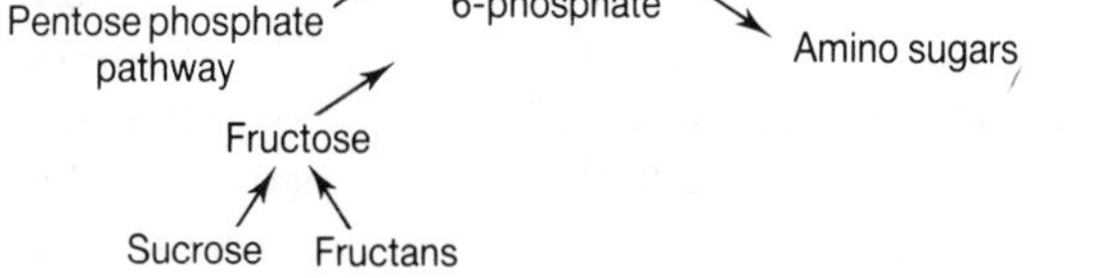

Sugar Nucleotide Formation and Interconversion

Nucleoside Diphosphate Sugar Formation via Pyrophosphorylase

The final stage in the provision of substrates for glycoside synthesis is formation of the appro-

Figure 4.9. The hexose monophosphate junction, showing some of the sources and sinks for carbon passing through it. 1, phosphoglucomutase; 2, phosphohexoisomerase.

priate nucleoside diphosphate sugars that act as donors of the monosaccharide units. At times a large proportion of the photosynthetic product is diverted towards synthesis of oligo- and polysaccharides, so that sugar nucleotide formation is an important feature of metabolism. The formation of nucleotide sugars occurs according to the general reaction

$$NTP + glycosyl\ 1\text{-phosphate} \rightarrow NDP\text{--glycose}$$
$$+ PP_i$$

(nucleotide   (nucleotide
triphosphate)   diphosphate sugar)

The reaction is catalysed by enzymes called pyrophosphorylases or nucleotidyltransferases. Many different pyrophosphorylases occur in plant tissues, each catalysing formation of a different nucleoside diphosphate sugar. Therefore, providing that the appropriate monosaccharide 1-phosphate and nucleotide triphosphate are available, plants can synthesize a wide range of nucleotide diphosphate sugars through the activity of these pyrophosphorylases. The different combinations of sugar and nucleotides that occur are not as extensive as that statement may at first make it appear. Certainly glucose is found in combination with a number of nucleotides, including ADP, GDP and UDP, but many of the sugars used in synthesis of polysaccharides are derivatives of uridine, and it is also the uridine derivatives that feature in the interconversion of the various sugars.

In plants a number of kinases exist that can phosphorylate free monosaccharides to form the monosaccharide 1-phosphate; they include kinases that act upon L-arabinose, D-galactose, L-fucose, D-galacturonic acid and D-glucuronic acid (Figure 4.10). Other sugars, for instance glucose and fructose, may be phosphorylated to form initially the 6-phosphates, and are converted to the 1-phosphates by the appropriate mutases. The activity of these hexose kinases may be quite high in some tissues and they may be especially important in bringing the free fructose and glucose released by the cleavage of sucrose into 'main-stream' metabolism (see earlier), but generally the concentration of free monosaccharides is very low in plants and so the formation of the nucleotide sugars via these routes is usually not very significant. Where the activity of these kinases takes on greater significance is in 'scavenging' monosaccharides produced from the hydrolysis of various polysaccharides, for instance cell wall and storage polysaccharides of seeds, and in the case of sucrose, which has already been referred to.

Sugar Nucleotide Oxidation Pathway and Associated Reactions

With the exception of ADP--glucose, which is used in the synthesis of starch, a large proportion of the nucleotide sugars employed in plant metabolism arise from a series of interlinked reactions in which substrates are acted upon by a series of epimerases, kinases, pyrophosphorylases and a decarboxylase, to produce a large range of nucleotide derivatives (Figure 4.10). The central part of this scheme (phosphoglucomutase, UDP--D-glucose pyrophosphorylase and UDP--glucose dehydrogenase) is referred to as the sugar nucleotide oxidation pathway, and a key enzyme in the scheme is UDP--D-glucose pyrophosphorylase, which catalyses formation of UDP--glucose from UTP and glucose 1-phosphate[2]. UDP--glucose is a nucleotide sugar that occupies a central position in sugar nucleotide metabolism. It serves both as a precursor in sucrose synthesis and in the formation of those sugar nucleotides required for the synthesis of the cell wall components, both cellulose and the non-cellulosic polysaccharides. At times, then, a considerable flux of carbon passes through the hexose monophosphate junction described earlier (Figure 4.9) and is directed towards sugar nucleotide synthesis through the formation of UDP--glucose. It is not surprising therefore to find that the activity of this pyrophosphorylase in some tissues may be several hundred times higher than that of pyrophosphorylases catalysing formation of sugar nucleotides that do not occupy such a central role. It is present in photosynthetic and non-photosynthetic

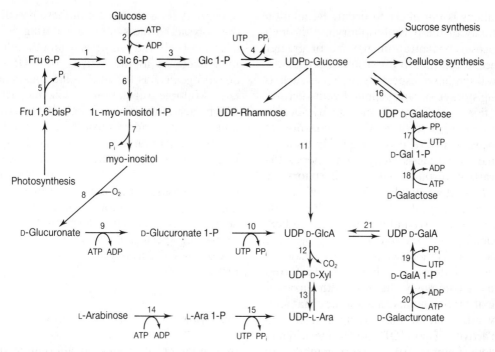

Figure 4.10. Sugar phosphate and sugar nucleotide interconversion, including the sugar nucleotide- and *myo*-inositol oxidation pathways. 1, phosphohexoisomerase; 2, hexokinase; 3, phosphoglucomutase; 4, UDP–glucose pyrophosphorylase; 5, fructose 1,6-bisphosphatase; 6, 1L-*myo*-inositol 1-phosphate synthase; 7, *myo*-inositol 1-phosphatase; 8, *myo*-inositol oxygenase; 9, glucuronokinase; 10, UDP–glucuronate pyrophosphorylase; 11, UDP–glucose dehydrogenase; 12, UDP–glucuronate decarboxylase; 13, UDP–xylose 4-epimerase; 14, arabinokinase; 15, UDP–arabinose pyrophosphorylase; 16, UDP–glucose 4-epimerase; 17, UDP–galactose pyrophosphorylase; 18, galactokinase; 19, UDP–galacturonate pyrophosphorylase; 20, galacturonokinase; 21, UDP-glucuronate-4-epimerase.

tissues and is particularly active in tissues such as developing leaf or storage tissues where sucrose and/or cell wall synthesis are prominent features of metabolism[2].

The UDP–glucose pyrophosphorylase reaction channels glucose 1-phosphate into the synthesis of a number of components that are derived from UDP–glucose, including sucrose, cellulose and non-cellulosic polysaccharides, the actual flux into each depending upon the particular developmental stage of the tissue. At such a 'branch' point in a metabolic pathway it is frequent to find controls that regulate the flux through the pathway with the demand for its product. In this particular sequence the activity of UDP–glucose pyrophosphorylase is regulated both by its own product and by metabolites of that product.

UDP–glucose has been found to be an allosteric inhibitor of the activity of the pyrophosphorylase at concentrations which appear to be physiological, at least in some tissues (K_i value for UDP–glucose, 0.1 mM), so that the flux of carbon through the reaction may be regulated by the rate at which the product (UDP–glucose) and those nucleotide sugars derived from it are used[2].

In the photosynthetically active leaf, inhibition of the pyrophosphorylase activity by UDP–glucose is unlikely to be a major regulatory influence, however. Under normal conditions synthesis of sucrose is rapid and the disaccharide is transported out of the leaf via the vascular system, or into the vacuole for temporary storage. Consumption of UDP–glucose by this reaction will be rapid and so its build-up to inhibitory concentrations is not likely to occur. Regulation of the activity of the pyrophosphorylase through UDP–glucose concentration may be of greater significance in relation to provision of precursors for cell wall synthesis, especially in tissues such as developing storage organs supplied with sucrose. In these tissues cleavage of sucrose by sucrose synthase leads to a substantial concentration of UDP–glucose which, if it is not quickly used for polysaccharide synthesis, may then restrict the flux of carbon through the pyrophosphorylase reaction. This effect may be important in storage tissue in regulating the flux of UDP–glucose between wall synthesis and its conversion to hexose monophosphate (see later).

The use of UDP–glucose as the precursor for more than one pathway has already been referred to and, as is commonly the case with enzymes operating at branch points, metabolites that subsequently arise out of the initial product also influence activity. Thus UDP–glucuronic acid and UDP–xylose (Figure 4.10) can exert an inhibitory effect on the pyrophosphorylase independently of each other, but they are more effective in combination, acting cumulatively to inhibit the enzyme and reinforce the effect of UDP–glucose[2]. The overall result of this feedback inhibition is to match the rate of formation of UDP–glucose with its use both in cellulosic and non-cellulosic polysaccharide synthesis and thus regulate flux of carbon into these polysaccharides.

UDP–glucose can be used directly in sucrose and cellulose synthesis, but further transformation of UDP–glucose is required to give rise to the nucleotide sugars that are the glycosyl donors for the non-cellulosic polysaccharides. The donor of D-glucuronic acid residues in polysaccharide synthesis is UDP–glucuronic acid, which is also the precursor of UDP–galacturonic acid, UDP–xylose and hence UDP–arabinose. All four nucleotide sugars are used in the synthesis of pectic and hemicellulosic polysaccharides of cell walls. UDP–glucuronic acid is formed through the activity of UDP–glucose dehydrogenase, an enzyme that is inhibited competitively by its product, UDP–glucuronic acid, and allosterically by UDP–xylose, a subsequent product in the scheme. A reduction in use of these precursors by the polysaccharide synthetases during periods of decreased cell wall synthesis would therefore lead to a decrease in the activity of UDP–glucose dehydrogenase as the nucleotide sugars accumulate. Multiple regulation of enzyme activity in which each of the end-products exerts a degree of influence allows for control to be exerted by all of the subsequent branches of the pathway. Maximum inhibition of enzyme activity occurs only when all of the end-products are present in excess concentrations. When only some of them are at inhibitory concentrations, partial inhibition of activity occurs. UDP–xylose regulates its own formation specifically by inhibiting the UDP–glucuronate decarboxylase responsible for its formation.

myo-*Inositol Oxidation Pathway*

In a number of tissues an alternative pathway leading to formation of UDP–glucuronic acid has been identified—the *myo*-inositol oxidation pathway (Figure 4.10). In many plants it is not possible to demonstrate activity of UDP–glucose dehydrogenase—the enzyme that catalyses oxidation of UDP–glucose to UDP–glucuronic acid— and yet it is possible to demonstrate the presence of substantial quantities of D-glucuronic acid in such tissues. Although some glucuronic acid may result from the breakdown of polysaccharides, the majority of the glucuronic acid that occurs in tissues lacking the dehydrogenase activity must arise from an alternative pathway, the *myo*-inositol oxidation pathway[3]. *myo*-Inositol is a cyclic hexitol that does not possess a carbonyl group; consequently the ring is formed from six carbon atoms. There being no ring oxygen, *myo*-inositol does not have an open chain form.

In the *myo*-inositol oxidation pathway *myo*-inositol is converted to UDP–glucuronic acid by the action of a series of enzymes which bypass the formation of UDP–glucose. This pathway can operate to use inositol reserves where they are available, for instance in seeds of a number of species or in pollen grains, where inositol is stored as the hexakisphosphate, phytic acid. Inositol is released from this stored form through the activity of a phytate-specific acid phosphohydrolase, termed phytase, which is active during germination and catalyses the sequential removal of phosphate groups from phytic acid to produce *myo*-inositol[4]. The free inositol produced has a multifunctional role in plant development and may be used in a variety of ways, including synthesis of pentose and uronide residues of the cell wall.

The sequential activities of three enzymes enable the conversion of free *myo*-inositol to UDP–glucuronic acid: they are *myo*-inositol oxygenase, which catalyses oxidative cleavage of *myo*-inositol to form glucuronic acid; glucuronokinase, which converts glucuronic acid to glucuronic acid 1-phosphate; and UDP–glucuronate pyrophosphorylase, the enzyme that forms UDP–glucuronic acid from glucuronic acid 1-phosphate and UTP. The addition of a further two enzymes—1L-*myo*-inositol 1-phosphate synthase, which catalyses the cyclization of glucose 6-phosphate to 1L-*myo*-inositol 1-phosphate; and *myo*-inositol 1-phosphatase, which catalyses formation of free *myo*-inositol from the phosphate derivative—links synthesis of free *myo*-inositol to the *myo*-inositol oxidation pathway and permits the conversion of D-glucose 6-phosphate to UDP–glucuronic acid. This scheme provides an alternative route to the sugar nucleotide oxidation pathway for UDP–glucuronate formation from glucose 6-phosphate, and in studies with D-glucose labelled specifically with both ^{3}H and ^{14}C a tritium isotope effect was found at C5, indicating that a major portion of the uronic acid and pentose residues of cell walls may be derived by way of a flux of carbon from glucose 6-phosphate passing down this pathway[5].

In rapidly growing tissues, for instance germinating lily pollen which synthesizes mostly noncellulosic polysaccharides, the activity of UDP–D-glucuronate pyrophosphorylase, an enzyme operational in the *myo*-inositol pathway, is much higher than that of UDP–glucose dehydrogenase (sugar nucleotide oxidation pathway), and in some tissues the dehydrogenase cannot be detected at all. Thus in these tissues the *myo*-inositol oxidation pathway may serve to provide hexuronic and pentose residues for cell wall synthesis from glucose. Synthesis of these residues would not then be dependent upon the supply of UDP–glucose for which synthesis of cellulose and other cellular components will be competing. In fact diversion of UDP–glucose towards sucrose, glucan and galactan synthesis may result from a large flux of carbon through the *myo*-inositol oxidation pathway since the UDP–xylose and UDP–glucuronate that the pathway produces are potent inhibitors of the dehydrogenase and will reduce UDP–glucose oxidation by the dehydrogenase.

The precise flux down either the *myo*-inositol oxidation or the sugar nucleotide oxidation pathway will depend upon such factors as the availability of *myo*-inositol from storage components, the availability of UDP–glucose from sucrose breakdown catalysed by sucrose synthase, and the stage of development of the tissue in terms of cell wall synthesis. Control over the *myo*-inositol pathway appears to operate at the glucuronokinase-catalysed step. The enzyme from lily pollen, for example, is inhibited both by its own product, glucuronic acid 1-phosphate, and the product of the subsequent step, UDP–glucuronate.

METABOLISM OF SUCROSE

General

In the leaf, when production of photosynthate exceeds its consumption the balance is exported to the rest of the plant. In the majority of plants the form in which photosynthate is transported is the disaccharide sucrose. A number of reasons have been advanced to explain the widespread role of this particular disaccharide as a transport molecule. These include its high solubility in water, its electrical neutrality, its apparent lack of inhibitory effect on the majority of biochemical processes, even at relatively high concentrations, and its non-reducing character which makes it less likely than other sugars, such as glucose or fructose, to interact with some of the other functional groups that occur in plants[6].

Whatever the particular advantages sucrose has as a transport molecule, it remains a fact that once formed it can be used metabolically in only a few ways. With the exception of fructan synthesis, as we shall see later, sucrose only achieves its full metabolic potential once it has been cleaved. Since only a limited number of enzymes can catalyse metabolic conversion of sucrose, its metabolism is markedly affected by the distribution of those enzymes within the plant.

Although some emphasis has been placed here on its role in transport, it should not be overlooked that sucrose metabolism plays a major part in the mechanism for fixation of carbon dioxide, and is a major storage carbohydrate in many plants[7]. Both aspects will be dealt with in later sections.

Synthesis

In the leaf most of the hexose that is used in sucrose biosynthesis is derived in the cytosol via gluconeogenesis (see earlier) from the triose phosphate exported from the chloroplast, though the ability to synthesize sucrose is widespread through plant tissues and any source of carbon that can feed into the hexose monophosphate junction can be used in sucrose synthesis.

There are two enzymic systems with the potential for catalysing production of sucrose:

1. UDP–glucose+fructose 6-phosphate

$$\xrightarrow{\text{sucrose phosphate synthase}} \text{UDP+sucrose 6-phosphate}$$

$$\text{sucrose 6-phosphate} \xrightarrow{\text{sucrose phosphatase}} \text{sucrose+P}_i$$

2. UDP–glucose+fructose

$$\xrightleftharpoons{\text{sucrose synthase}} \text{UDP +sucrose}$$

While these two routes are theoretically possible, however, the available evidence points to a role in synthesis only for the sucrose phosphate synth-

ase : sucrose phosphatase pair. The activity of sucrose synthase appears to be restricted to the cleavage of sucrose in certain situations where the concentration of sucrose is elevated[6].

This view is based upon a consideration both of the kinetic properties of the enzymes concerned and of their distribution within plant tissues. Thus values in excess of 3000 have been reported for the equilibrium constant, K_{eq} ([UDP] [sucrose 6-phosphate]/[UDP–glucose] [fructose 6-phosphate]) at pH 7.5 for the reaction catalysed by sucrose phosphate synthase. That sort of value for K_{eq} indicates that at the physiological concentration of the components the reaction is essentially irreversible in the direction of sucrose synthesis. In contrast, K_{eq} ([UDP] [sucrose]/[UDP–glucose] [fructose]) for the reaction catalysed by sucrose synthase has been reported to be in the region of 1.3–2.0 at pH 7.5. Reactions with a K_{eq} close to unity are sensitive to fluctuations in concentration of any of the components of the reaction, and a small increase in the prevailing sucrose concentration would readily lead to its conversion to UDP–glucose and fructose in the presence of sucrose synthase. Essentially, then, the reaction catalysed by sucrose synthase is readily reversible and whether the flux operates from UDP–glucose and fructose in the direction of sucrose or vice versa will depend very much on the concentrations of the substrates available. Since sucrose synthase is frequently found in association with tissues with a high concentration of sucrose it is likely to be catalysing the cleavage of sucrose rather than its synthesis[6].

Both of the reactions described above involve UDP–glucose, so that it will be the concentrations of fructose 6-phosphate and free fructose that markedly influence the synthesis of sucrose by the sucrose phosphate synthase or sucrose synthase-catalysed reactions, respectively. Normally the concentration of fructose in leaf tissue is too low to detect, so that any flux of sucrose generated through the sucrose synthase-catalysed reaction would be extremely small. In contrast, millimolar concentrations of fructose 6-phosphate can occur in leaves, resulting from the triose phosphates exported from the chloroplast, and in the presence of sucrose phosphate synthase a substantial flux of sucrose could be generated from it.

The proposal that the sucrose phosphate synthase-catalysed reaction predominates in sucrose synthesis is supported by studies of the distribution of the enzymes catalysing the two reactions. Although all three enzymes are located in the cytosol, sucrose phosphate synthase appears always to be present in leaves, where substrate in the form of hexose phosphates and UDP–glucose for sucrose synthesis is abundant, its activity being particularly high in the developing leaf. In contrast, its activity in non-photosynthetic tissues such as root is usually very low. In comparison the activity of sucrose synthase is usually found to be at its highest in non-photosynthetic tissues, especially storage tissues where photosynthate in the form of sucrose is imported and converted to a polymeric, storage carbohydrate following its cleavage to monomers[6].

Thus both the kinetic properties and distribution indicate different roles for the two enzymes. The K_{eq} of the sucrose phosphate synthase together with its occurrence in parts of the plant where ample carbon supplies are available point to a key role in sucrose synthesis. Sucrose synthase, on the other hand, catalyses a reaction that is freely reversible and hence very sensitive to fluctuations in substrate supply. Since it is frequently found in 'sink' tissues where elevated concentrations of sucrose are available, a role in cleavage of sucrose appears to be consistent with its properties.

For the conversion of sucrose phosphate to sucrose a phosphatase is also required. Sucrose phosphatase is a specific enzyme that removes the phosphate group from sucrose 6-phosphate. Its distribution is widespread and parallels the distribution of the sucrose phosphate synthase.

Sucrose phosphate synthase plays an important role in regulation of sucrose synthesis and this will be discussed in the later section on 'Regulation of Sucrose Metabolism'.

Storage

Experiments in which isolated chloroplasts have been fed $^{14}CO_2$ have shown that they are unable to form sucrose. This fact, combined with the rather low permeability of the chloroplast envelope to sucrose, indicates that sucrose synthesis takes place in the cytosol. While some sucrose is retained by the cytosol, the majority of it is exported via the phloem to other parts of the plant (see later). When the rate of photosynthesis exceeds the rate of sucrose export, however, sucrose accumulates in the leaf, where it is stored in the vacuole of the cell. In many plants sucrose is the major component of storage tissues and the size of this store may be substantial, as is the case in sugar-cane, sugar-beet and many fruits, where sucrose is an important reserve carbohydrate and is stored in the vacuole of parenchymatous tissue. The size of the store varies with species and environmental conditions. In leaves of spinach, for instance, storage of sucrose is transient, serving to buffer photosynthesis, and so the vacuolar concentration of sucrose is found to rise rapidly during illumination but fall during the subsequent dark period. In specialized storage organs build-up of sucrose occurs over a much longer period and mobilization takes place usually in the next growing period.

Both the influx of sucrose into and the efflux of sucrose out of the vacuole are mediated by the same carrier, which is located in the tonoplast membrane. The process is not energy dependent and direction therefore depends presumably upon the relative concentrations of sucrose in the cytosol and vacuole[8]. Evidence is accumulating, however, that transport may be linked to an H^+ antiport system. There has also been some suggestion that sucrose stored in the vacuole may be hydrolysed by the acid invertase of the vacuole and that the resulting hexoses are then released to the cytoplasm during mobilization of stored sucrose.

Transport of Sucrose

When the leaf becomes a net exporter of material it is referred to as a 'source', a term that may also be applied to storage tissues in which the reserve material is mobilized and exported. Material is transported from the source to non-photosynthetic 'sink' tissues such as meristems, roots or the endosperm of developing seeds, and even to leaf tissue in the early stages of development, where it is used to synthesize new cellular components or reserve material. In the majority of plants sucrose is the form in which carbohydrate is transported, and two main transport pathways exist—the symplast and the apoplast.

The Symplast

Symplastic transport refers to the movement of material through the cytoplasm of cells (Figure 4.11) and in the case of sucrose occurs from cell to cell in the mesophyll tissue of the leaf. Movement through this route is possible because the cytoplasm of adjacent cells is connected via membrane-lined pores called plasmodesmata. Experiments with exogenous [^{14}C]sucrose applied to leaf tissue have shown that it is rapidly accumulated in the leaf veins, and micro-injection of fluorescent tracer dyes has revealed that the plasmodesmata are indeed operative in transport of material between cells. In the leaf the direction of flux is from the chloroplast-containing cells via the mesophyll to the phloem cells located in the leaf veins. The sieve tubes of the phloem form an integral part of the symplast route. They are composed of sieve elements—elongated cells that are arranged end-to-end and form a network extending throughout the plant. It is through the sieve tubes of the phloem that transport of solute occurs over long distances.

The Apoplast

The cell walls of plants form a continuum that can also accommodate transport of solutes. This extracellular pathway of transport is referred to as the apoplastic route (Figure 4.11). It has been established as an important route for the transport and distribution of solutes delivered from the root by way of the xylem, but evidence supporting its function as a major route for transport of sucrose from the mesophyll cells to the phloem is ambiguous. There is evidence, however, to indi-

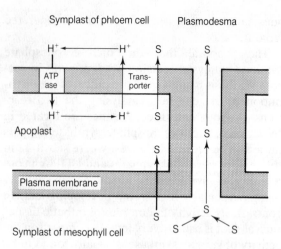

Figure 4.11. Phloem loading and sucrose transport showing the symplastic and apoplastic routes of sucrose movement. The movement of sucrose (S) through a plasma membrane-lined plasmodesma connecting a mesophyll and a phloem cell occurs down a concentration gradient. This is not considered to be a major route for phloem loading but is of greater significance with respect to transport between mesophyll cells. Movement of sucrose through the apoplastic route requires transport of sucrose from the mesophyll cell into the cell wall by a mechanism which is as yet unknown. Sucrose is transferred from the apoplast through the activity of an H$^+$/sucrose cotransporter. This route appears to be the major one for phloem loading.

cate that during transfer of sucrose from the adjacent mesophyll cells to the phloem—a process referred to as phloem loading—the apoplastic route is an important one[9].

Overall, then, movement of sucrose from the chloroplast to the leaf veins appears to take place via the symplastic route, the direction of movement being down the concentration gradient that exists between the source cells of the mesophyll to the sink cells of the phloem. In terms of sucrose transport the apoplastic route appears to have its major significance in the process of loading sucrose into the sieve elements.

Phloem Loading

The process of phloem loading is an active one that can take place against a considerable concentration gradient (Figure 4.11). From the fact that the process is selective (few sugars other than sucrose can be loaded) and shows saturation kinetics (transport velocity is concentration dependent until V_{max} is reached) it has been deduced that it is carrier mediated. In fact the carrier is a co-transporter which is located in the plasma membrane of the sieve elements and couples uptake of sucrose from the apoplast to uptake of protons[9]. A necessary part of the scheme is generation of a proton gradient in such a way that protons are actively transported into the apoplast. The gradient is generated and maintained by an ATPase of the plasma membrane. The energy of the proton gradient is subsequently used to transport sucrose from the apoplast by an obligatory coupling of the inward movement of protons, mediated by the proton symporter, to the inward flux of sucrose. Proton symporters appear to be a feature of the plasma membrane of all plant cells where different symporters mediate uptake of a number of different solutes. The mechanism by which the flux of sucrose from cytoplasm into the apoplast of the cells adjacent to the sieve elements is achieved has not yet been established, but it is a necessary part of the scheme described for phloem loading. Within the sieve tube the concentration of sucrose can be in the range 500–1000 mM, compared to 1–2 mM in the cytoplasm of mesophyll cells.

Unloading and Cleavage

Once loaded into the sieve elements sucrose may be transported through the sieve tubes throughout the plant. Unloading of the sieve tubes occurs wherever a 'sink' tissue is adjacent to the phloem, a sink being defined as any tissue which consumes photoassimilate. Sucrose is released to the cells of the sink along the concentration gradient that is maintained between phloem and 'sink' tissue. It is possible that sucrose could enter the cell via either the symplast or apoplast but the majority of tissues appear to receive sucrose via the symplast, with the significant exception of developing seeds.

A number of mechanisms are used to maintain a low concentration of sucrose in sink tissues and thereby maintain the unloading pressure. One mechanism is the storage of sucrose in the vacuole of adjacent parenchyma cells (see earlier), where it can no longer contribute to the concentration gradient. As an alternative to sequestration of sucrose in the vacuole, cleavage of the disaccharide can also reduce the cytoplasmic concentration of sucrose and give rise to carbon skeletons that can be used in a number of metabolic pathways. Cleavage can be achieved through the activity of two enzymes: invertase (β-fructofuranosidase) or sucrose synthase. Both types of enzyme may be found in sink tissues and their activities give rise to hexose products which are not taken back up into the phloem:

$$\text{sucrose} \xrightarrow{\text{invertase}} \text{glucose} + \text{fructose}$$

$$\text{sucrose} \xrightarrow{\text{sucrose synthase}} \text{UDP–glucose} + \text{fructose}$$

Sucrose synthase is a cytoplasmic enzyme, whereas two types of invertase occur—acid (pH optimum $c.$ 5.0) and alkaline ($c.$ 7.5). Alkaline invertase occurs in the cytoplasm, while acid invertase is located in the vacuole and the wall. The effect of these enzymes is to keep the concentration of sucrose at a minimum and thereby maintain a steep concentration gradient of sucrose between phloem and surrounding cells. In tissues that import sucrose via the apoplastic route the possibility arises that sucrose is hydrolysed by the acid invertase of the wall before entering the cell. In fact some studies have indicated that hydrolysis occurs, while in others sucrose is not hydrolysed prior to uptake. As mentioned previously, acid invertase may also be responsible for hydrolysis of the sucrose in the vacuole during mobilization, producing hexoses that are taken up by the cytoplasm.

In storage tissues such as potato, sucrose synthase appears to be the enzyme catalysing cleavage of the majority of the imported sucrose. Activities of invertase are insufficient to account for the observed rates of sucrose metabolism[10]. In other tissues, however, the activities of either enzyme could account for the observed rate of sucrose breakdown, and the precise contribution of each is difficult to establish. Sucrose synthase can catalyse formation of both UDP– and ADP–glucose *in vitro*. However, *in vivo* the available evidence indicates that UDP–glucose is the major sugar nucleotide produced by the enzyme. The UDP–glucose may be used in the synthesis of cell wall components directly or by way of the sugar nucleotide oxidation pathway. The supply of UDP–glucose is likely to greatly exceed the demands of these pathways, however, and much of it must enter the hexose phosphate pool described earlier. This is especially true in tissues where the activity of invertase is insufficient to generate hexose monophosphates for respiration, etc. Entry is possible through the reaction catalysed by UDP–glucose pyrophosphorylase operating in the direction of glucose 1-phosphate formation. The K_{eq} for the reaction in the direction of UDP–glucose formation is *c*. 0.8, so that providing pyrophosphate is available a flux of UDP–glucose from sucrose would lead to glucose 1-phosphate formation. Measurements have shown that significant amounts of pyrophosphate are present in the cytoplasm and that pyrophosphatases that might otherwise hydrolyse pyrophosphate are located in the vacuole and plastids, not the cytoplasm.

The role of alkaline invertase in sucrose degradation appears to be significant in some tissues. In soya bean leaf, for instance, its activity accounts for about half of the sucrose degradation occurring in older leaves. The free glucose and fructose produced by these reactions can enter the hexose monophosphate pool through the activities of the appropriate kinases.

METABOLISM OF STARCH

Starch is the major storage polysaccharide in plants and is present in all the major organs. In cereal grains and potato tubers it can reach 80% of the dry weight, and it is an important product of photosynthesis.

Structure

Two types of polymer are present in starch (Figure 4.12). Amylose is a straight-chain glucan in which the α-D-glucose residues are joined via 1→4 linkages, while amylopectin, in addition to the α(1→4) linkages, has a number of residues substituted at the 6 position with linear glucan chains. The degree of polymerization (DP) of each type of molecule will depend on the source tissue and its physiological state, but a DP of 1000 or more is not unusual for either class of molecule and for amylopectin may reach 20 000 or more. In amylopectin the α(1→6) linkages comprise only about 4–5% of the total linkages but they impart a branched structure to the molecule, the average length of the chain between branches being approximately 25.

Obviously, then, synthesis of starch requires the formation of two types of glucosidic linkage. The mechanism involved in the formation of each is distinctly different.

Starch in the Chloroplast

Accumulation of sucrose in leaves in situations where the rate of photosynthesis exceeds the rate

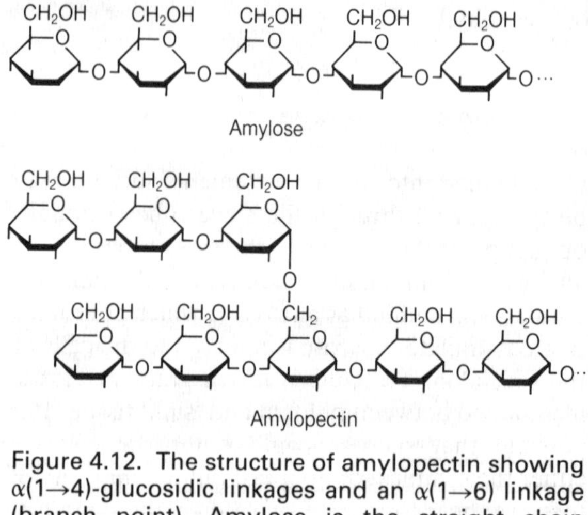

Figure 4.12. The structure of amylopectin showing α(1→4)-glucosidic linkages and an α(1→6) linkage (branch point). Amylose is the straight chain α(1→4)-linked glucan component of starch which lacks the branch points.

of sucrose export has already been referred to. The leaf has a finite capacity for storage, however, and if that condition continues the photosynthate is retained within the chloroplast and converted to starch. Thus it is normal for starch to accumulate within the chloroplast in the light and for it to be degraded in the dark and the products of its breakdown to be exported from the chloroplast. Under such circumstances storage of the photosynthate in the form of starch in the chloroplast is temporary and starch synthesis acts to buffer the photosynthetic capacity of the chloroplast, allowing for maximal rates of photosynthesis to be maintained even when sucrose synthesis in the cytoplasm is unable to absorb its full output.

Synthesis

The predominant route for starch synthesis within the chloroplast is by way of transfer of glucose from a nucleoside diphosphate donor to the non-reducing end of a pre-existing $\alpha(1\rightarrow4)$-glucan primer in a reaction catalysed by starch synthase (ADP–D-glucose:$(1\rightarrow4)$-α-D-glucan 4-α-glucosyl-transferase).

$$NDP\text{–glucose}+(\text{glucose})_n\rightarrow NDP+(\text{glucose})_{n+1}$$
$$\text{primer}$$

In leaf tissue starch synthase is specific for ADP–glucose. Neither the starch granule-bound or soluble synthase, both of which are present in leaves, has significant activity with UDP–glucose, a glucose donor that shows some activity with the starch synthase of some reserve tissues. Within the chloroplast ADP–glucose is provided from the carbon fixed by photosynthesis through the activity of ADP–D-glucose pyrophosphorylase (ATP:α-D-glucopyranosyl phosphate adenylyl-transferase), an enzyme that plays a key role in control of chloroplast starch metabolism[11].

$$\text{glucose 1-phosphate}+ATP\rightarrow ADP\text{–glucose}+PP_i$$

Both starch synthase and ADP–glucose pyro-phosphorylase are located within the chloroplast, and in tissues such as spinach leaf their activities are sufficient to account for the observed rates of starch synthesis.

The relationship between the granule-bound and soluble synthases is not yet entirely clear, though both contribute to formation of the starch granule, and in addition to these two classes of synthase multiple forms of both the granule and soluble synthase have been shown to exist. In spinach leaf, for instance, two forms of the soluble synthase have been recognized, differing in their activities and primer requirements[11]. The precise function of these two synthases is unknown, though one has a very high affinity for endogenous primer and may play a role as an 'initiator' synthase similar to that claimed for the synthase of reserve tissues. In developing maize endosperm two soluble and two granule-bound synthases have been identified, and these have been shown to be distinctly different forms of the enzyme, differing in their substrate specificities and other characteristics. The relationship between the different multiple forms and between the granule and soluble forms is unclear, though it has been suggested that each form may play different roles in the construction of the starch granule.

Formation of the branch points in amylopectin results from the activity of a branching enzyme, $(1\rightarrow4)$-α-D-glucan:$(1\rightarrow4)$-α-D-glucan 6-glycosyl transferase (sometimes known as Q enzyme). The enzyme catalyses a transglycosylation reaction in which a section of $(1\rightarrow4)$-α-D-glucan chain from amylose is hydrolysed from the main chain and is transferred onto the 6 position of a glucose residue in another part of the molecule. Transfer appears to be predominantly interchain, i.e. to adjacent glucan chains, rather than intrachain (transfer to the same glucan chain) and the enzyme uses an acceptor molecule of 30–40 glucose units or longer. Once the branch has been formed elongation can occur by glycosyl transfer from the nucleotide sugar through the activity of starch synthase.

Evidence that this is the principal route for starch synthesis, as opposed to, for instance, reversal of the starch phosphorylase-catalysed reaction, comes from a number of experiments, including observations that activities of ADP–glucose pyrophosphorylase and starch synthase are sufficient for the rates of starch accumulation observed. The starchless mutant of *Arabidopsis* that lacks ADP–glucose pyrophosphorylase but is normal with regard to starch phosphorylase adds proof that the route involving ADP–glucose pyrophosphorylase is the major one[12]. In leaf the source of carbon for starch synthesis is of course carbon dioxide fixation during photosynthesis.

Degradation

In contrast to the situation in reserve tissues, metabolism of starch in the chloroplast is characterized by rapid alterations in the direction of flux, between net synthesis and degradation. Breakdown of starch occurs when the rate of photosynthesis is low, or has ceased, and, because the sequence takes place within the chloroplast, events are closely related to the transport capabilities of the chloroplast. Starch of the leaf is frequently referred to as 'transitory starch' because of its temporary nature and its role as a buffer to photosynthesis described earlier. Its metabolism is intimately connected with that of sucrose[13] in the leaf (see later).

General Features of Starch Degradation

In general, breakdown of starch requires the cooperative activity of a number of enzymic activities, including hydrolases that cleave glycosidic linkages by addition of water, and phosphorylases which function by addition of phosphate and produce a monosaccharide phosphate:

$$G_n1{\rightarrow}4G_m + H_2O \xrightarrow{\text{hydrolase}} G_n + G_m$$

$$G1{\rightarrow}4G_m + P_i \xrightleftharpoons{\text{phosphorylase}} G1\text{–}P + G_m$$

Phosphorylase is an 'exo' enzyme that catalyses phosphorolytic cleavage at the non-reducing end of the glucan chain, while hydrolases may be exo enzymes, cleaving from the non-reducing end of the chain, or 'endo' enzymes that can hydrolyse linkages both at exterior and interior points in the chain and so bring about a rapid depolymerization of the molecule. A number of hydrolytic enzymes are necessary for starch breakdown, including α-amylase, an endo-hydrolase that cleaves $\alpha(1{\rightarrow}4)$ linkages, and β-amylase, which is an exo-hydrolase that can only work on the exterior parts of the chains to release maltose units and cannot work past the branch points. The initial activity of α-amylase on starch is to bring about a rapid depolymerization to form a series of linear and branched oligosaccharides called maltodextrins, as internal linkages are cleaved. This is followed by a rather slower hydrolysis of the maltodextrins. The major product of starch hydrolysis by α-amylase is maltose, since α-amylase cannot cleave the $\alpha(1{\rightarrow}4)$ linkage in maltose, with small quantities of glucose, maltotriose and oligosaccharide of six to eight units. The products of hydrolysis with β-amylase are maltose and 'limit-dextrin', a high-molecular-weight structure in which hydrolysis has been stopped at the branch points. The $\alpha(1{\rightarrow}6)$ linkages of the branch points may be hydrolysed by amylopectin 6-glycanohydrolase (also called R enzyme or limit-dextrinase) which releases short oligosaccharides from the branch points. Complete hydrolysis to glucose of maltose and the oligosaccharides released through the activity of the amylases results from the activity of α-glucosidase, an exo-hydrolase which hydrolyses the $\alpha(1{\rightarrow}4)$ linkages of maltose and maltodextrins from the non-reducing end, to give glucose.

The Chloroplast

There is much evidence to indicate that in the chloroplast starch breakdown occurs predominantly via the action of phosphorylase. Chloroplasts of pea do not contain significant amounts of α- or β-amylase but substantial quantities of phosphorylase, phosphoglucomutase, phosphoglucoisomerase and phosphofructokinase have been detected in the chloroplasts of these plants. Thus the product of starch breakdown via phosphorylase, glucose 1-phosphate, may be converted to fructose 1,6-bisphosphate and be incorporated into the reactions of the pentose phosphate pathway via the activity of these enzymes. From there of course it may be transported via the phosphate translocator (see earlier) to the cytoplasm. Results from studies of the mobilization of starch in chloroplasts isolated from spinach, however, show that glucose and maltose are formed as products in addition to triose phosphate and glycerate 3-phosphate. This result indicates that amylases also play a part in starch breakdown, and it now appears that in addition to phosphorylase activity endoamylases may play a significant part in starch degradation. It must be taken into account, however, that the specificity of such endoamylases in spinach could mean that maltose would not be a major product of hydrolysis and the maltose may arise instead through the activity of maltose phosphorylase, which has been identified in isolated pea chloroplasts and which catalyses the reaction

$$\text{glucose} + \text{glucose 1-phosphate} \rightarrow \text{maltose} + P_i$$

Transport of maltose and glucose out of the chloroplast is made possible by the activity of a transport system which, though less active than the phosphate translocator, is capable of transporting significant quantities of carbon[21].

Starch in Reserve Tissue

In contrast to the situation in chloroplasts, starch accumulation in reserve tissues such as the endoplasm of seeds, tubers, etc. takes place over longer periods of time, sometimes several months. It may then remain unchanged until the next growing season, when it is mobilized to support new growth. In tissues such as tubers and in some seeds the starch is deposited in the form of water-insoluble granules within amyloplasts—structures surrounded by a double membrane—and, like chloroplast starch, it is composed of a mixture of the two polymers amylose and the branched-chain amylopectin. The ratio of these two polymers varies with the species but for most plants lies within the ranges 20–25% amylose and 75–80% amylopectin. There is also considerable variation in the shape and final size of starch granules, those of potato reaching 100 μm in diameter, but a range of 20–30 μm would represent the 'average' size.

Just as in the case of the chloroplast, the double membrane surrounding the plastid imposes limitations on which metabolites have access to the granule. Import of carbon from the cytoplasm is made possible by the presence of a transport system which is described in the following section.

Synthesis

In reserve tissues the major carbon source for starch synthesis is usually sucrose translocated from source tissues via the phloem. Before it can be converted to starch within the plastid it must be processed in the cytoplasm into a form compatible with the substrate characteristics of the plastid transport system. Until quite recently much of what was known concerning uptake of metabolites by amyloplasts came from studies of the rates of conversion into starch of substrates supplied to isolated plastids, or from observations of the enzymic activities of such amyloplasts. Hence studies with amyloplasts from soya bean indicated that triose phosphate is taken up and converted to starch, while in plastids from cauliflower phosphoenolpyruvate appeared to be taken up in addition to triose phosphate. Such results were taken to indicate the presence of a phosphate translocator, analogous to the one in the inner membranes of chloroplasts, and a pathway in which the hexose monophosphates released by cleavage of sucrose are converted to triose phosphate in the cytosol prior to uptake by the plastid. A number of observations have indicated, however, that hexose phosphates may be taken up directly by amyloplasts. For instance, amyloplasts isolated from maize were able to synthesize starch from glucose 6-phosphate, while experiments with isolated wheat amyloplasts showed that only glucose 1-phosphate gave substantial synthesis of starch, the pattern of ^{14}C labelling in the synthesized starch indicating that the hexose was not cleaved prior to incorporation. In addition, amyloplasts of wheat endosperm lack the fructose 1,6-bisphosphatase activity necessary for resynthesis of hexose phosphate for starch synthesis from triose phosphates. In fact, isolation of a phosphate translocator from pea root plastids has now taken place and, like the phosphate translocator of the inner chloroplast membrane, it can transport P_i, DHAP and glycerate 3-phosphate[23]. It differs from the chloroplast translocator, however, in that it can also transport glucose 6-phosphate, and this finding has been extremely significant to an understanding of the origin and nature of the flux of carbon into the plastid for starch synthesis (Table 4.1) (see also Chapter 2).

Hence, the sucrose imported into the cytoplasm may be cleaved either by sucrose synthase or alkaline invertase, though in tissues such as potato tuber the activity of invertase is not sufficient to account for the observed rates of starch synthesis and so the sucrose synthase route must predominate. In many tissues both enzymes are present in activities sufficient to account for the observed rates of sucrose breakdown, however, and in such cases it has rarely proved possible to determine the relative roles of the two enzymes. This aspect of sucrose metabolism was referred to earlier in relation to phloem unloading, when conversion of the free glucose and fructose

Table 4.1. Kinetic constants for the phosphate translocator of pea root plastids.

	K_m (mM)	K_i (mM)
P_i	0.18	
Glucose 6-phosphate		0.33
Glycerate 3-phosphate		0.31
PEP*		0.20
DHAP		0.11

Results taken from Borchert et al.[26].
*Phosphoenol pyruvate.

released to the hexose monophosphates and the fate of the UDP–glucose was also discussed. These considerations taken together with the specificities of the plastid translocator indicate that the bulk of carbon from the sucrose imported into such tissues is taken up into the amyloplast in the form of hexose monophosphate.

Once imported into the plastid the hexose monophosphate must be converted to ADP–glucose. The ADP–glucose pyrophosphorylase of non-photosynthetic tissues is similar to that of leaf in being activated by glycerate 3-phosphate (maximum stimulation with 2–10 mM) and inhibited by P_i (50% inhibition with 1–3 mM). These figures indicate that the enzyme is less sensitive to the effectors than the leaf enzyme, and the role of these metabolites in controlling formation of ADP–glucose in reserve tissues is difficult to assess, especially if, as seems to be the case in some tissues, carbon enters the plastid in the form of hexose monophosphate rather than triose phosphate.

The mechanism of starch synthesis in the amyloplast is similar to that described for the chloroplast, and it has already been indicated that two classes of starch synthase (ADP–glucose:1→4-D-glucose 4-α-glucosyltransferase) are present in reserve tissue—one which is tightly associated with the amylose component of the starch granule, and a second which appears in the cytosol. The soluble synthases are active only with ADP–glucose, while some of the granule-bound synthases can utilize UDP– as well as ADP–glucose. The rate of transfer from ADP–glucose is some three- to tenfold higher than from UDP–glucose, however, and the K_m for the latter sugar nucleotide at c. 60 mM is 15- to 30-fold higher than for ADP–glucose. Taken together with the fact that UDP–glucose occurs in the cytoplasm and the plastid envelope is impermeable to UDP–glucose, these considerations mean that it is unlikely that UDP–glucose could make a significant contribution to starch synthesis. It has been suggested that adsorption of the soluble synthase to the surface of the granule alters the substrate specificity of the enzyme. In a situation similar to that in chloroplasts, multiple forms of the soluble synthase have been identified, differing in metal ion requirements and primer requirements. Branching enzyme also exists in multiple forms in developing maize endosperm but their significance is unknown.

Degradation

Reserve starch in cereal seeds is stored in the endoplasm and the enzymes for its breakdown are either not present in the seed at the time of germination or are stored in an inactive form. In this type of seed maltose and glucose appear to be the major products of degradation of the starch, so that hydrolysis rather than phosphorolysis appears to be the major mechanism involved. Because of its endo-activity α-amylase hydrolyses amylose to produce, in the initial stages, maltodextrins—high-molecular-weight oligosaccharides of glucose. α-Amylase is not present in resting cereal seeds, however; it is synthesized and secreted into the endoplasm at the time of germination, from either the scutellum or aleurone layer, depending on the species. Synthesis occurs in response to release of gibberellic acid (GA) within the seed and the response includes *de novo* synthesis of the α-amylase by an effect at the transcriptional level. It would appear that GA is the principal regulator of all α-amylase mRNAs in cereal seeds[22].

The effect of α-amylase on starch is biphasic, achieving a rapid production of maltodextrins in the first phase, followed by a second phase of hydrolysis which is somewhat prolonged and in which maltose is the major product together with oligosaccharides of five to six units. β-Amylase, which is an exoenzyme that hydrolyses starch at the non-reducing end of the chain to produce maltose, is present in these seeds at the onset of germination. The majority of it, however, is stored as an inactive form that is released in an active form through the action of proteolytic enzymes that are synthesized in response to gibberelins at the onset of germination. It is very effective at hydrolysing maltodextrins to produce maltose. The activities of both α- and β-amylase have been observed to increase during germina-

tion. Maltase, an enzyme that can cleave maltose, is also present in seeds; its activity leads to the production of glucose. R enzyme, the debranching enzyme, and α-glucosidase are also necessary components of the degradation system that occurs in seeds. Glucose released by the action of these enzymes is absorbed by the tissues of the scutellum, converted to sucrose and transported to the embryo.

In contrast to the situation in cereal seeds, phosphorylase activity appears to play a major role in starch degradation in potato, though initial degradation of the starch granule appears to require endoamylase activity. A critical role for endoamylases in initiating breakdown emerged following studies in which phosphorylase was apparently unable to hydrolyse intact starch grains. These and other studies suggested that initial breakdown of the starch granule requires the activity of an endoamylase, but the soluble oligosaccharides released may be degraded by the activity of amylase, phosphorylase or both. However, demonstrations that spinach leaf phosphorylase can degrade starch grains from chloroplasts of spinach and similar observations with pea have called into question the proposed role for endoamylases as the initiators of degradation. Whatever the situation regarding initiation of breakdown, phosphorylase of potato, in combination with debranching enzyme, is effective in degrading amylose and amylopectin to produce glucose 1-phosphate. This product can be converted to UDP–glucose and fructose 6-phosphate to be used in synthesis of sucrose for export to other parts of the growing plant or may be used for the synthesis of cellular components or in respiration, growth, etc.

Phosphorolysis appears also to be the major route for degradation of starch in pea cotyledons, where starch breakdown is an intracellular process in contrast to the cereal grains.

CONTROL OF STARCH AND SUCROSE METABOLISM IN THE LEAF

The metabolism of chloroplast starch and sucrose are intimately related through their positions as primary products of the photosynthetic process. It is not surprising then that starch and sucrose metabolism are both coordinated with the rate of photosynthetic carbon assimilation.

The use of fixed carbon for synthesis of starch in the chloroplast and sucrose in the cytoplasm must be balanced against the requirement for sufficient ribulose 1,5-bisphosphate (RuBP) to be available for optimal operation of the photosynthetic reduction cycle. Provided RuBP is regenerated at a rate sufficient to match the intake of carbon dioxide and production of ATP and NADPH, then carbon in excess of that requirement may be diverted towards synthesis of starch and sucrose. Although these two processes— sucrose and starch synthesis—occur in different metabolic compartments they are linked by the phosphate translocator (chapter 2) and so in effect use the same pool of triose phosphate as precursors.

A number of controls exist to coordinate these processes and avoid a conflict in the fate of the carbon between synthesis of sucrose, synthesis of starch and maintenance of the photosynthetic reduction cycle. They operate to ensure that as far as possible the highest rates of photosynthesis are matched by a capacity to process the fixed carbon. Wherever possible most of the triose phosphate is exported via the chloroplast P_i translocator and is used to synthesize sucrose in the cytosol. When, however, carbon fixation exceeds the rate at which triose phosphate can be converted to sucrose and the sucrose exported from the cytoplasm, an increasing proportion of photosynthate is converted to starch within the chloroplast.

Cytosolic inorganic phosphate has an important influence on the rate of photosynthesis (chapter 2). It is produced in the cytosol during sucrose synthesis and if the rate of synthesis is too high its concentration in the cytosol will force an exchange of triose phosphate out of the chloroplast. By depleting the intermediate of the cycle in this way the rate of photosynthesis will be decreased. If sucrose synthesis is low, insufficient P_i will be returned to the chloroplast, so that ATP synthesis in the chloroplast and hence photosynthesis will be inhibited.

Regulation of Starch Synthesis in the Chloroplast

ADP–Glucose Pyrophosphorylase

A key component regulating starch synthesis in the chloroplast is the enzyme ADP–glucose pyrophosphorylase, which catalyses formation of ADP–glucose from ATP and glucose 1-phosphate, and the response of this enzyme to glycerate 3-phosphate and phosphate appears to be the main element of the regulatory process:

$$ATP + \text{glucose 1-phosphate} \rightarrow ADP\text{–glucose} + PP_i$$

The enzyme from spinach leaf has an M_r of 210 000 and is composed of four subunits. Its activity is subject to allosteric control by a number of effectors, some of which are positive while others are negative. A number of glycolytic intermediates are able to activate the enzyme but the most effective of them is glycerate 3-phosphate, a concentration of 20 μM achieving 50% of the maximal activation compared to, for example, the 250 μM concentration of phosphoenol pyruvate required to achieve 50% maximal activation [11]. The effect of glycerate 3-phosphate is to stimulate V_{max} between eight- and 100-fold and to lower the K_m for ATP by about tenfold, the degree of change depending upon the source of the enzyme. The result is that the activity of ADP–glucose pyrophosphorylase is raised when the concentration of glycerate 3-phosphate is elevated by increases in photosynthetic rate and there is therefore a plentiful supply of reduced carbon available for starch synthesis.

The most effective inhibitor of leaf ADP–glucose pyrophosporylase is inorganic phosphate, a concentration of 22 μM bringing about a change to sigmoidal kinetics for the enzyme and a 50% inhibition of activity in the absence of glycerate 3-phosphate. In the presence of 1.0 mM glycerate 3-phosphate, however, the concentration of P_i required to bring about 50% inhibition of the spinach enzyme is, at 1.3 mM, c. 60 times higher, so that the presence of the activator glycerate 3-phosphate has the effect of decreasing the sensitivity of the enzyme to the inhibitor. Conversely, the presence of P_i at 0.5 mM greatly increases the concentration of glycerate 3-phosphate required for activation [11].

Clearly there is an interaction between glycerate 3-phosphate, a primary product of photosynthesis, and P_i, a key metabolite required in the Calvin cycle, and the reponse of ADP–glucose pyrophosphorylase to these two effectors is a major factor regulating starch synthesis. Both components serve as signal molecules providing information on the state of the metabolic fluxes within the stroma and, because both are also transported across the chloroplast membrane by the phosphate translocator and form a connection with the cytoplasm, on events in that metabolic compartment.

Experimentally lowering the concentration of P_i in the medium of isolated chloroplasts leads to a stimulation of starch synthesis by relieving inhibition of ADP–glucose pyrophosphorylase. A similar decrease in stromal P_i concentration can be observed when the illumination is increased, or when recycling of P_i to the chloroplast from sucrose synthesis in the cytosol is restricted through the inhibitory effects of cytosolic fructose 2,6-bisphosphate upon the enzyme fructose 1,6-bisphosphatase (see later). Conditions that raise the stromal concentration of glycerate 3-phosphate also stimulate starch synthesis. In this respect it has been shown that when P_i becomes limiting during photosynthesis the concentration of glycerate 3-phosphate is raised in the chloroplast, most probably because the lowered concentration of ATP that occurs when P_i is limiting decreases the rate at which glycerate 3-phosphate is converted to triose phosphate. Hence the stromal glycerate 3-phosphate : P_i ratio reflects both the availability of carbon for starch synthesis within the chloroplast and the status of carbohydrate metabolism in the cytoplasm, allowing the rate of starch synthesis to be coordinated with the photosynthetic rate and the rate of sucrose metabolism in the cytosol. A consequence of starch synthesis within the chloroplast is recycling of P_i to the intermediates of the Calvin cycle, making it less probable that the P_i concentration will become limiting for photosynthesis.

Regulation of Sucrose Metabolism

Fructose 2,6-Bisphosphate

A further link between the chloroplast and cytosol exists in the form of the regulatory metabolite fructose 2,6-bisphosphate (Figure 4.11). First identified as a key factor in regulating glycolysis and gluconeogenesis in mammalian liver, fructose 2,6-bisphosphate is now known to play an important role in regulating photosynthetic carbon metabolism, particularly with respect to the use of photoassimilate in sucrose synthesis[24].

In order to understand how this metabolite can play a role as a signal and affect carbohydrate metabolism in the leaf, it is necessary to understand how it is synthesized and the factors that affect its synthesis. Fructose 2,6-bisphosphate is synthesized from fructose 6-phosphate through the activity of a specific fructose 6-phosphate-2-kinase located in the cytosol. An associated fructose 2,6-bisphosphate phosphatase activity removes the phosphate to regenerate fructose 6-phosphate. In spinach both of these activities are present on a single bifunctional protein. This substrate cycle has an important part to play in regulating carbohydrate metabolism because the activities of the kinase and phosphatase that determine the concentration of the bisphosphate *in vivo* are regulated by metabolic effectors that are also key metabolites in the synthesis of sucrose[24]. Some of these effectors of the two enzymes are listed in Table 4.2. Fructose 6-phosphate and P_i are both activators of the kinase but inhibitors of the phosphatase and so increases in the concentration of either of these effectors will, through their effects upon both enzyme activities, raise the concentration of fructose 2,6-bisphosphate in the cytosol. On the other hand, the three-carbon phosphates, dihydroxyacetone phosphate and glycerate 3-phosphate, are inhibitors of the kinase, though they are without effect upon the phosphatase. Thus increases in the concentration of these intermediates will lead to a decrease in the concentration of fructose 2,6-bisphosphate. The concentration of fructose 2,6-bisphosphate depends on the relative activities of the kinase and phosphatase, and normally the

Table 4.2. Effectors that regulate the intracellular concentration of fructose 2,6-bisphosphate.

Effector	Fructose 6-phosphate-2-kinase activity	Fructose 2,6-bisphosphate phosphatase activity
DHAP	Inhibitor	No effect
3-PGA	Inhibitor	No effect
P_i	Activator	Inhibitor
Fructose 6-phosphate	Activator	Inhibitor

concentration of this effector is within the range 1–10 μM and it is confined to the cytosol.

Feed-forward Control of Sucrose Synthesis

The significance of changes in concentration of fructose 2,6-bisphosphate lies in the inhibitory effect of this signal metabolite upon the cytosolic fructose 1,6-bisphosphatase, the enzyme catalysing removal of P_i from fructose 1,6-bisphosphate, the committed step on the pathway to sucrose (Figure 4.13)[24]. At concentrations in the nanomolar to micromolar range it decreases the affinity of the enzyme for fructose 1,6-bisphosphate some 80–100-fold and induces sigmoidal kinetics. Sensitivity of the enzyme to inhibitors such as AMP and to product inhibition by P_i is also increased. Therefore, because its own concentration is affected by the availability of cytosolic glycerate 3-phosphate and P_i, fructose 2,6-bisphosphate provides a link between availability of fixed carbon from photosynthesis and the rate of sucrose synthesis, through its effect on fructose 1,6-bisphosphatase activity.

Through the phosphate translocator an increase in photosynthetic rate leads to an increase in the cytosolic concentration of both triose phosphate and glycerate 3-phosphate, as well as a decrease in concentration of P_i. In terms of effect upon the concentration of fructose 2,6-bisphosphate the ratio of glycerate 3-phosphate : P_i is the critical signal and an increase in that ratio will lead to an inhibition of the fructose 6-phosphate-2-kinase and result in a lowering of the fructose 2,6-bisphosphate concentration. Indeed increasing the light intensity or carbon dioxide concentration experimentally leads to a three- to fivefold decrease in fructose 2,6-bisphosphate concentration in spinach, whereas inhibitors that lower the concentration of glycerate 3-phosphate increase the concentration of fructose 2,6-bisphosphate three- to fivefold. Lowering the P_i concentration will also relieve the inhibition of the fructose 2,6-bisphosphate phosphatase and decrease the concentration of fructose 2,6-bisphosphate.

Lowering the fructose 2,6-bisphosphate concentration relieves the inhibition of the fructose 1,6-bisphosphatase in the cytosol and so increases the flux of photosynthate towards sucrose. In addition to the effect of fructose 2,6-bisphosphate on its activity, fructose 1,6-bisphosphatase is subject to control by its own substrate, fructose 1,6-bisphosphate. In this case the control is feed-forward activation and, in addition therefore to an effect upon fructose 2,6-bisphosphate concen-

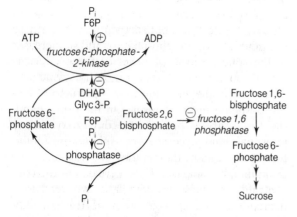

Figure 4.13. Fructose 2,6-bisphosphate and its effects on sucrose synthesis.

tration, the increased concentration of cytosolic triose phosphate will, by giving rise to fructose 1,6-bisphosphate, exert a feed-forward control over the pathway to sucrose. The combined effect of a decrease in concentration of fructose 2,6-bisphosphate and an increase in fructose 1,6-bisphosphate is a powerful feed-forward control relating availability of photosynthate to sucrose synthesis.

If, however, the ratio of glycerate 3-phosphate : P_i in the cytosol decreases, as photosynthesis is reduced, an increase in fructose 2,6-bisphosphate concentration will result as the inhibition of the fructose 6-phosphate-2-kinase is relieved. Inhibition of the fructose 1,6-bisphosphatase will result from the increased concentration of fructose 2,6-bisphosphate, and the use of triose phosphate in sucrose synthesis will be restricted.

These alterations in fructose 2,6-bisphosphate concentration are thus a central element in a feed-forward mechanism to coordinate sucrose synthesis with the rate of carbon dioxide fixation. Increased production of photosynthate within the chloroplast leads to feed-forward activation of the synthetic sequence to sucrose, the signal between chloroplast and cytoplasm being in the form of triose phosphate and P_i carried by the translocator. The increased activity of the fructose 1,6-bisphosphatase leads not only to an increase in sucrose synthesis to absorb the fixed carbon, but to an increase in P_i recycling to the Calvin cycle and thus maintains photosynthetic efficiency.

Phosphorylation of Sucrose Phosphate Synthase

Fructose 1,6-bisphosphatase is not the only site of control relating the rate of triose phosphate production in the chloroplast to the rate of sucrose synthesis in the cytosol. Sucrose phosphate synthase is also a key element in regulating sucrose synthesis. The enzyme catalyses a reaction which in vivo is irreversible, and two types of control operate to regulate its activity. Fine control refers to inhibition or activation of activity by metabolic effectors, while in coarse control the quantity of extractable enzyme activity is modified.

In general terms sucrose phosphate synthase is activated by glucose 6-phosphate and inhibited by P_i, so that when the concentration of the hexose monophosphate pool increases with increasing photosynthetic rate, sucrose synthesis will increase. The hexose monophosphate pool derives, under these circumstances, from the activity of the fructose 1,6-bisphosphatase which is itself, through fructose 2,6-bisphosphate, regulated by the three-carbon products of photosynthesis. This dependence of the sucrose phosphate synthase activity upon glucose 6-phosphate means therefore that there is a coordinated regulation of the pathway to sucrose synthesis.

Sucrose phosphate synthase is also subject to a so-called coarse control mechanism, which involves protein phosphorylation and gives rise in spinach leaves to two forms of the enzyme that have different kinetic properties[25]. When leaves are illuminated, or the P_i is reduced by mannose treatment, sucrose phosphate synthase activity is raised while treatments that raise the sucrose concentration decrease the activity. Monoclonal antibodies have been used to show that the amount of enzyme protein was the same in both cases and subsequently inactivation was shown to result from phosphorylation of the active protein. Phosphorylation gave an enzyme protein of lower activity with a lower affinity for substrate and which required a greater concentration of glucose 6-phosphate for activation and was more sensitive to inhibition by P_i. Experiments in which partially purified sucrose phosphate synthase has been subjected to phosphorylation and dephosphorylation in vitro have confirmed that deactivation of the enzyme is associated with phosphorylation. The protein kinases and phosphatases that are involved with this regulatory mechanism have not yet been identified, nor is it known how factors such as availability of P_i and sucrose concentration can influence the degree of phosphorylation of the enzyme. The effect of the transition, however, is to make sucrose synthesis more active and less subject to control by glucose 6-phosphate and P_i in high light intensities when the cytosol sucrose concentration is not excessive.

Feedback Control

Fructose 2,6-bisphosphate is also involved in a feedback regulation that relates production of triose phosphate in the chloroplast to the capacity of cytosol to synthesize sucrose and 'process' it. This regulation may lead to starch synthesis within the chloroplast when the rate of carbon fixation exceeds the capacity of the leaf to synthesize and export sucrose. Under circumstances where the rate of synthesis exceeds the rate of transport, sucrose accumulates in the leaf. While sucrose can accumulate in the vacuole and have little effect on the export of triose phosphate from the chloroplast, at some point an increase in the cytosolic concentration of sucrose occurs which leads to the deactivation of sucrose phosphate synthase described above. As a result the concentration of fructose 6-phosphate rises, activating the fructose 6-phosphate-2-kinase and increasing the concentration of fructose 2,6-bisphosphate (Figure 4.13). Experimentally increasing the sucrose content of leaves by applying exogenous sucrose has been shown to achieve a similar increase in effector concentration. The resultant increase in the concentration of fructose 2,6-bisphosphate restricts the activity of the fructose 1,6-bisphosphatase, so reducing the quantitiy of sucrose synthesized[24]. This feedback regulation leads to increased rates of starch synthesis in the chloroplast as the raised concentrations of glycerate 3-phosphate in the stroma activate the ADP–D-glucose pyrophosphorylase. Thus when photosynthetic rate exceeds the rate at which sucrose can be exported from the leaf, or stored in the vacuole, the flux of carbon towards sucrose is restricted and increasing amounts of photosynthate are stored within the chloroplast in the form of starch.

Starch Degradation in the Chloroplast in the Dark

The mobilization of starch during periods of low illumination has already been referred to. When chloroplasts isolated from spinach are incubated in the dark a mixture of glucose, maltose, triose phosphates and glycerate 3-phosphate is produced, indicating that starch is degraded both hydrolytically and phosphorolytically. High concentrations of P_i in the cytosol stimulate the exchange of triose phosphates of the chloroplast with P_i and are likely to lead to a stimulation of phosphorolysis of starch as the concentration of P_i is raised in the stroma. High cytosolic P_i concentrations would occur when the rate of sucrose synthesis is high and/or the rate of respiration is increased, both situations demanding a high rate of carbon export from the chloroplast. Glucose and maltose are produced in significant quantities by isolated chloroplasts and may be transported to the cytosol *in vivo* by the hexose transporter. Their production is increased when P_i is lowered, so that under such conditions hydrolysis rather than phosphorolysis may be the major route for starch degradation in the chloroplast.

FRUCTAN METABOLISM

While starch may be regarded as the major storage form of carbohydrate, it is not the only one and a number of plants accumulate substantial quantities of sucrose or the related polymers of fructose (fructans) in their leaves and stems[14]. Storage of sucrose in the vacuole has been referred to earlier and the vacuole is also the storage location of the water-soluble fructans.

Storage

Fructan accumulation, like starch storage, shows two different metabolic patterns. In both situations where it occurs the accumulated fructan represents a reserve of carbohydrate, but in heterotrophic 'sink' organs such as roots, stems and grains, fructan synthesis is based upon conversion of carbon imported from a source tissue via the phloem. Like the case of starch, storage in these tissues is more long term in nature. Fructan accumulation in leaf tissue, however, is closely related to the synthesis and export of sucrose, and so is closely connected with photosynthetic carbon reduction. In both types of tissue the site of fructan accumulation is the vacuole.

Structure

Fructans consist of homologous series of non-reducing oligo- and polysaccharides, each member of the series containing one more fructose residue than the previous member. Because of the mechanism by which they are synthesized, every fructan contains one glucose moiety per chain. Fructans from different sources show quite different structures, though a basic pattern can be recognized related to the structures of three isomers of monofructosyl sucrose (Figure 4.14).

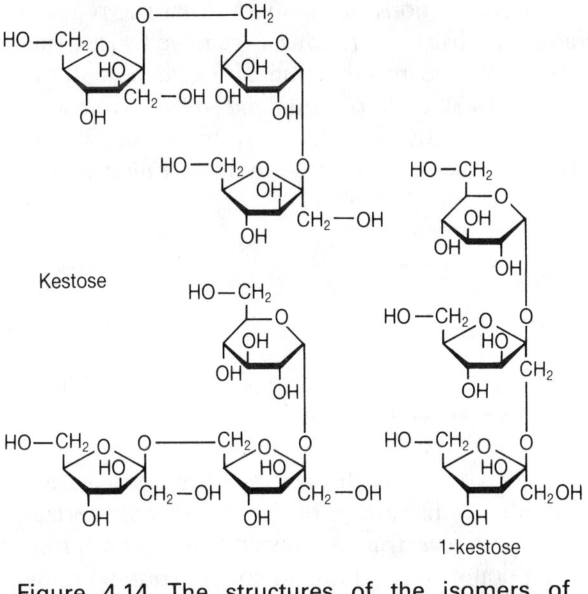

Each isomer forms the basis of a different homologous series and each series is characterized by a different linkage pattern.

Members of the 1-kestose series are based on the trisaccharide 1-kestose (Figure 4.12). These polymers consist of a linear chain of β-2,1-linked fructose residues (F) with a single terminal (i.e. a residue linked into the end of the chain through its 1 position) glucose residue (G) and so the molecule is non-reducing. The general form of the molecule is

$$G\text{-}1,2\text{-}F\text{-}1,(2\text{-}F\text{-}1)_n,2\text{-}F$$

Where n has a maximum in the region of 35. The fructans from tubers of Jerusalem artichoke *(Helianthus tuberosus)* are typical of this series. Frequently such fructans are referred to as inulins and they are accumulated in a number of species.

The family of fructans based on the trisaccharide 6-kestose is characterized by β-2,6 linkages between adjacent fructose residues. They have the general form

$$G\text{-}1,2\text{-}F\text{-}6,(2\text{-}F\text{-}6)_n,2\text{-}F$$

These polymers are of a characteristically high M_r and n has a maximum close to 250. Several grasses possess this type of fructan, which are often referred to as levans or phleins.

The third series of fructans originates from the trisaccharide neokestose, an oligosaccharide in which the glucose residue is linked to two fructose residues through its 1 and 6 positions. This means that subsequent elongation of the chain can occur on both fructose residues, giving rise to the following general structure:

$$F\text{-}2,(1\text{-}F\text{-}2)_m,\ 1\text{-}F\text{-}2,6\text{-}G\text{-}1,2\text{-}F\text{-}1,(2\text{-}F\text{-}1)_n,2\text{-}F$$

In fructans of this type both m and n reach a maximum around 10.

Branched fructans have also been isolated from grasses. Such fructans appear to have backbones based upon both the 1-ketose and 6-ketose series.

Figure 4.14. The structures of the isomers of monofructosyl sucrose that give rise to the three fructan series.

Synthesis

Some similarities between the storage patterns of starch and fructans were referred to in the previous section but, in contrast to the synthesis of starch, sugar nucleotide intermediates are not directly involved in the synthesis of fructans. Instead their synthesis occurs through the activity of glycosyl transferases which can catalyse transfer of fructose residues from one oligosaccharide that acts as donor to a second oligosaccharide that represents the acceptor.

Formation of 1-Kestose

At least two distinct enzymes are involved in the synthesis of fructans. In the initial reaction sucrose acts as the primary donor of fructose residues and the acceptor is a second sucrose molecule. The product of the transfer is the trisaccharide 1-kestose, which occurs in the inulin-type fructan (Figure 4.14):

$$\underset{\text{sucrose}}{\text{G-1,2-F}} + \underset{\text{sucrose}}{\text{G-1,2-F}} \xrightarrow{\text{SST}} \underset{\text{1-kestose}}{\text{G-1,2-F-1,2-F}} + \underset{\text{glucose}}{\text{G}}$$

The reaction is catalysed by the enzyme sucrose–sucrose fructosyl transferase (SST), and this reaction appears to be the major route by which the flux of carbon from sucrose into fructans occurs[15]. The K_m values for the enzyme from a number of tissues have been reported and tend to be relatively high. The K_m for the enzyme purified from roots of asparagus *(Asparagus officinalis)* for instance is 0.11 M, a figure which indicates that a substantial build-up of sucrose has to occur before the enzyme will begin to operate in the region of its maximum velocity. Indeed it has been shown that in some grasses a threshold concentration of sucrose has to be accumulated before fructan synthesis will occur[15].

The substrate specificity of the enzyme from tubers of *H. tuberosus* indicate that the reaction is essentially irreversible in the direction of 1-kestose formation, the enzyme failing to catalyse transfer of a fructose residue from 1-kestose to glucose. However, the enzyme from *A. officinalis* has been observed to catalyse the reverse reaction but only *in vitro* and only at high concentrations of 1-kestose and glucose.

Chain Elongation

SST from *H. tuberosus* cannot catalyse transfer of a fructose residue from sucrose to 1-kestose and so cannot extend the fructan oligosaccharide beyond the original trisaccharide product. Chain elongation is achieved by a second enzyme, fructan:fructan fructosyl transferase (FFT), which catalyses transfer of the terminal β-2,1-linked fructose residue from an appropriate donor to an acceptor to form β-2,1-linked fructan chains:

$$\underset{\text{acceptor}}{\text{G-F-(F)}_x} + \underset{\text{donor}}{\text{G-F-(F)}_y} \xrightarrow{\text{FFT}} \text{G-F-(F)}_{x+1} + \\ \text{G-F-(F)}_{y-1}$$

The enzyme from tubers of *H. tuberosus* can catalyse transfer between oligosaccharides of various sizes, but sucrose and the fructans of higher M_r are the most effective acceptors, while 1-kestose, the product of the reaction catalysed by SST, can act as an acceptor or donor. When 1-kestose functions as a donor, sucrose remains after the fructosyl residue is removed. This sucrose molecule may then function as an acceptor in the FFT-catalysed reaction and so can give rise to fructan directly (2). Alternatively the sucrose can be converted to 1-kestose by means of the reaction catalysed by SST (1):

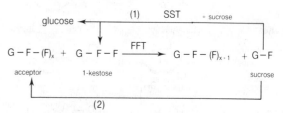

The activity of FFT leads to synthesis of oligosaccharides with a range of degrees of polymerization (DPs) as fructose residues are transferred from donor to acceptor. In combination with the enzyme SST, the activity of which ensures continued production of 1-ketose to be used as a fructosyl donor in this chain-elongation reaction, FFT leads to net accumulation of fructan oligosaccharides of the 1-kestose series. The distribution of DPs within the group of oligosaccharides synthesized by individual plant species is often characteristic of that species.

A. officinalis accumulates fructans of both the 1-kestose and neokestose series, and in tissues of this plant two fructosyltransferase activities have been identified. One, 1-F-fructosyl transferase, has an activity equivalent to the FFT from *H. tuberosus* and its activity leads to the production of β-2,1 chains of fructose. Unlike the enzyme from *H. tuberosus*, however, the *A. officinalis* enzyme can also use both neokestose and neokestose–oligosaccharides as acceptors, extending the chains at both of the terminal fructose residues attached to the central glucose residue, to produce fructan polymers of the neokestose series.

The second transferase activity is 6-G-fructosyl transferase (6GFT), an enzyme that catalyses transfer of a terminal β-2,1-linked fructose residue from a donor to the 6 OH group of the glucosyl residue of the acceptor[16]. 1-Kestose is used as donor in the reaction and when sucrose is the acceptor neokestose is produced:

$$\text{G-1,2-F-1,2-}\overset{\text{X}}{\overset{\bullet}{\text{F}}} + \overset{\bullet}{\text{G}}\text{-1,2-}\overset{\bullet}{\text{F}} \xrightarrow{\text{6GFT}}$$

isokestose   sucrose

$$\xrightarrow{\text{6GFT}} \text{G-1,2-}\overset{\text{X}}{\overset{\bullet}{\text{F}}} + \text{F-2,6-}\overset{\bullet}{\text{G}}\text{-1-}\overset{\bullet}{\text{F}}$$

sucrose   neokestose

When larger oligosaccharides of the isokestose series are used as acceptors the products are the corresponding oligosaccharides of the neokestose series. The neokestose and higher homologues of neokestose produced through this reaction may then be extended by the activity of the 1-F-fructosyl transferase.

Fructan Degradation

At various times the fructan reserves are mobilized. While the timing of mobilization will depend on the tissue under consideration—root, stem, grain or leaf tissue—the mechanism involved is the same and involves depolymerization of the stored fructan through the activity of a fructan exo-hydrolase (FEH). The enzyme is a β-fructofuranosidase that catalyses sequential removal of terminal fructose residues from the chain. Such enzymes have been isolated from a number of plant species and, in common with many other glycosidases, they have been shown to have pH optima in the range 4–5.5. They are effective in degrading oligofructosides of various degrees of polymerization, releasing fructose as they do so. The enzyme from *H. tuberosus* cannot hydrolyse sucrose and its activity is actually inhibited by sucrose. The fructose released can be used in a number of pathways but, depending upon the tissue in question, the majority of it appears to be phosphorylated by hexokinase and is then converted by the activity of sucrose phosphate synthase and sucrose phosphatase to sucrose for export to sink tissues. Thus the inhibition of the hydrolase by sucrose may be an important factor relating the breakdown of fructan to synthesis and export of sucrose.

THE CELL WALL

Plant cells are surrounded by a wall which is laid down external to the plasma membrane and which is composed of polysaccharides, glycoproteins and phenolic compounds. The thickness generally lies within the range 0.1–1.0 μm but both size and thickness can vary with the cell type. In most cell walls three layers can be identified: the middle lamella, and the primary and the secondary wall. Each is laid down at a different stage in development of the cell, although some overlap may occur in the phases of wall synthesis, so that material characteristic of the primary wall may be synthesized at the same time as the middle lamella.

The middle lamella is a thin layer of matrix polysaccharides that occurs between adjacent cells. It is the first layer to be synthesized, being deposited at the cell plate at the time of cell division. Once cell division has been completed the two 'daughter' cells deposit primary wall material at the surface of the plasma membrane. Deposition occurs over the entire cell surface and continues throughout the period of cell expansion. When surface growth has ceased the final phase of synthesis occurs—deposition of the secondary wall. Again deposition of secondary wall continues over the entire surface, but the secondary wall differs markedly from the primary wall both in the types and composition of its polysaccharides and in its overall characteristics. The primary wall is laid down at a time when the cell is expanding, so whereas it must be sufficiently rigid to support the cell it must also be flexible enough to allow cell expansion. The secondary wall, on the other hand, is laid down when no further expansion will occur. It has different types of polysaccharide, with a greater proportion of cellulose and hemicellulosic polysaccharides present, and overall it is a thicker structure than the primary wall, with a greater rigidity and strength. Synthesis and assembly of the wall polymers take place at the plasma membrane, so those parts of the wall synthesized first are most remote from the plasma membrane and the secondary wall is internal to the primary.

Classification and Structure of Wall Polysaccharides

The Microfibrillar Phase

All cell walls are composed of two phases: the microfibrillar phase and the matrix. As its name suggests, the microfibrillar phase is a highly crystalline phase which consists of long, thin structures called microfibrils (average width 10 nm, depending upon the tissue). Each microfibril is composed of between 30 and 100 molecules of cellulose, arranged in a direction parallel to the long axis of the microfibril[17]. The microfibrils themselves lie within the plane of the cell surface and in some cells they are deposited with an orientation perpendicular to the major axis of cell

elongation, an arrangement resembling that of hoops around a barrel.

Structure of Cellulose

Cellulose is an unbranched, β-4,1-linked polymer of D-glucose residues (Figure 4.15). In secondary walls the DP of cellulose may be as high as 15 000, equivalent to an M_r of 2.4×10^6, but in primary walls the DP is considerably less, in the region of 3–5000. Within the microfibril the cellulose chains are arranged into an ordered crystalline lattice which is stabilized by hydrogen bonds formed between (intermolecular) and within (intramolecular) chains. X-ray diffraction patterns indicate the presence of hydrogen bonds between the ring oxygen of one glucose residue and the C3 hydroxyl of an adjacent residue in the same chain (intramolecular bonding). Bonds between chains (intermolecular bonding) occur

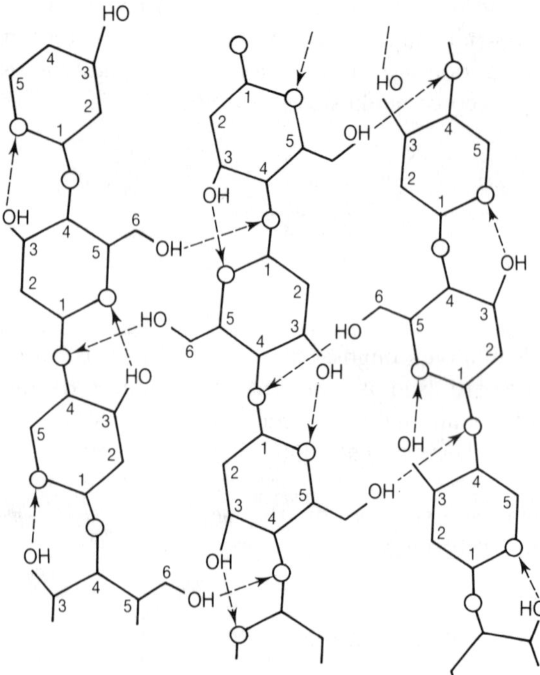

Figure 4.15. The structure of cellulose.

between hydroxyl groups at C6 and the oxygen of the glycosidic linkage in adjacent chains. Not all of the microfibril is so ordered, however, and X-ray diffraction and electron microscope studies indicate the presence of a much less ordered arrangement of cellulose molecules at the surface of the microfibril. In this region also it appears that traces of the non-cellulosic sugars, mannose and xylose, may occur.

The Matrix Phase

The matrix is non-crystalline and in contrast to the microfibrillar phase, where only one polymer—cellulose—is present, is composed of a variety of polysaccharides, proteins and phenolic compounds. Walls vary with respect to the ratio of microfibrillar to matrix phase present but typically cellulose represents c. 20–30% of the dry weight of the primary wall but between 45% and 90% of the secondary wall.

The major polysaccharide types present in the matrix are described in Table 4.3. The proportion of each polymer present will vary with tissue and the species, though it will be apparent that some are more typical of the primary wall and others of the secondary[18].

Matrix polysaccharides can be classified into one of two groups based upon their solubilization from the wall with different solvents. Because their ease of solubilization by a particular solvent relates to their chemical composition and structure, some generalizations regarding the structure of the polymers within the groups can be made.

Treatment of the wall with aqueous chelating agents, such as ethylenediaminetetraacetate (EDTA) or oxalate, extracts Ca^{2+} from the wall and leads to solubilization of the pectic polysaccharides. Subsequent treatment of the wall with solutions of alkali extracts the hemicellulose polysaccharides and those pectic polysaccharides not removed by chelating agents. Treatment with alkali is effective in solubilizing hemicelluloses—polymers in which xylose, glucose and mannose predominate. Different types of hemicellulose are extracted by different concentrations of alkali, allowing for some fractionation of this group of polymers.

This operational classification is not, however, clear cut and some galacturonic acid-rich polysaccharides appear in the alkali fraction, while some polymers more typical of the hemicellulosic fraction may be extracted together with the pectic polysaccharides. Pectic polysaccharides are the polymers of the middle lamella and primary wall

Table 4.3. Main polysaccharides of the plant cell wall showing structure of the interior chains.

Polysaccharide	Interior chain
Cellulose	→4)-Glc-(1→4)-Glc-(1→4)-Glc-(1-
Hemicelluloses	
Xyloglucan	→4)-Glc-(1→4)-Xyl-(1→4)-Glc-(1-
Xylan	→4)-Xyl-(1→4)-Xyl-(1→4)-Xyl-(1-
Mannan	→4)-Man-(1→4)-Man-(1→4)-Man-(1-
Glucomannan	→4)-Man-(1→4)-Glc-(1→4)-Man-(1-
Callose	→3)-Glc-(1→3)-Glc-(1→3)-Glc-(1-
Arabinogalactan	→3)-Gal-(1→3)-Ara-(1→3)-Gal-(1-
Pectins	
Homogalacturonan	→4)-GalA-(1→4)-GalA-(1→4)-GalA-(1-
Rhamnogalacturonan	→4)-GalA-(1→2)-Rha-(1→4)-GalA-(1-
Arabinan	→5)-Ara-(1→5)-Ara-(1→5)-Ara-(1-
Galactan	→4)-Gal-(1→4)-Gal-(1→4)-Gal-(1-

of dicotyledonous plants, where they may constitute up to 50% of the wall. In monocots the
proportion of pectic polysaccharides is normally
less than this and in secondary walls the proportion of hemicellulosic polysaccharides greatly
exceeds the amount of pectic polysaccharides.

Structure

Unlike proteins, polysaccharides are not synthesized against a molecular template, and while a
basic structure can be recognized for each type
considerable variation occurs in the fine structure
of both the pectic and hemicellulosic polysaccharides from different sources. It is only possible
therefore to describe the more general features
here. Polysaccharides are also polydisperse, i.e.
individual molecules of a single polysaccharide
type are not all the same size, so that it is only
possible to quote an average for M_r or DP—the
number of monosaccharide units per molecule.

Pectic Polysaccharides

The main structural characteristic of this group of
polysaccharides is a 1,4-α-linked chain of D-
galacturonic acid residues forming the backbone
of the molecule (Table 4.3). These polymers are
rarely simple homogalacturonans containing only
galacturonic acid, however, and neutral sugars
are normally present as integral components of
the polysaccharide. Only one other type of sugar
residue appears in the backbone—L-rhamnose
(Table 4.3), which is linked via the hydroxyl at its
C2 position. A rhamnose residue generally occurs
between approximately every 20 galacturonic acid
residues. Because of the occurrence of rhamnose
as a feature of the molecule they are referred to as
rhamnogalacturonans. Frequently pectic polysaccharides are found to contain regions in which the
ratio of rhamnose to galacturonic acid is low, c.
1 : 20, connected to regions of rhamnogalacturonan where the ratio is much higher, at c. 1 : 1. In
the latter region the backbone is an alternating

sequence of galacturonic acid and rhamnose
residues, and side chains appear to be attached to
nearly all the rhamnose residues.

All rhamnogalacturonans possess side chains
made up of neutral sugars attached at the C3 of
galacturonic acid and C4 of rhamnose. Considerable variation has been detected in the side chains
of rhamnogalacturonans from different sources,
but the regions of low rhamnose content appear
to be the point of attachment of neutral polymers
of arabinose (arabinans), galactose (galactans)
and arabinose plus galactose (arabinogalactans).
They occur attached to the C4 of rhamnose and
the DP of these polysaccharides may be quite
high. In that part of the molecule where rhamnose and galacturonic acid alternate, side chains
are so frequent that these sections of the chain are
referred to as the 'hairy' regions. Two rhamnogalacturonans in which the regions have been particularly well characterized are RGI[18] and RGII[19],
both of which have been isolated from the walls of
cultured sycamore cells.

RGI is a major component of the middle
lamella and primary wall of dicotyledonous plants
and features the alternating rhamnose, galacturonic acid type of backbone described above.
The side chains attached to C4 of rhamnose
residues are all rich in arabinose and/or galactose
and display considerable variation in structure.
At least thirty different types have been
identified[18].

RGII is a minor component of dicot cell walls.
It too is characterized by a backbone rich in
galacturonic acid residues that possess side chains
of complex but precise structure. Arabinose and
galactose make up the side chains together with a
number of sugars rather less commonly found in
plant polysaccharides, for example fucrose, aceric
acid, apiose and 3-deoxy-manno-octulosonic
acid[19].

It is not altogether clear to what extent highly
branched rhamnogalacturonans such as RGI and
RGII exist as individual polysaccharides, or
whether they occur covalently linked to the less-
branched galacturonans, described earlier to form
very long molecules. *In vivo* many of the carboxyl
groups of the galacturonic acid residues are

methylated, which will prevent aggregation of the galacturonan chains.

Three other polysaccharides found in association with the pectic fraction have already been referred to—an arabinan, a galactan and an arabinogalactan (Table 4.3). Generally these polymers have a much higher M_r than the side chains of the rhamnogalacturonans RGI and RGII described above. There is some evidence, however, that *in vivo* in addition to occurring as covalently linked to the pectic polysaccharides they occur as individual polysaccharides of the matrix.

Hemicelluloses

Those polysaccharides associated with the hemicellulosic fraction include glucans, xylans, xyloglucans and the mannans (Table 4.3).

Xylans are major components of the walls of monocots, representing about 20% of each of the primary and secondary walls. In dicots they make up a similar percentage of the secondary wall but represent only 5% of the primary wall. This group of polysaccharides is characterized by a backbone of 1,4-β-linked xylose residues carrying side chains of single glucuronic acid residues on some C2 positions and arabinose residues on some C2 and C3 positions. The degree to which the C2 and C3 hydroxyls are substituted and the proportions of glucuronic acid and arabinose carried vary with the source of the xylan. Generally, however, monocot xylans tend to have a higher proportion of arabinose than the xylans of dicots, whereas in the xylans of secondary walls the proportion of glucuronic acid is increased compared to the xylans of primary walls. The nature of the side chain gives rise to the name arabinoxylan for the former and glucuronoarabinoxylan for the latter type of polysaccharide.

A number of hemicellulosic polysaccharides contain glucose as a substantial component, including the xyloglucans, which make up about 20% of the primary wall in the tissues of dicots. In monocots it represents only *c.* 5% of the primary wall and it appears to be absent from most secondary walls. Like cellulose, xyloglucan chains have a backbone of 1,4-β-linked D-glucose residues but in contrast to cellulose *c.* 70% of the glucose residues in xyloglucans are substituted at the C6 position. The majority (*c.* 80%) of the substituents are single xylosyl residues but more complex side chains containing galactose and fucose have been identified.

Cell walls of grasses typically contain a mixed linkage glucan in which the backbone is composed of a chain of D-glucose residues connected by 1,3 linkages (*c.* 30%) and 1,4 linkages. A sequence of between two and four 1,4-linked residues appears to be separated by a single 1,3-linked residue.

Another wall polymer featuring β-linked D-glucose residues is callose, a polysaccharide that primarily occurs at the sieve plate of phloem vessels and in pollen tubes. The linkage in callose is 1,3 and there appear to be no side chains present. Callose may be present in small amounts in all cell walls but it is known to accumulate in large amounts in a number of tissues in response to wounding, and its presence in most tissues may result primarily from damage.

One further group of polysaccharides containing D-glucose is the glucomannans. As the name implies, these polysaccharides also contain mannose, which is present together with the glucose residues in the backbone of the polymer. Glucose and mannose are joined by β-1,4 linkages and occur in the ratio of *c.* 1 : 3. Glucomannans are a major hemicellulose of the secondary wall, where they are known to carry single galactose side chains on the C6 of some mannose residues.

In addition to carbohydrate polymers a number of other components are present in the cell wall, including glycoproteins. Chief amongst these are extensins, a series of basic glycoproteins which have peptide backbones extremely rich in hydroxyproline and with substantial quantities of serine, lysine, valine and tyrosine. These amino acids contribute the bulk of the polypeptide backbone which is characterized by a repeating oligopeptide sequence. Most of the hydroxyproline residues carry tri-or tetrasaccharide side chains composed of arabinose. The majority of the

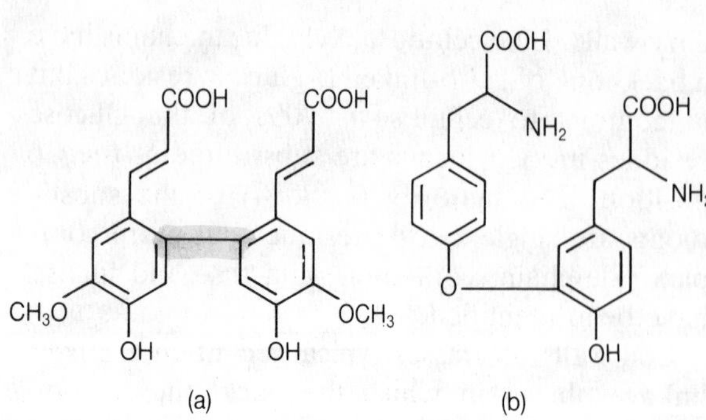

Figure 4.16.

extensin present in the cell wall is covalently bound and it has been suggested that it acts as a 'skeleton' giving coherence to the wall, or that it functions in control of extension growth.

The polysaccharide components of the wall are frequently cross-linked by phenolic compounds, traces of which are present in all cell walls. The most common are ferulic-, p-coumaric- and p-hydroxybenzoic acid, and in walls of cells that have ceased growth there are substantial (up to 35%) quantities of lignin, a hydrophobic polymer of p-hydroxycinnamyl, coniferyl and sinapyl alcohols, some of which is bound to the polysaccharide of the wall. Ferulic acid can form an ester linkage between its carboxylic group and a specific OH group of a polysaccharide, while lignin may be bound via ether bonds. The phenolic side chains of matrix polysaccharides such as ferulic acid are particularly significant because they can be cross-linked via biphenyl bonds to form diferulate bridges and thereby cross-link the polysaccharides (Figure 4.16a). Tyrosine residues of adjacent extension chains can form isodityrosine and so cross-link the glycoprotein (Figure 4.16b).

Synthesis of Polysaccharides

Pathways of Formation

Nucleoside diphosphate sugars are used as the donor of sugar residues, in most cases the evidence indicating involvement of the UDP derivatives. The sugar nucleotides are generated by the reactions of the sugar nucleotide and *myo*-inositol oxidation pathways and associated reactions described earlier (Figure 4.8). The available evidence suggests that sugar residues are added directly from the sugar nucleotide to the growing polysaccharide chain rather than by way of a lipid intermediate, a mechanism that occurs in the synthesis of glycoproteins where polyprenyl phosphate derivatives are used as the direct donors of sugar residues to the oligosaccharide chain.

Experiments with ^{3}H-labelled sugars show that synthesis of the matrix polysaccharides occurs within the endomembrane system, i.e. the endoplasmic reticulum, Golgi bodies and vesicles, and not in the cell wall. The reactions that give rise to the nucleoside diphosphate sugars occur in the cytosol, however, and since membranes are generally impermeable to sugar nucleotides there must be present in the endomembrane system of plants a system for their transport similar to the one identified in mammalian cells. Pulse-chase experiments, combined with cell fractionation studies and autoradiography at the electron microscopic level, have shown that the major site of assembly of the matrix polysaccharides is the Golgi body. Transition of polysaccharide material occurs from elements of the endoplasmic reticulum, through the Golgi bodies from where vesicles are generated to carry the polymers to the plasma membrane. On their arrival at the plasma membrane the vesicles fuse with it and the newly synthesized polysaccharides that are released are integrated into the cell wall, the transit time from endoplasmic reticulum to cell wall being of the order of 20–30 minutes. Little is known concerning the composition of the polysaccharides within the vesicles in terms of whether a single type of polysaccharide is present at any one time or whether a mixture representing the entire range of matrix of polysaccharides is present.

Glycosyltransferases

Formation of the polysaccharides occurs through the activity of glycosyltransferases, enzymes that catalyse transfer of a sugar residue from one molecule to another. In the case of the polysaccharide-synthesizing transferases they are called polysaccharide synthases and they catalyse the reaction

$$NDP\text{-glycose} + (glycan)_n \rightarrow NDP + (glycan)_{n+1}$$

The reaction is essentially irreversible because of the large change in free energy that occurs when the glycosyl residue is transferred from the nucleotide sugar to the polysaccharide chain.

A variety of polysaccharide synthase activities have been identified, all of them membrane bound and all, with the exception of cellulose synthase and glucan synthase II, located in the Golgi bodies. Because they are membrane bound they are difficult to solubilize and purify. Most experiments therefore have been performed with membrane preparations where more than one activity has been present. This, combined with the difficulties of providing a suitably defined oligo- or polysaccharide molecule to act as acceptor, access of the NDP–sugar to the enzyme and subsequent characterization of the product, has made study of these enzymes extremely difficult. Nevertheless a number of polysaccharide synthase activities have been identified and studied, including those involved in synthesis of arabinan, galactan, galacturonan, glucomannan, glucan, xylan and xyloglucans.

Little is known, however, concerning details of the reactions catalysed by the synthases. *In vitro* synthesis is enhanced by addition of a 'primer', an oligosaccharide or polysaccharide onto which the sugar residues can be transferred. It is not known whether such a carbohydrate primer, or a protein primer similar to the one that functions in starch synthesis, is required *in vivo*. There has been some evidence for involvement of a protein primer in xyloglucan synthesis, however, and in maize the polysaccharide which is secreted from root caps is synthesized initially as a glycoprotein from which the protein is subsequently cleaved.

Other problems that must also be considered in polysaccharide synthesis arise from the fact that synthesis of polysaccharides does not involve a molecular template in the way that protein synthesis takes place against the sequence of nucleotides in mRNA. There is, for example, the question of chain initiation and termination. In the absence of a recognized template, what intracellular signal exists to initiate synthesis or to indicate that it is complete? It may not be, of course, that the size of the polysaccharide is 'measured' by the cell directly but final size of the polymer may simply be a reflection of a number of factors, including the activities of the enzymes involved, the availability of substrates and the time the growing polysaccharide spends in transit within the endomembrane system. Termination of synthesis would occur as the polysaccharide in its passage through the endomembrane system moved away from the enzymes and the supply of substrate.

The absence of a template also raises questions with regard to the mechanism by which heteropolymers are assembled by the synthases. While a homopolymer such as starch or cellulose may be assembled by the repeated action of a single glycosyltransferase, and assembly of the backbone of a polymer such as xylan could occur in the same way, at some stage in synthesis of a heteropolymer a variety of different glycosyltransferases must come into play to add the side chains that are a feature of many of the wall polymers. Although it has not yet been clearly established it is probable that each combination of sugar residue and linkage will require a specific separate transferase, so that the activities of a number of glycosyltransferases will be required for assembly of a heteropolymer. In the case of the side chains of the rhamnogalacturonans, for example, or the backbone of the glucomannans, these activities must be precisely coordinated to yield structures that have a precise definition. For other polymers where the structure is more variable, a rather lower degree of coordination would suffice. There is no information available to indicate how the activities of the various glycosyltransferases may be coordinated.

Cellulose

Synthesis of cellulose takes place at the plasma membrane, with UDP–glucose apparently being the donor of glucose residues. Despite at one time being implicated in cellulose synthesis, GDP–glucose appears to be used mainly in the synthesis of glucomannans. Synthesis of cellulose *in vitro* using tissue homogenates has been extremely difficult to achieve and it has never been possible to obtain a glucosyltransferase activity that synthesizes β-glucan chains at a rate comparable to synthesis of cellulose chains *in vivo*[17]. Frequently the polymer produced in such experiments has turned out to be a β-1,3-glucan rather than cellulose. An explanation for this problem appears to be related to the structure of the cellulose synthase. Electron microscopic studies with freeze-fractured preparations of plasma membrane have revealed the presence of hexagonal arrays of groups or rosettes of particles, some of which are associated with the ends of microfibrils. Each rosette appears to be composed of six particles, the whole rosette representing the cellulose synthase complex which is capable of synthesizing an entire microfibril. Since each microfibril is an aggregate of between 30 and 100 cellulose chains, assembling the microfibril must be a coordinated process which relies on the structural integrity of the synthase complex on the membrane. Failure to observe significant rates of synthesis *in vitro* may result from the inevitable loss of integrity that will occur during tissue homogenization. There is some suggestion that the complex responsible for cellulose synthesis may, when it is disrupted, synthesize callose, the β-1,3-glucan which is synthesized very rapidly at the plasma membrane in response to wounding[20]. Such a model would explain the failure to demonstrate significant β-1,4-glucan synthesis *in vitro*.

REFERENCES

1. Pigman, W. and Horton, D. (eds) (1972) *The Carbohydrates: Chemistry and Biochemistry.* Academic Press, New York and London.
2. Feingold, D.S. (1982) Aldo (and keto) hexoses and uronic acids. In *Plant Carbohydrates I, Encyclopedia of Plant Physiology*, New Series, Vol. 13A (eds F.A. Loewus and W. Tanner), pp. 3–76. Springer-Verlag, Berlin.
3. Loewus, F.A., Everard, J.D. and Young, K.A. (1990) Inositol metabolism: precursor role and breakdown. In *Inositol Metabolism in Plants*, Vol. 9 in Plant Biology Series (eds D.J. Marré, W.F. Boss and F.A. Loewus), pp. 21–45. Wiley–Liss, New York.
4. Raboy, V. (1990) Biochemistry and genetics of phytic acid synthesis. In *Inositol Metabolism in Plants*, Vol. 9 in Plant Biology Series (eds D.J. Marré, W.F. Boss and F.A. Loewus), pp. 55–76. Wiley–Liss, New York.
5. Loewus, M.W. and Loewus, F.A. (1980) The C-5 hydrogen isotope-effect in 1L-*myo*-inositol 1-phosphate synthase as evidence for the *myo*-inositol oxidation pathway. *Carbohydrate Research*, **82**, 333–342.
6. Avigad, G. (1982) Sucrose and other disaccharides. In *Plant Carbohydrates I, Encyclopedia of Plant Physiology*, New Series, Vol. 13A (eds F.A. Loewus and W. Tanner), pp. 217–347. Springer-Verlag, Berlin.
7. ap Rees, T. (1984) Sucrose metabolism. In *Carbohydrates in Vascular Plants* (ed. D. H. Lewis), pp. 53–73. Cambridge University Press, Cambridge.
8. Kaiser, G. and Heber, V. (1984) Sucrose transport into vacuoles isolated from barley mesophyll protoplasts. *Planta*, **161**, 562–568.
9. Delrot, S. (1989) Loading of photoassimilates. In *Transport of Photoassimilates* (eds D.A. Baker and J.A. Milburn), pp. 167–205. Longman, Harlow.
10. ap Rees, T. (1988) Hexose phosphate metabolism by non-photosynthetic tissues of higher plants. In *The Biochemistry of Plants*, Vol. 14 (ed. J. Preiss), pp. 1–33. Academic Press, New York.
11. Preiss, J., Robinson, N., Spilatro, S. and McNamara, K. (1985) Starch synthesis and its regulation. In *Regulation of Carbon Partitioning in Photosynthetic Tissue* (eds R. Heath and J. Preiss), pp. 1–26. American Society of Plant Physiology, Rockville.
12. Lin, T.P., Caspar, T., Somerville, C. and Preiss, J. (1988) Isolation and characterisation of a starchless mutant of *Arabidopsis thaliana* (L.) Heynh lacking ADP glucose pyrophosphorylase activity. *Plant Physiology,* **86**, 1131–1135.
13. Stitt, M.N., Huber, S.C. and Kerr, P. (1987) Regulation of photosynthetic sucrose synthesis. In *The Biochemistry of Plants*, Vol. 13 (eds M.D. Hatch and N.K. Boardman), pp. 327–409. Academic Press, New York.
14. Pollock, C.J. and Cairns, A.J. (1991) Fructan metabolism in grasses and cereals. *Annual Review of Plant Physiology and Plant Molecular Biology,* **42**, 77–101.
15. Pollock, C.J. (1986) Fructans and the metabolism

of sucrose in vascular plants. *New Phytologist,* **104,** 1–24.

16. Shiomi, N. (1981) Purification and characterization of 6G-fructosyltransferase from the roots of asparagus (*Asparagus officinalis* L.). *Carbohydrate Research,* **96,** 281–292.

17. Delmer, D.P. (1987) Cellulose biosynthesis. *Annual Review of Plant Physiology,* **38,** 259–290.

18. McNeil, M., Darvill, A.G. and Albersheim, P. (1980) Structure of plant cell walls. X. Rhamnogalacturonan I, a structurally complex pectic polysaccharide in the walls of suspension-cultured sycamore cells. *Plant Physiology,* **66,** 1128–1134.

19. Melton, L.D., McNeil, M., Darvill, A.G., Albersheim, P. and Dell, A. (1986) Structural characterization of oligosaccharide isolated from the pectic polysaccharide rhamnogalacturonan II. *Carbohydrate Research,* **146,** 279–305.

20. Jacob, S.R. and Northcote, D. (1985) *In vitro* glucan synthesis by membranes of celery peptides: the role of the membrane in determining the linkage formed. *Journal of Cell Science, Supplement,* **2,** 1–11.

21. Herold, A., Leegood, R.C., McNeil, P. and Robinson, S.P. (1981) Accumulation of maltose during photosynthesis in protoplasts isolated from spinach leaves treated with mannose. *Plant Physiology,* **67,** 85–88.

22. Beck, E. and Ziegler, P. (1989) Biosynthesis and degradation of starch in higher plants. *Annual Review of Plant Physiology and Plant Molecular Biology,* **40,** 95–117.

23. Heldt, H.W., Flügge, U. and Borchert, S. (1991) Diversity of specificity and function of phosphate translocators in various plastids. *Plant Physiology,* **95,** 341–343.

24. Stitt, M. (1990) Fructose 2,6-bisphosphate as a regulatory molecule in plants. *Annual Review of Plant Physiology and Plant Molecular Biology,* **41,** 153–185.

25. Stitt, M. (1991) Rising CO_2 levels and their potential significance for carbon flow in photosynthetic cells. *Plant, Cell and Environment,* **14,** 741–762.

26. Borchert, S., Grosse, H. and Heldt, H.W. (1989) Specific transport of phosphate, glucose 6-phosphate, dihydroxyacetone phosphate and 3-phosphoglycerate into amyloplasts from pea roots. *FEBS Letters,* **253,**183–186.

FURTHER READING

Baker, D.A. and Milburn, J.A. (1981) *Transport of Photoassimilates.* Longman, Harlow.

Fry, S.C. (1988) *The Growing Plant Cell Wall: Chemical and Metabolic Analysis.* Longman, Harlow.

Loewus, F.A. and Tanner, W. (eds) (1982) *Plant Carbohydrates I, Encyclopedia of Plant Physiology,* New Series, Vol. 13A. Springer-Verlag, Berlin.

Loewus, F.A. and Tanner, W. (eds) (1982) *Plant Carbohydrates II, Encyclopedia of Plant Physiology,* New Series, Vol. 13B. Springer-Verlag, Berlin.

Preiss, J. (ed) (1988) *The Biochemistry of Plants: A Comprehensive Treatise, Vol. 14: Carbohydrates.* Academic Press, San Diego/London.

Stoddart, R.W. (1984) *The Biosynthesis of Polysaccharides.* Croom Helm, London and Sydney.

5

PLANT LIPIDS: THEIR METABOLISM, FUNCTION AND UTILIZATION

D.J. Murphy

John Innes Centre for Plant Science Research, Norwich, UK

INTRODUCTION

The term lipid is difficult to define satisfactorily. In the past, lipids have often been defined as being compounds which are insoluble in water but which are soluble in organic solvents such as ether or chloroform. While this definition covers the majority of common lipids, it excludes many lipids, particularly the lysolipids, which are relatively soluble in water and less soluble in organic solvents. There are also numerous substances, such as hydrophobic proteins, which are obviously not lipids and which are soluble in organic solvents but relatively insoluble in aqueous media.

Probably the best-known examples of physiologically important lipids are the phospholipids and galactolipids, which are the major components of

---

Plant Biochemistry and Molecular Biology. Edited by P.J. Lea and R.C. Leegood.

all biological membranes. Another important lipid class is the neutral lipids such as triacylglycerols, which act as an energy reserve in many organisms. Many other small hydrophobic molecules, however, are also included in the definition of lipids. These include sterols and the various hydrophobic pigments such as chlorophylls, carotenoids and xanthophylls, which play such an important role in light-harvesting processes during photosynthesis. Pigments are discussed in more detail in chapter 8. For the purposes of this chapter, the discussion will be restricted to acyl lipids, i.e. those lipids containing long-chain fatty acids and their derivatives. Three important reasons for taking a particular interest in plant lipids are their unusual nature, their dietary significance and their increasingly useful role as agricultural products:

1. Plant lipids are highly unusual and differ considerably from the lipids found in most animal cells. For example, the major plant membrane system is the thylakoid membranes of chloroplasts. The lipidic phase of these membranes is made up mostly of galactolipids. Galactolipids are relatively rare in animal membranes, but make up about three-quarters of the lipid phase of chloroplast membranes. Furthermore, these chloroplast galactolipids are highly polyunsaturated, normally containing up to six double bonds per galactolipid molecule. In contrast, the phospholipids of most animal membranes normally contain only two to three double bonds per lipid molecule. The reason for the unusual nature of the major plant lipids is at present unknown. It is possible that these polyunsaturated galactolipids play an important role in the unique function of photosynthetic membranes.

2. Plant lipids are essentially dietary components for most mammals. This is because animal cell membranes require the presence of phospholipids containing polyunsaturated fatty acids for their optimal function. Mammals are unable to synthesize polyunsaturated fatty acids *de novo* and must therefore obtain them from their diet. Hence linoleic and α-linolenic acids are

regarded as essential fatty acids and must be present in the diet of any mammal. Deficiencies in the dietary intake of polyunsaturated fatty acids lead to membrane abnormalities and hormonal imbalances, and will often result in the eventual death of the animal.

3. Finally, plant lipids are important agricultural commodities. Seed storage lipids, which are normally made up of triacylglycerols, are the basis of most cooking oils and margarines. Such lipids are also used for many non-edible applications and may eventually come to replace much of the oil which is presently extracted from non-renewable fossil reserves.

We have seen, therefore, that many plant lipids are highly unusual. They are essential components of our diets and, furthermore, are of great and expanding economic significance. These factors make plant lipids an intriguing class of compounds for further study. In the subsequent sections of this chapter, the detailed structure and metabolism of plant lipids will be examined. Their role in membrane organization and function will be explored. Finally, their agricultural and biotechnological significance will be outlined.

STRUCTURE

Fatty acids

Many hundreds of fatty acids have been identified in plants, but only a very small number of them are of quantitative importance. Fatty acids are carboxylic acids attached to a hydrocarbon carbon chain. In most plant membranes the length of a typical fatty acid chain is 16 or 18 carbons. The fatty acids found in storage lipids are somewhat more diverse and may range from 8 to 24 carbons in length. Fatty acids are often described by means of trivial names, such as oleic, palmitic or stearic acids. More recently, a systematic nomenclature has been established for describing different fatty acids in terms of their carbon chain length and the number and location of double bonds on the carbon chain. A third way of

Table 5.1. The major fatty acids in plants.

Common name	Symbol	Systematic name	Plant source[a]	Major uses
Lauric	12:0	Dodecanoic	S	Cosmetics, detergents
Myristic	14:0	Tetradecanoic	S	Detergents
Palmitic	16:0	Hexadecanoic	S, M	Edible
Stearic	18:0	Octadecanoic	S	Edible
Oleic	$18:1\Delta_{9c}$	*cis*-9-Octadecanoic	S, M	Edible
Ricinoleic	$18:1\text{-OH}\Delta_{9c}$	13-Hydroxy-*cis*-9-octadecanoic	S	Pharmaceuticals, plastics
Linoleic	$18:2\Delta_{9,12c}$	*cis,cis*-9,12-Octadecadienoic	S, M	Edible
α-Linolenic	$18:3\Delta_{9,12,15c}$	*all cis*-9,12,15-Octadecatrienoic	S, M	Paints, drying agents
γ-Linolenic	$18:3\Delta_{6,9,12c}$	*all cis*-6,12,12-Octadecatrienoic	S	Health products
Gadoleic	$20:1\Delta_{9c}$	*cis*-9-Eicosenoic	S	Lubricating oils
Erucic	$22:1\Delta_{13c}$	*cis*-13-Docosenoic	S	Plastics, lubricating oils
Nervonic	$24:1\Delta_{15c}$	*cis*-13-Tetracosenoic	S	Industrial oils

[a]S, seed oils; M, membrane lipids.

describing fatty acids is to use a numerical symbol, whereby the chain length is represented first, followed by the number of double bonds in the fatty acid. For example, 18:1 represents a fatty acid with 18 carbons and a single double bond. This fatty acid is known by the trivial name of oleic acid. Its systematic name is *cis*-9-octadecenoic acid. A list of the major fatty acids occurring in plants is given in Table 5.1.

Glycerolipids

Glycerolipids are the most important constituents of plant membranes and oils. Glycerolipids are esters of the trihydric alcohol glycerol, and fatty acids. Three major categories of glycerolipids are found in plants, i.e. acylglycerols, phospholipids and glycolipids. The structures of these lipids are given in Figure 5.1. Acylglycerols are the major components of seed storage oils. The most common acylglycerol in seed oils is triacylglycerol. The molecule is made up of a single glycerol to

which three fatty acids are esterified. Mono- and diacylglycerols are also found in plants and, while these are important metabolites, they are not normally present in large quantities in either seed reserves or in plant membranes. The major glycolipids in plants are the galactolipids. These lipids consist of a glycerol backbone with fatty acid groups attached to the *sn1* and *sn2* positions and one or more galactose residues attached to the *sn3* position. The galactolipids comprise 70–80% of the total lipid of photosynthetic membranes in plants. Since green matter from plants accounts for the overwhelming majority of the biomass on earth, galactolipids are the most common membrane lipids in the biosphere. Another important class of plant glycolipid is the sulpholipid, which consists of a single sulphonated hexose, sulphoquinovose, esterified to a diacylglycerol moiety. Phospholipids are important components of all non-chloroplast membranes in plants. These lipids are made up of a glycerol backbone, with fatty acid groups attached to the *sn1* and *sn2* positions and a phosphate ester group attached

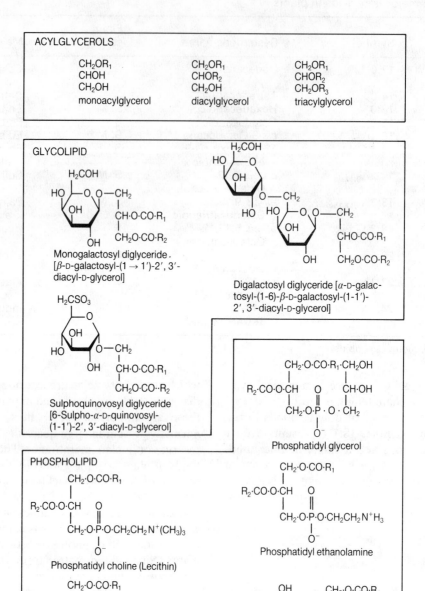

Figure 5.1. The major plant glycerolipids.

to the *sn3* position. The most common phosphate esters found in plant phospholipids are choline, glycerol, ethanolamine, serine, glycerol and inositol.

BIOSYNTHESIS

Generation of Acetyl-CoA

The ultimate precursor of all carbon compounds in plants is photosynthetically fixed carbon dioxide. Carbon dioxide is fixed in the stroma of chloroplasts to form triose phosphates (Chapter 2). Triose phosphates can then be converted to pyruvate and hence to acetyl-CoA by glycolytic enzymes. There is some controversy as to whether the acetyl-CoA used for fatty acid synthesis in plants is indeed generated in chloroplasts directly from triose phosphates. The presence of glycolytic enzymes in chloroplasts has been confirmed but it is unclear as to whether these enzymes are sufficiently active to account for known rates of fatty acid biosynthesis. It is known that virtually all plant fatty acid synthesis occurs in plastids. If the activity of the intra-plastid glycolytic enzymes is not sufficient to generate enough acetyl-CoA for fatty acid biosynthesis, acetate must be imported from outside the plastid. Large pools of free acetate have been found in plant cells. It is possible that the acetate is generated in plant mitochondria and then diffuses into the plastids. Plastids contain an enzyme—acetyl-CoA synthetase—which can esterify a CoA group onto free acetate, therefore forming acetyl-CoA. Another suggestion is that the source of plastid acetyl-CoA is acetyl carnitine. This would require the presence of an acetyl carnitine carrier on the plastid envelope membrane. At present it is not possible to choose between any of these three possible mechanisms for acetyl-CoA generation on plastids. It is possibly fair to say that the current consensus favours the operation of a glycolytic pathway in plastids, i.e. that triose phosphates are converted directly to acetyl-CoA inside the plastids. In the case of plastids that are unable to fix carbon dioxide, e.g. in non-green tissues such as roots and tubers, the carbon source

for fatty acid biosynthesis is sucrose imported from photosynthetic organs of the plant. Sucrose cannot enter plastids and is therefore converted to hexose phosphates in the cytosol of the non-green tissues (chapter 4). The hexose phosphates diffuse across the plastid envelope membranes and are then converted by the glycolytic pathway to acetyl-CoA. Hence in both green and non-green plastids it is likely that acetyl-CoA arises from a glycolytic pathway. These reactions are summarized in Figure 5.2.

Fatty Acid Biosynthesis

The vast majority of fatty acid biosynthesis in plants occurs in the plastids. There have been some recent reports of fatty acid biosynthetic activity in plant mitochondria, but the rates of this activity are very small in comparison to those found in plastids. The synthesis of fatty acids *de novo* requires the presence of two enzyme complexes, i.e. acetyl-CoA carboxylase and fatty acid synthetase. Acetyl-CoA carboxylase consists of a protein complex of between 220 and 240 kDa. There have been several reports that the activity of acetyl-CoA carboxylase may be rate limiting for fatty acid biosynthesis in plastids. This has stimulated a great deal of interest in this enzyme system, particularly in the case of oil seeds, where the introduction of extra copies of the acetyl-CoA carboxylase gene may result in the accumulation of greater quantities of commercially useful storage oils.

The fatty acid synthetase from plants is a dissociable type II complex made up of six subunits. This is different from the animal fatty acid synthetase, which consists of a single protein molecule with six different active sites. The plant fatty acid synthetase resembles more the dissociable prokaryotic enzyme system. This fact has been adduced as further evidence for the endosymbiotic origin of plastids in plants (see chapter 9). Acetyl-CoA is converted into malonyl-CoA by the acetyl-CoA carboxylase complex. Another molecule of acetyl-CoA serves as the primer for fatty acid synthesis. The primer acetyl-CoA combines with a malonyl-CoA with the loss of one molecule of carbon dioxide to form a four-carbon fatty acid. Successive condensations with further malonyl-CoA groups give rise to the fatty acids containing six, eight, ten, etc. carbon molecules, until a final chain length of 16 carbons is reached.

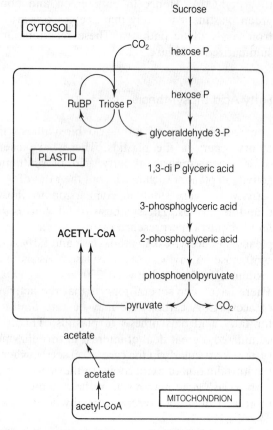

Figure 5.2. Generation of acetyl-CoA in plastids. Three pathways have been proposed for the generation of acetyl-CoA in plastids:
1. the operation of a partial intra-plastid glycolytic sequence via pyruvate,
2. the import of free acetate generated in mitochondria,
3. an acetate shuttle from mitochondria via acetyl carnitine.
In chloroplasts, acetyl-CoA is derived from photosynthetically-fixed carbon dioxide via triose phosphate P. In non-green plastids, acetyl-CoA is derived from sucrose imported from photosynthetic tissues, via hexose phosphate and triose phosphate.

At this stage the fatty acid dissociates from the fatty acid synthetase complex. The full reaction scheme is shown in Figure 5.3. The normal product of the plant fatty acid synthetase complex is the C_{16} fatty acid, palmitic acid, esterified to an acyl carrier protein (ACP), i.e. palmitoyl-ACP. Fatty acids with chain lengths of less than 16 carbons are relatively rare in most plant cells, but are sometimes found as components of seed oils. In this case the fatty acid synthetase probably contains an extra activity leading to premature chain termination, resulting in the accumulation of medium- and short-chain fatty acids, such as C_8, C_{10} and C_{12} fatty acids. These short- and medium-chain fatty acids are normally excluded from plant membranes owing to their detergent-like activity, which can result in severe membrane disruption and eventual lysis. Instead, they are sequestered in the specialized storage organelles, termed oil bodies, that are found in the endosperm and embryo tissues of most oil seeds.

Formation of Oleoyl-CoA

The further elongation of palmitoyl-ACP requires the presence of an extra enzyme, palmitoyl elongase. This enzyme is a complex similar to the fatty acid synthetase. Its function is to convert palmitoyl-ACP into stearoyl-ACP. Stearoyl-ACP is an 18-carbon saturated fatty acid, i.e. it contains no double bonds. At this stage it is possible to introduce double bonds into the molecule by means of a desaturase enzyme. Stearoyl-ACP desaturase introduces a double bond into the $\Delta 9$ position of stearic acid to form oleoyl-ACP. The final reaction in this part of the plastid fatty acid synthesis system is the transacylation of oleoyl-ACP to oleoyl-CoA. The result is the accumulation of oleoyl-CoA in plastids. Oleoyl-CoA is often regarded as a central metabolite in plant lipid metabolism. While all the reactions on the pathway from acetyl-CoA to oleoyl-CoA are present in the plastid, the subsequent metabolism of oleoyl-CoA occurs mainly outside the plastid. Oleoyl-CoA can also undergo a variety of fates, such as further elongation, desaturation or the

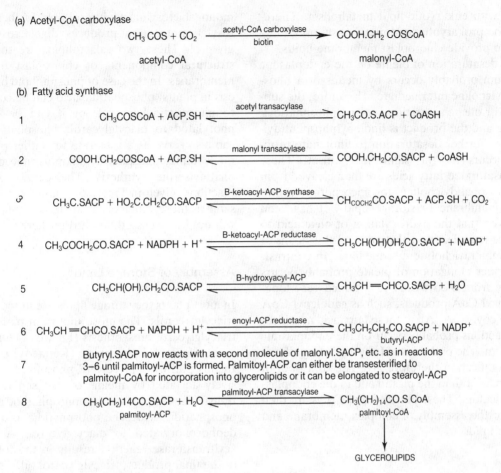

(a) Acetyl-CoA carboxylase

$$CH_3 COS + CO_2 \xrightarrow[\text{biotin}]{\text{acetyl-CoA carboxylase}} COOH.CH_2 COSCoA$$

acetyl-CoA   malonyl-CoA

(b) Fatty acid synthase

1    $CH_3COSCoA + ACP.SH \xrightleftharpoons{\text{acetyl transacylase}} CH_3CO.S.ACP + CoASH$

2    $COOH.CH_2COSCoA + ACP.SH \xrightleftharpoons{\text{malonyl transacylase}} COOH.CH_2CO.SACP + CoASH$

3    $CH_3C.SACP + HO_2C.CH_2CO.SACP \xrightleftharpoons{\text{B-ketoacyl-ACP synthase}} CH_{COCH2}CO.SACP + ACP.SH + CO_2$

4    $CH_3COCH_2CO.SACP + NADPH + H^+ \xrightleftharpoons{\text{B-ketoacyl-ACP reductase}} CH_3CH(OH)OH_2CO.SACP + NADP^+$

5    $CH_3CH(OH).CH_2CO.SACP \xrightleftharpoons{\text{B-hydroxyacyl-ACP}} CH_3CH=CHCO.SACP + H_2O$

6    $CH_3CH=CHCO.SACP + NAPDH + H^+ \xrightleftharpoons{\text{enoyl-ACP reductase}} CH_3CH_2CH_2CO.SACP + NADP^+$
                                                                 butyryl-ACP

7   Butyryl.SACP now reacts with a second molecule of malonyl.SACP, etc. as in reactions 3–6 until palmitoyl-ACP is formed. Palmitoyl-ACP can either be transesterified to palmitoyl-CoA for incorporation into glycerolipids or it can be elongated to stearoyl-ACP

8    $CH_3(CH_2)14CO.SACP + H_2O \xrightleftharpoons{\text{palmitoyl-ACP transacylase}} CH_3(CH_2)_{14}CO.S \, CoA$
     palmitoyl-ACP   palmitoyl-CoA

GLYCEROLIPIDS

Figure 5.3. Fatty acid biosynthesis. All of the fatty acid biosynthetic reactions shown above occur in the plastids of plants. Acetyl-CoA is converted by a series of reactions into oleoyl-CoA, which is the central fatty acid intermediate for further modification reactions, many of which occur outside the plastid.

insertion of groups such as hydroxyl moieties to produce a variety of acyl-CoAs. Alternatively, the oleoyl-CoA can be incorporated directly into glycerolipids in the plastid.

Modification of Oleoyl-CoA

There are two major categories of plant which differ only in their utilization of oleoyl-CoA. In the case of plants with so-called prokaryotic lipid metabolism, oleoyl-CoA can be desaturated inside the plastid to form linoleoyl and linolenoyl fatty acid derivatives. In the case of plants with so-called eukaryotic lipid metabolism, the oleoyl-CoA is exported from the plastid and subsequent desaturations or other modifications occur on the endoplasmic reticulum. There is no desaturation of oleoyl-CoA inside the plastids in such plants. The prokaryotic lipid metabolism leads to the formation of chloroplast membrane lipids exclusively within plastids. However, in order to supply oleate for non-plastid membrane systems or for seed storage oils, the oleoyl-CoA must be exported from the plastid and modified further in the endoplasmic reticulum in a similar way to plants with eukaryotic lipid metabolism. Therefore, the prokaryotic lipid metabolism only functions to provide chloroplast membrane lipids.

The desaturation of oleate on the endoplasmic reticulum probably occurs by means of a phosphatidylcholine intermediate. Therefore, the substrate for oleate saturation is oleoylphosphatidylcholine and the product is linoleoylphosphatidylcholine. Further desaturation to linolenate probably occurs by the same mechanism. These polyunsaturated fatty acids are then cleaved from the phosphatidylcholine for incorporation into either membrane or storage lipids. It has been suggested that the hydroxylation of oleic acid to produce ricinoleic acid also proceeds by means of a phosphatidylcholine intermediate. In contrast, the further elongation of oleate probably occurs directly from oleoyl-CoA to produce very long-chain acyl-CoA products, such as gadoleoyl-CoA and erucoyl-CoA. All of these fatty acid modification reactions probably occur on the endoplasmic reticulum. The pathways of fatty acid biosynthesis and modification therefore give rise to a pool of acyl-CoAs both in the plastid and in the endoplasmic reticulum. These pools of acyl-CoAs are then used for the assembly of complex membrane and storage lipids.

Assembly of Membrane Lipids

The pathways for the assembly of the major membrane phospholipids and galactolipids are shown in Figure 5.4. In both cases, glycerol 3-phosphate is successively acylated to form lysophosphatidic acid and phosphatidic acid. In the case of galactolipid synthesis, the phosphatidic acid is dephosphorylated to produce diacylglycerols. The diacylglycerols accumulate transiently on the chloroplast envelope membrane. The galactose moiety is derived from cytoplasmic pools of carbohydrate via glucose 1-phosphate and UDP–glucose. The immediate precursor for galactolipid synthesis is cytoplasmic UDP–galactose. This combines with the diacylglycerol on the chloroplast envelope by means of a UDP–galactose diacylglycerol galactosyltransferase to produce monogalactosyldiacylglycerol. A further galactosylation reaction produces digalactosyldiacylglycerol. These two galactolipids are the major structural components of chloroplast thylakoid membranes. In the case of phospholipid biosynthesis in plants, phosphatidic acid can be converted to phosphatidylglycerol, or it can be dephosphorylated to diacylglycerol. Phosphatidic acid can also serve as a substrate for either phosphatidylcholine, phosphatidylethanolamine or phosphatidylserine synthesis. The details of these reactions, given in Figure 5.4, are essentially the same as those which occur in animal systems, and will not be discussed any further here.

Assembly of Storage Lipids

In most plants the storage lipids are in the form of triacylglycerols. The most important pathway for triacylglycerol biosynthesis (Figure 5.4) in plants is the glycerol 3-phosphate (Kennedy) pathway. In this pathway, glycerol 3-phosphate is successively acylated by means of two separate acyltransferases to produce lysophosphatidic acid and phosphatidic acid. The phosphatidic acid is then dephosphorylated to diacylglycerol. A further acyltransferase reaction results in the formation of the final product, triacylglycerol oil. The three acyltransferase reactions involved in this pathway often have different specificities for their acyl-CoA substrates in different plant species. This results in the accumulation of seed storage oils with very different fatty acid compositions. The acyltransferases therefore play a key role in the regulation of the final fatty acid composition of the seed oil. In a few plant species, such as jojoba, the major seed storage lipid is a liquid wax, rather than an oil. This wax is made up of a long-chain fatty acid, esterified to a long-chain alcohol, and is widely used in the cosmetics industry.

Seed storage oils are synthesized on a component of the endomembrane system, probably the endoplasmic reticulum, in developing seed tissues. It is likely that triacylglycerols accumulate in the middle of the lipid bilayer of the endoplasmic

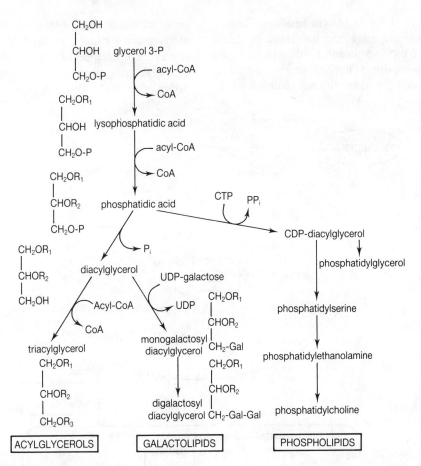

Figure 5.4. Assembly of glycerolipids. Glycerolipid assembly begins with the acylation of glycerol 3-phosphate. The central intermediate is phosphatidic acid. Phospholipids are synthesized directly from phosphatidic acid. Galactolipids and acylglycerols are formed from diacylglycerols, the dephosphorylated form of phosphatidic acid.

reticulum membrane. Eventually the triacylglycerols will cause the membrane to swell up and bud off as a triacylglycerol droplet bounded by a monolayer of phospholipid. Such nascent oil droplets do not appear to contain large amounts of protein. The small oil droplets coalesce rapidly to form large oil bodies in the cytoplasm of the developing seed tissues. Eventually, these oil bodies become coated with a layer of specific protein termed oleosin. These proteins serve to encapsulate the oil and probably have a role in protecting it during seed dehydration and subsequent imbibition. The oleosins also serve to

sequester potentially cytotoxic oils, e.g. those containing short-chain fatty acids, from other cell compounds. The ontogeny of storage oil bodies is summarized in Figure 5.6.

MOBILIZATION

Membrane Lipid Turnover

Membrane lipids, like other active cell components, are subject to continual degradation and resynthesis. It has been found that the head-groups of plant membrane lipids are turned over

122   D.J. MURPHY

more rapidly than the fatty acid residues. Therefore the phosphate ester and galactose moieties are removed from membrane lipids and replaced by new head-groups. These reactions are performed by specific phospholipases and galactoli-

pases, the activities of which have been found in many plant extracts. It has been estimated that the average lifetime of a galactolipid molecule in the thylakoid membrane chloroplast is of the order of several days.

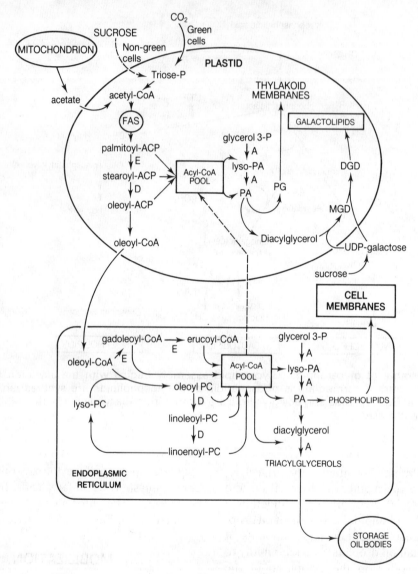

Figure 5.5. Biosynthesis of plant membrane and storage lipids. Acetyl-CoA may be formed in plastids by a variety of reactions depending on the tissue type and developmental stage of the plant. Acetyl-CoA is then converted into oleoyl-CoA within the plastid. Subsequent metabolism of oleoylCA occurs on the endoplasmic reticulum. The plastid lipids may be assembled within the plastid or on the endoplasmic reticulum, depending on the plant species. Most phospholipids and all storage lipids are assembled on the endoplasmic reticulum. These reactions are mediated by a series of acyltransferases (A), desaturases (D), and elongases (E). DGD, digalactosyldiacylglycerol; FAS, fatty acid synthetase; MGD, monogalactosyldioxylglycerol; PA, phosphatidic acid; PC, phosphatidylcholine; PG, phosphatidylglycerol.

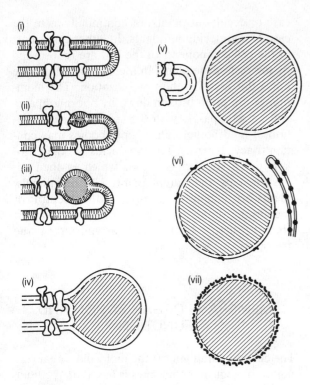

Storage Lipid Mobilization

Lipids are a convenient form of storing energy and storage lipids are found in most organisms. Lipids are relatively compact, anhydrous and have a much higher calorific value than proteins or carbohydrates. Virtually all seeds contain some storage lipids. In some seeds, such as rapeseed or sunflower, the proportion of stored lipid is as high as 50% of total seed weight. Storage lipids are mobilized by lipases, which cleave the fatty acids from all three positions of the triacylglycerol molecule. Lipases are not present in most dry seeds and are probably synthesized *de novo*, several days after seed germination. It is found that when oil seeds such as rapeseed germinate it is the storage proteins which are mobilized first. The amino acids derived from the storage proteins then serve as precursors for the synthesis of enzymes such as lipases, which are required for further mobilization of seed storage lipids. Lipase activity and storage lipid mobilization in rapeseed occurs several days after the mobilization of the seed storage proteins. The lipase successively releases fatty acids from the *sn1*, *sn2* and *sn3* posi-

Figure 5.6. Ontogeny of storage oil bodies in plants. This hypothetical scheme is based on a number of recent studies of the mechanism of oil body formation in oil seeds. (i) It is believed that all the enzymes of triacylglycerol biosynthesis are associated with the endoplasmic reticulum in cells which are synthesized in storage oils. (ii) The activity of these enzymes will result in the deposition of increasing amounts of triacylglycerols. Since the latter are neutral hydrophobic lipid molecules they will tend to accumulate in the middle of the phospholipid bilayer of the endoplasmic reticulum membrane. (iii) As more triacylglycerol accumulates, the two leaflets of the phospholipid membrane will be forced apart, causing a swelling in the endoplasmic reticulum membrane. (iv) The accumulating triacylglycerol is now in the form of an oil droplet bounded by a monolayer of phospholipid, but still attached to the endoplasmic reticulum. Note that intrinsic membrane proteins are unable to partition into such a region and the nascent oil body will therefore tend not to contain endoplasmic reticulum proteins. (v) Eventually the oil body will bud off from the endoplasmic reticulum membrane, forming a nascent oil droplet bounded by phospholipid monolayer. Experiments *in vivo* and *in vitro* have demonstrated that these nascent oil droplets do not contain appreciable amounts of protein. (vi) At a later stage of seed development, the biosynthesis of oleosins begins. Oleosins may be formed on rough endoplasmic reticulum, which is often seen adjacent to immature oil bodies at this stage of seed development. (vii) Eventually the oil body becomes coated with a layer of oleosin and its maturation is complete. Such oil bodies are extremely stable and are able to withstand the rigours of dehydration and rehydration during seed dormancy and germination.

tions of the triacylglycerol molecule. Since free fatty acids are deleterious to cell membranes, and never accumulate in healthy cells, lipolysis is probably coordinated with an activation of the fatty acid by an acetyl-CoA synthetase to produce acyl-thioesters. The acyl-CoAs are then available for β-oxidation, followed by entry into the glyoxylate cycle in glyoxysomes to produce compounds such as sucrose for the developing seedling. In germinating castor bean seedlings there is a quantitative, gram for gram, conversion of storage triacylglycerol to sucrose. Sucrose is a main product of storage lipid mobilization in most oil seed tissues but is not normally the exclusive

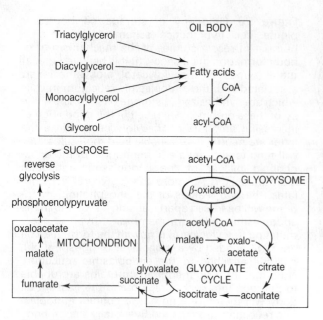

Figure 5.7. Compartmentation of lipid biosynthesis in plants. Four subcellular compartments are involved in storage lipid mobilization. The initial breakdown of the triacylglycerol to fatty acids and glycerol is carried out by a single lipase present in storage oil bodies. The fatty acids are converted to acyl-CoAs, which are then transported to glyoxysomes for further breakdown to acetyl-CoA via the β-oxidation pathway. The acetyl-CoA enters the glyoxylate cycle and this results in the export of succinate to mitochondria, where it is converted to phosphoenolpyruvate. Finally, the phosphoenolpyruvate is converted in the cytosol to hexoses via reverse glycolysis and thence to sucrose.

product as in castor beans. The sucrose is generally transported to the axis to support growth of the new seedling. The pathway of storage lipid mobilization is shown in Figure 5.7 (see also chapter 1).

Membrane Lipid Mobilization

While the membrane lipids of active cells are continually being turned over, they are not normally mobilized. In most perennial plants, however, organs such as leaves are continually being replaced. This may occur annually as in the case of deciduous plants, or continually as in the case of non-deciduous plants. In either case, most of the cell components of the leaf are reabsorbed by the rest of the plant. In the case of membrane lipids this involves their mobilization. The membranes are often broken down by phospholipases and galactolipases, with the transient accumulation of lipid bodies. These lipid bodies may contain triacylglycerols. The fatty acids are then broken down to form sucrose, which is then reabsorbed by the parent plant. Unlike animals, plants are unable to transport either neutral or membrane lipids and must therefore convert them to sucrose for intercellular and inter-tissue transport.

MEMBRANE ORGANIZATION AND FUNCTION

The current paradigm for the molecular organization of biological membranes is based on the fluid mosaic model, initially proposed by Singer *et al.* in 1971. According to this model, biological membranes are visualized as consisting of a planar lipid bilayer formed of more or less cylindrically shaped lipid molecules. Into this lipid bilayer are embedded, to varying extents, the different membrane-associated proteins. When applied to plant membranes such as those of the chloroplast thylakoids, as shown in Figure 5.8, this model must be elaborated in order to explain such factors as transverse and lateral asymmetries in lipid distribution across the membranes and the presence of large amounts of bulky pigment protein complexes embedded in the membranes. It must also account for the presence of large quantities of non-bilayer-forming lipids, such as the monogalactolipids which are such an important feature of thylakoid membranes.

The essential features of the fluid mosaic model for plant membranes are as follows.

1. The membrane is made up of a bimolecular leaflet of phospholipids and/or glycolipids. The lipid bilayer serves both as a solvent for in-

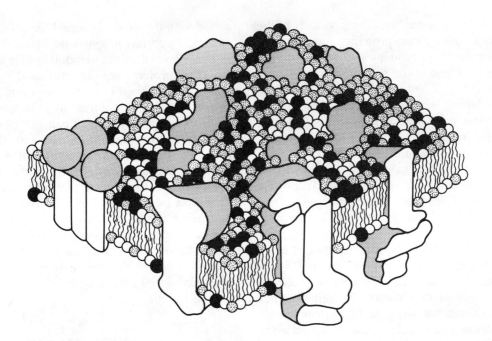

Figure 5.8. Molecular organization of plant membranes. Plant membranes, like other biological membranes, exist as bimolecular leaflets of amphipathic polar lipids into which are embedded the integral membrane proteins. Virtually all integral membrane proteins characterized to date span the entire lipid bilayer and CoA can only be extracted by means of detergents or organic solvents. Extrinsic (non-integral) membrane proteins are attached to one of the external faces of the membrane and are normally associated with intrinsic or other extrinsic membrane proteins by means of ionic interactions. Such extrinsic membrane proteins can be removed by washing with agents such as salts or EDTA. Note that the lateral and transverse distribution of the bilayer lipids is not random. For example, in the transverse plane some lipid classes are more common in the upper monolayer (hatched areas), some are found mainly in the lower monolayer (unshaded areas), while others are found equally in both layers (dark shaded areas). In the lateral plane, some lipids occur more frequently adjacent to integral membrane proteins (hatched areas), while others are more often found in the bulk bilayer phase (unshaded areas). The basic principles of this 'fluid mosaic' model of biological membrane organisms were outlined by Singer and Nicholson (1972) and are presented in more detail by Alberts *et al* (1989).

tegral membrane proteins and as a general permeability barrier.

2. There are two types of protein associated with biological membranes. Integral membrane proteins extend through the lipid bilayer and have polar domains exposed to the aqueous phase at either face of the membrane. The core of the protein is embedded in the hydrophobic core of the membrane. Such proteins are able to diffuse laterally with the membrane, but are unable to undergo rotation and can only be extracted from membrane preparations following the use of chaotropic

agents such as detergents. Non-integral, or extrinsic membrane proteins are adsorbed onto the surface of the membrane and do not extend into the hydrophobic core of the bilayer. Such proteins are almost always associated with intrinsic proteins or with other extrinsic proteins which are themselves associated with the projecting polar domains— intrinsic proteins.

3. While most membrane lipids are in a fluid bilayer phase, there is a population of the lipids which is associated specifically with intrinsic membrane proteins. Such lipids tend to

have long residence times in the vicinity of the intrinsic membrane proteins and tend to accompany these proteins as the latter diffuse along the plane of the membrane.

Lipid molecules within the bilayer phase of bio-·logical membranes are able to diffuse laterally within the membrane at relatively high rates. Transverse diffusion of lipid molecules, i.e. from one bilayer to another (also known as flip-flop), is however, a much slower process. Transverse asymmetries of lipid distribution between the two bilayer leaflets will therefore tend to be relatively stable, and numerous examples have now been well documented in both animal and plant membrane systems. It has been proposed that factors such as lipid-packing properties, relative hydrophobicity/hydrophilicity and different pH/ionic environments give rise to a spontaneous, thermodynamically stable transbilayer lipid asymmetry. For example, it has been shown in several studies that most of the phosphatidylglycerol in the thylakoid membranes of chloroplasts occurs on the outer leaflet of the bilayer, whereas most of the phosphatidylcholine occurs on the inner membrane leaflet.

More recently, it has been found that there are also lateral asymmetries of lipid distribution in thylakoid membranes. These lateral asymmetries are probably mainly due to the secondary effects of lipid–protein associations. Many thylakoid protein complexes are distributed in a laterally asymmetric manner; for example, the ATP–synthetase complex is found only in non-appressed (non-stacked) regions of thylakoid membranes, while the photosystem II complex tends to occur mostly in the appressed regions of the membranes (chapter 1). Since each of these protein complexes will tend to have different lipid populations associated with them, it follows that these lipids will also tend to be distributed asymmetrically in the lateral plane of the membrane.

One intriguing feature of chloroplast thylakoid membranes is that they are largely made up of the non-bilayer-forming lipid, monogalactosyldiacylglycerol. This lipid will not form stable bilayers under any physiological conditions. Even when mixed with the other thylakoid lipids in the physiologically correct proportions, extensive regions of non-bilayer lipid are found. It is only where the thylakoid proteins are present that the monogalactosyldiacylglycerol is able to form a stable bilayer. This is in contrast to other membrane lipids such as most phospholipids, which will form stable bilayers, even in the absence of membrane proteins. It has been proposed that the function of monogalactosyldiacylglycerol in the thylakoid membranes is to facilitate the incorporation and operation of the very bulky protein complexes containing the light-harvesting and photosynthetic electron transport systems. The optimal function of these protein complexes may depend upon their presence within an environment of highly polyunsaturated galactolipids.

AGRICULTURAL AND BIOTECHNOLOGICAL USES OF PLANT LIPIDS

Oil Seeds

Oil seeds can be defined as a class of plant in which relatively large amounts of lipid are stored in the seed tissue. The amount of lipid found in oil seeds ranges from 10–20% in species such as maize and soya bean, to 50% in rapeseed and sunflower. Seed storage oils are normally in the form of triacylglycerols, but in some plants, such as jojoba, they may be liquid waxes. Oil seeds are among the most ancient crops domesticated by mankind. For example, there is evidence for the cultivation of oil-bearing varieties of linseed in the Middle East from over 8000 years ago. From the very earliest times, oil seed products were utilized for a variety of edible and non-edible applications. The ancient Persians used sesame oil in cooking, as a body massage, for illumination, in cosmetics and as a lubricant in simple machines. During the twentieth century, the non-edible uses of oil seed products declined substantially, due to the availability of relatively inexpensive oil derived from fossil reserves. It is now recognized that the fossil reserves are not renewable and interest is growing once more in the possibility of using seed oils for non-edible applications. For example, seed oils can be used as renewable sources of such products as detergents, plastics, solvents, lubricants, paints and even fuels. Some seed oils can also be used in the cosmetic and pharmaceutical industries. As the price of fossil-derived oils continues to mount, the possibility of utilizing oil seed crops as sources of industrial oils becomes ever more attractive.

Utilization of Seed Oils

Edible products

The most important edible oils are those containing palmitic, stearic, oleic and linoleic acids. Of these four fatty acids, oleic and linoleic are the most valuable. Palmitic and stearic acids are saturated lipids and recent publicity concerning the possible adverse nutritional consequences of diets high in saturated lipid has reduced the appeal of oils high in such fatty acids. Oil seed crops containing high levels of oleic and linoleic acids include rapeseed, sunflower, soya bean and olive, to name but a few. Linoleic acid is particularly important, since this is an essential dietary component for humans. The monounsaturated oleic acid is also regarded as a desirable although not essential dietary component.

Non-edible Oils

Almost any type of fatty acid is of potential use for non-edible oil products. For example, very short-chain fatty acids can be used as fuels, or as detergents. Highly polyunsaturated fatty acids are used as drying agents. Hydroxylated fatty acids can be used for cosmetics and plastics manufacture. Very long-chain fatty acids are often used as lubricating oils. Alternatively, the very long-chain fatty acid erucic acid is used as a raw material for the manufacture of valuable high-grade engineering plastics such as nylon 1313, as shown in Figure 5.9.

Designer Oil Seed Crops

At present, most oil seed crops are utilized for edible purposes. In many developed countries, however, the continued accumulation of surpluses of edible products, including oil seeds, has focused attention on the possible non-edible uses of such crops. While most seed oils are of potential use in industrial processes, it is desirable to have a homogeneous fatty acid composition in the oil. Many efforts are now underway to manipulate the seed oil quality using such techniques as plant breeding, induced mutation and genetic engineering. The objective is to produce a whole series of new seed varieties, each of which will contain a different but relatively homogeneous oil composition targeted towards a specific end use.

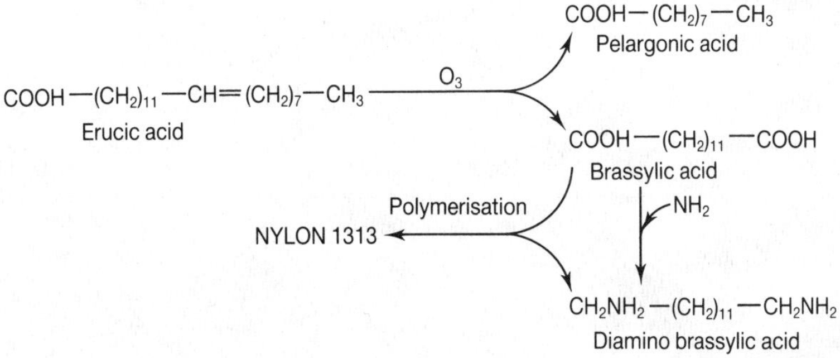

Figure 5.9.

For example, the major European oil seed crop, rapeseed, can in principle be engineered to produce a whole series of varieties with different oil compositions. One variety could be suitable for the plastics industry, one for the synthesis of pharmaceuticals, another for the manufacture of detergents and yet another for cosmetics.

FURTHER READING

Harwood, J.L. and Russell, N.J. (1984) *Lipids in Plants and Microbes*. Allan and Unwin, London.

Staehelin, A. and Arntzen, C.J. (eds) (1986) *Encyclopedia of Plant Physiology, Vol 19: Photosynthesis*. Springer-Verlag, Berlin.

Christie, W.W. (1982) *Lipid Analysis*. Pergamon Press, Oxford.

Harwood, J.L. (1989) Lipid metabolism in plants. *Critical Reviews in Plant Sciences*, **8,** 1–43.

Murphy, D.J. (1991) Storage lipid bodies in plants and other organisms. *Progress in Lipid Research*, **29,** 299–324.

Singer, S.J. and Nicholson, G.L. (1972) *Science*, **175,** 720–731.

Roughan, P.G. and Slack, C.R. (1982) Cellular organisation of lipid metabolism. *Annual Review of Plant Physiology*, **33,** 97–132.

Alberts, B., Bray, D., Lewis, J., Raff, M. and Roberts, K. (1989) *Molecular Biology of the Cell*, 2nd edn. Garland, New York.

6

NITROGEN FIXATION

R. John Smith

Institute of Environmental and Biological Sciences, Lancaster University, UK

John R. Gallon

School of Biological Sciences, University of Wales, Swansea, UK

INTRODUCTION

The inclusion of a chapter on nitrogen fixation in a book about plant biochemistry and molecular biology may at first sight seem strange since nitrogen fixation is not, strictly speaking, a plant process. The ability to fix nitrogen is confined to a relatively small number of prokaryotes (bacteria). On the other hand, nitrogen-fixing cyanobacteria may be thought of as primitive plants and the well-documented and much studied symbiotic associations between nitrogen-fixing bacteria (diazotrophs) and higher plants is of such agricultural importance that it would seem equally strange not to mention it here.

Plants require a source of inorganic nitrogen. Of the nitrogen available to the biosphere, 99.95% (4×10^{15} tonnes) is present at atmo-

Plant Biochemistry and Molecular Biology. Edited by P.J. Lea and R.C. Leegood.
© 1993 John Wiley & Sons Ltd

spheric or dissolved nitrogen, yet this is directly available to only a handful of plants. Most plants obtain their cellular nitrogen from nitrate and ammonium in the soil or dissolved in water and, in order to maintain and improve agricultural productivity, billions of pounds are spent annually to provide fertilizer nitrogen. As well as being expensive, overuse of fertilizer nitrogen has given rise to concern because of its polluting effects on drinking water. Small wonder therefore that there is an intensive research effort devoted to the better exploitation of nitrogen fixation.

In this chapter it is impossible to present more than a general overview of the process of nitrogen fixation. Readers who would like a more detailed account should consult one or more of the following recent books on the topic: Dixon and Wheeler (1986); Postgate (1987); Gallon and Chaplin (1987); Sprent and Sprent (1990).

Sources of Nitrogen: The Nitrogen Cycle

Nitrogen can exist in a variety of oxidation states, ranging from +5 (nitrate) to −3 (ammonium), and biological systems can transform nitrogen between these states. Thus, assimilatory nitrate reduction converts nitrate (+5) to ammonium (−3), dissimilatory nitrate reduction (denitrification) converts nitrate (+5) to nitrous oxide (+1) and nitrogen (0), while, in the opposite direction, nitrification converts ammonium (−3) to nitrite (+4) and nitrate (+5). A cycling of nitrogen therefore occurs in the biosphere, with nitrogen fixation, the reduction of nitrogen (0) to ammonium (−3), an integral part of this process. Figure 6.1 shows a simplified version of the nitrogen cycle.

For incorporation into organic material, nitrogen must be in the form of ammonium. Any plant that assimilates nitrogen in one of its more oxidized forms is therefore faced with the problem of first reducing that compound to the level of ammonium. How plants reduce nitrate to ammonium, and how they assimilate ammonium into organic material, is discussed in chapter 7; here we confine ourselves to the process of nitrogen fixation itself.

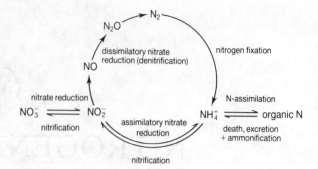

Figure 6.1. A simple nitrogen cycle: the biological interconversion of different forms of inorganic nitrogen.

Nitrogen Fixation

Nitrogen (N_2) is a very unreactive molecule. The reason for its chemical stability lies in the electronic structure of the molecule, but nitrogen is not totally inert. For example, at room temperatures it forms complexes with the salts of transition metals and, in some cases, the nitrogen bound in these complexes can be chemically reduced. Figure 6.2 shows a scheme for reduction of nitrogen bound as a metal complex. Industrial reduction of nitrogen to ammonia (principally by the Haber-Bosch process) is also extensively practised.

The simplest equation for nitrogen fixation is

$$N_2 + 3H_2 \rightarrow 2NH_3$$

This equation describes the Haber–Bosch process. The standard free energy (ΔG^0) for this reaction is −16.6 kJ mol^{-1}. The negative value of ΔG^0 indicates that energy is released during nitrogen fixation, i.e. the reaction is exergonic. This in turn implies that, in the presence of a suitable catalyst, the reaction should proceed spontaneously at room temperature. Unfortunately, no catalyst exists that allows a reasonable rate of nitrogen reduction at room temperature, so in practice the Haber-Bosch process is carried out at high temperatures (400–600 °C) and pressures (100–200 atm), with concomitant input of energy.

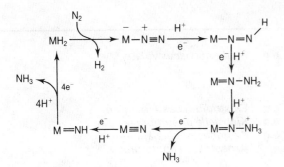

Figure 6.2. Reduction of nitrogen in metal complexes. This scheme is based on studies with model complexes containing molybdenum or tungsten centres. Electrons (e⁻) for reduction of nitrogen may be supplied by the metal centre, in which case the metal becomes oxidized and the cycle stops after one turn, following liberation of ammonia. Alternatively, if electrons are provided externally, for example from an electrode, the cycle can operate indefinitely. Postulated intermediates in the enzymic reduction of nitrogen (Figure 6.4) are based on intermediates of this scheme. Redrawn from R.L. Richards, 1988, *Chemistry in Britain*, **24**, 133–136, by permission of the Royal Society of Chemistry.

Industrial nitrogen fixation is therefore an expensive process.

Biological nitrogen fixation, exemplified by the equation

$$N_2 + 10H^+ + 8e^- \rightarrow 2NH_4^+ + H_2$$

is also an exergonic reaction but in practice requires an input of energy (16 molecules of ATP are hydrolysed per molecule of nitrogen reduced). In addition, since protons replace hydrogen (the reaction takes place in an aqueous environment), a source of electrons (e⁻) is required. Note also that hydrogen is a reaction product of biological nitrogen fixation.

The Organisms

Although the total number of organisms is relatively small, a wide range of different bacterial species can fix nitrogen. Sprent and Sprent (1990) give an excellent account of these and Tables 6.1 and 6.2 serve only to provide a general guide. Diazotrophs are usually divided into free-living and symbiotic forms, but the position is not always clear-cut. Some organisms (notably certain cyanobacteria) appear to be equally able to fix nitrogen independently and in symbiotic association.

The free-living diazotrophs include representatives of both archaebacteria and eubacteria. Among the latter there is an almost random distribution of the ability to fix nitrogen among the various families. Diazotrophy also spans a range of physiological behaviours, covering autotrophy and heterotrophy and also anaerobiosis, facultative anaerobiosis and aerobiosis (Table 6.1). As organisms capable of oxygenic photosynthesis, cyanobacteria (blue-green algae) have the distinction of being the only diazotrophic plants, although as prokaryotes they are more properly classified as bacteria.

Symbiotic diazotrophs (Table 6.2) include the various genera of rhizobia that form the well-documented symbiosis with legumes, though one non-legume, *Parasponia*, is known to become infected by a strain of *Rhizobium*. Infection of legumes by rhizobia results in the formation of nodules, usually on roots. However, some rhizobia, for example *Azorhizobium caulinodans*[1] and the newly discovered photosynthetic *Photorhizobium thompsonianum*[2], form stem nodules on certain plants.

A number of non-legume plants also produce nitrogen-fixing nodules. However, in this case the microsymbiont is not *Rhizobium* but an actinomycete, *Frankia*. The macrosymbionts, referred to as actinorhizal plants, are mainly woody species, and belong to a variety of plant families. For a detailed description of infection and nodule development in both legume and actinorhizal systems, see Sprent and Sprent (1990).

Finally, cyanobacteria participate in a variety of diazotrophic symbioses (see Gallon and Chaplin, 1987; Sprent and Sprent, 1990; Rai, 1990). The co-symbiont may be an angiosperm, a gymnosperm, a pteridophyte, a bryophyte, a fungus (some lichens can fix nitrogen), a diatom or, in

Table 6.1. Some free-living diazotrophs.

Archaebacteria	
Methanogens	e.g. *Methanococcus volate*
Eubacteria	
a. Heterotrophs	
Anaerobes	e.g. *Clostridium pasteurianum*
Facultative anaerobes/microaerobes	e.g. *Klebsiella pneumoniae*
Aerobes	e.g. *Azotobacter vinelandii*
	Azospirillum lipoferum[a]
b. Autotrophs	
Chemotrophic bacteria	e.g. *Thiobacillus ferrooxidans*
Photosynthetic bacteria	e.g. *Rhodospirillum rubrum*
Cyanobacteria:	
Unicells	e.g. *Gloeothece*
Filamentous forms	e.g. *Oscillatoria, Plectonema*
Heterocystous forms	e.g. *Anabaena, Nostoc*

[a]*Azospirillum* spp. are often found in association with roots of grasses, where they fix nitrogen microaerobically.

the case of the coral reef sponge *Siphonochalina*, even an animal.

NITROGENASE

The nitrogen-fixing enzyme is known as nitrogenase. Between 1960, when the enzyme was first isolated, and 1980 only one type of nitrogenase was known. However, we now recognize three genetically distinct nitrogenase enzymes: the original enzyme, which contains molybdenum and iron (Mo-nitrogenase), an enzyme that contains vanadium and iron (V-nitrogenase) and an enzyme that contains iron but only very low levels of

Table 6.2. Some symbiotic diazotrophs.

		Host
Rhizobiaceae		Legumes and *Parasponia*
Azorhizobium	(fixes N_2 *ex planta*)	
Bradyrhizobium	(usually slow growing, some strains fix N_2 *ex planta*)	
Photorhizobium	(photosynthetic)	
Rhizobium	(usually fast growing, most strains do not fix N_2 *ex planta*)	
Sinorhizobium	(fast growing)	
Actinomycetales		Non-legumes (actinorhizal plants)
Frankia		*Alnus, Casuarina, Elaeagnus, Myrica*
Cyanobacteria		Various, e.g.
		Gunnera (angiosperm)
		Macrozamia (cycad: gymnosperm)
		Azolla (pteridophyte)
		Blasia (bryophyte)
		Rhizosolenia (diatom)
		Certain lichens (fungi)
		Siphonochalina (sponge)

Table 6.3. A comparison of molybdenum nitrogenase, vanadium nitrogenase and 'alternative' nitrogenase.

	Mo-nitrogenase	V-nitrogenase	'Alternative' nitrogenase
	Fe-protein	Fe-protein	Fe-protein (2)
Molecular weight	57–72 kDa	63 kDa	65 kDa
Subunit structure	γ_2	γ_2	γ_2
Metal content	4Fe : 4S	4Fe : 4S	4Fe : 4S
	MoFe-protein	VFe-protein	Fe-protein (1)
Molecular weight	220 kDa	200–210 kDa	216 kDa
Subunit structure	$\alpha_2\beta_2$	$\alpha_2\beta_2\delta_2$	$\alpha_2\beta_2\delta_2$
	$\alpha = 50$ kDa	$\alpha = 50$ kDa	$\alpha = 50$ kDa
	$\beta = 60$ kDa	$\beta = 60$ kDa	$\beta = 60$ kDa
		$\delta = 14$ kDa	$\delta = 15$ kDa
Metal content	2Mo : 24–32Fe : 24–30S	2V : 17–21Fe : 18–20S	24Fe : 18S
Product of C_2H_2 reduction	C_2H_4	$C_2H_4 + C_2H_6$	$C_2H_4 + C_2H_6$
% electron flux to NH_3 (balance to H_2)	75	50	50

Note: These properties of the enzymes are typical. Precise values vary from organism to organism.

molybdenum and vanadium (Fe-nitrogenase or, more usually, 'alternative' nitrogenase, since the possibility that it contains some other transition metal, though unlikely, has not rigorously been excluded). All three enzymes catalyse the reduction of nitrogen to ammonium, and are related. However, there are some differences in their properties, as shown in Table 6.3.

Molybdenum Nitrogenase

Isolation and Properties

Nitrogen fixation by cell-free extracts was first successfully demonstrated by Carnahan et al.[3] Because these workers selected the obligate anaerobe Clostridium pasteurianum as the source of their extracts, they were careful to exclude all traces of oxygen during their preparative procedures. This proved to be very wise since it is now known that, regardless of source, all nitrogenases are rapidly and irreversibly inactivated by oxygen. In addition, nitrogenase is cold labile, so the standard enzymological practice of working at

4 °C can also result in loss of activity. However, providing anaerobic conditions are maintained, and providing work is carried out at room temperature, isolation and purification of nitrogenase is a routine procedure, and active preparations have been obtained from a variety of sources. For details of the preparative procedure, see Eady[4].

The properties of Mo-nitrogenase are remarkably constant, regardless of source. For example, the enzyme consists of two proteins: a molybdenum–iron protein (protein 1 or dinitrogenase) and an iron protein (protein 2 or dinitrogenase reductase). The MoFe-protein is an $\alpha_2\beta_2$ tetramer of molecular weight around 220 kDa, with subunits of molecular weight 50 kDa (α) and 60 kDa (β). It contains molybdenum, iron and acid-labile sulphur in the ratio 2Mo : 24–32Fe : 24–30S per molecule. Studies using advanced physical techniques such as electron paramagnetic resonance (EPR), Mössbauer spectroscopy, X-ray absorption spectroscopy and magnetic circular dichroism (MCD) spectroscopy (for details, see Gallon and Chaplin, 1987) suggest that the metals are organized into two iron–molybdenum cofactor (FeMoco) centres, each of stoichiometry $MoFe_{6-8}S_{4-10}$, four Fe_4S_4 'P' clusters (probably one per subunit), similar to those found in bacterial ferredoxins, and an 'S' centre containing two Fe atoms. FeMoco can be extracted from nitrogenase by treatment with dilute acid followed by N-methyl formamide. It may then be studied in isolation.

The Fe-protein of nitrogenase is a γ_2 dimer of molecular weight 57–72 kDa, depending on source. It contains a single Fe_4S_4 cluster that is shared between the two subunits.

Together, the MoFe- and Fe-proteins form an active nitrogenase complex that, given a suitable source of ATP and reducing power, can catalyse nitrogen fixation. *In vitro*, sodium dithionite ($Na_2S_2O_4$) is used as reductant (the active reducing species is $SO_2^{\cdot-}$). ATP is supplied at low concentrations in the presence of a regenerating system, such as creatine phosphate/creatine phosphokinase. This is because the product of ATP hydrolysis, ADP, is an inhibitor of nitrogenase. The active species of ATP and ADP that are involved in nitrogen fixation are $MgATP^{2-}$ and $MgADP^-$ respectively.

The reaction catalysed by nitrogenase under these conditions is therefore:

$$N_2 + 10H^+ \xrightarrow[16MgATP^{2-} + 16H_2O \rightarrow 16MgADP^- + 16H_2PO_4^-]{8(SO_2^{\cdot-}) + 8OH^- \rightarrow 8HSO_3^-} 2NH_4^+ + H_2$$

Many heterologous nitrogenases, consisting of MoFe-protein from one diazotroph incubated with Fe-protein from another, can catalyse nitrogen fixation. This demonstrates the high degree of structural homology among the nitrogenases of different organisms.

In addition to reducing nitrogen to ammonium, nitrogenase can reduce a variety of other substrates (Table 6.4). These substrates have in common their small size and, in many cases, their possession of a triple bond. For more details, see Gallon and Chaplin (1987). Notable among these substrates is acetylene (ethyne) because this forms the basis of a widely used and convenient assay for nitrogenase[5,6]. Mo-nitrogenase reduces acetylene to ethylene (ethene) but no further.

Table 6.4. Some reducible substrates of nitrogenase.

Substrate	Products
Dinitrogen ($N \equiv N$)	NH_3, H_2
Azide ($[N \equiv N—N]^-$)	N_2, N_2H_4, NH_3
Nitrous oxide ($N \equiv N—O$)	N_2
Cyanide ($[C \equiv N]^-$)	CH_4, NH_3, CH_3NH_2
Alkyl cyanides ($R—C \equiv N$)	$R—CH_3$, NH_3
Cyanamide ($N \equiv CNH_2$)	CH_4, NH_3, CH_3NH_2
Acetylene ($HC \equiv CH$)	$H_2C = CH_2$
Alkynes ($R—C \equiv CH$)	$R—CH = CH_2$
Allene $H_2C = C = CH_2$	$H_3C—CH = CH_2$
Proton (H^+)	H_2

Ethylene formed in this way can readily be measured by gas chromatography. In contrast, measurement of nitrogen fixation itself, by monitoring reduction of $^{15}N_2$ by mass spectrometry, is laborious and time consuming. However, the acetylene reduction technique gives only an indirect measurement of nitrogen fixation and, when a quantitative assessment of absolute amounts of nitrogen fixation is needed, should be used with caution.

Unlike the reduction of nitrogen, no hydrogen is generated by nitrogenase as a by-product of the reduction of acetylene or any other substrate. However, in the absence of an alternative substrate, nitrogenase catalyses the reduction of protons H^+ to hydrogen. This is referred to as the ATP-dependent (H_2) evolution reaction catalysed by nitrogenase.

As mentioned above, nitrogenase is inhibited by oxygen, cold and ADP. The oxygen sensitivity of the enzyme lies in both constituent proteins, though the Fe-protein is particularly sensitive to inactivation by oxygen. Carbon monoxide is also a potent inhibitor of all nitrogenase-catalysed reactions except ATP-dependent hydrogen evolution. In addition, hydrogen is an inhibitor of nitrogen reduction, though it does not interfere with the reduction of any other substrate. Finally, it should be mentioned that the various substrates of nitrogenase interfere with each other's reduction. The kinetics of these inhibitions are, in many cases, less simple than might be expected, reinforcing the idea that nitrogenase is a very complex enzyme.

Reaction Mechanism

The way in which the components of nitrogenase cooperate in order to catalyse reduction of nitrogen is now understood and, as far as is known, is the same in all diazotrophs. Nevertheless, studies have largely been confined to a single organism, *Klebsiella pneumoniae*. The Fe-protein of nitrogenase functions to supply electrons to the MoFe-protein, which in turn transfers these electrons to the substrate, nitrogen. It is convenient therefore to divide the reaction catalysed by nitrogenase into three steps: reduction of Fe-protein by an external electron donor such as dithionite; reduction of MoFe-protein by reduced Fe-protein; and reduction of substrate by reduced MoFe-protein. These reactions are summarized in Figures 6.3 and 6.4.

The Fe-protein has two binding sites for $MgATP^{2-}$ (or $MgADP^-$). Binding of $MgATP^{2-}$ or $MgADP^-$ to the Fe-protein results in conformational changes and a lowering of the protein's redox potential by about 0.1 V (to around -0.45 V). Reduction of the Fe-protein by an external

donor results in the acquisition of one electron by the single Fe_4S_4 centre. This reduction is accompanied, or closely followed, by ATP/ADP exchange. Reduced Fe-protein then complexes with oxidized MoFe-protein. There are two binding sites for Fe-protein on each molecule of MoFe-protein, but it is assumed that the MoFe-protein actually functions as two non-cooperative halves. Each half consists of an $\alpha\beta$ dimer and a single FeMoco centre, and forms a complex with a single molecule of Fe-protein. Following complex formation, a single electron is transferred to the MoFe-protein and, simultaneously, ATP hydrolysis occurs. The catalytic cycle is completed by dissociation of the complex. During the operation of this cycle (Figure 6.3), a single electron is transferred from electron donor to the MoFe-protein, and two molecules of ATP are hydrolysed.

Dissociation of the complex, following electron transfer to the MoFe-protein, is a fairly slow process and constitutes the rate-limiting step of nitrogen reduction. Nitrogenase is, at best, a sluggish enzyme, so diazotrophs usually produce relatively large amounts in order to ensure a reasonable rate of nitrogen fixation. Nitrogenase can amount to as much as 10% of total cellular protein.

Nitrogen binds to the FeMoco centre of the

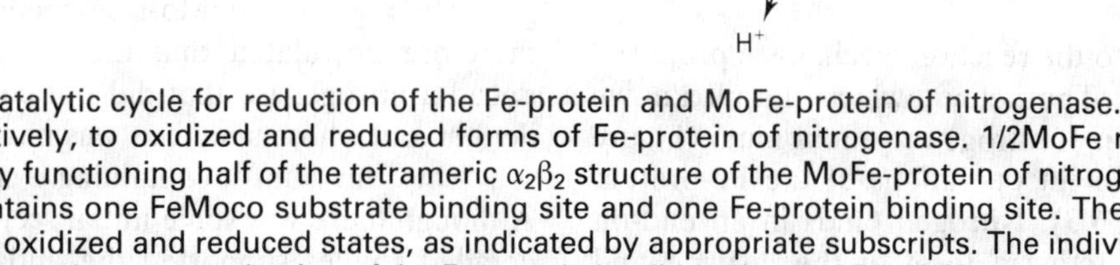

Figure 6.3. Catalytic cycle for reduction of the Fe-protein and MoFe-protein of nitrogenase. Fe_{ox} and Fe_{red} refer, respectively, to oxidized and reduced forms of Fe-protein of nitrogenase. 1/2MoFe represents one independently functioning half of the tetrameric $\alpha_2\beta_2$ structure of the MoFe-protein of nitrogenase (i.e. $\alpha\beta$). Each half contains one FeMoco substrate binding site and one Fe-protein binding site. The MoFe-protein also exists in oxidized and reduced states, as indicated by appropriate subscripts. The individual reactions of the catalytic cycle are: 1, reduction of the Fe-protein by dithionite; 2, ADP : ATP exchange; 3, association with the MoFe-protein; 4, ATP cleavage; 5, electron transfer; and 6, dissociation of the Fe-protein : MoFe-protein complex.

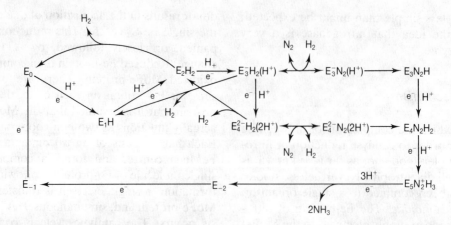

Figure 6.4. A theoretical scheme for the reduction of nitrogen on the MoFe-protein of nitrogenase. E_0 represents 1/2MoFe-protein in the dithionite reduced state. E_n refers to states of the protein reduced by n electron equivalents. Each electron (e^-) is delivered to E through the operation of the catalytic cycle shown in Figure 6.3. (H^+) and ($2H^+$) represent protons bound at a site adjacent to, but separate from, the metal binding site of E. These protons can react with metal-bound hydrogen or nitrogen but cannot react with the metal centre itself to yield a tri- or tetrahydride. For further detail of the possible structure of the metal-bound intermediates shown in this scheme, see Figure 6.2. Note that forms E_2H_2, E_3H_3 and E_4H_4 can decay to more oxidized states of the enzyme, liberating hydrogen. These reactions explain ATP-dependent hydrogen evolution catalysed by nitrogenase. Hydrogen evolution that accompanies nitrogen fixation arises from nitrogen : hydrogen exchange at the level of E_3 and/or E_4.

MoFe-protein. Mutants of *K. pneumoniae* that produce a defective FeMoco show altered substrate specificity[7]. By extrapolation from studies on complexes between nitrogen and transition metals, it has been assumed that nitrogen binds in an end-on manner to molybdenum in FeMoco. However, there is no direct evidence for this, and the discovery of Mo-independent nitrogenases has stimulated a re-evaluation of this assumption. Complexes between iron and nitrogen are known, so it is not impossible that nitrogen binds to Fe in FeMoco, with Mo involved in electron transport and/or H^+ donation to nitrogen. Alternatively, nitrogen may form a bridging complex between Fe and Mo.

According to the reaction mechanism proposed by Lowe and Thorneley[8], which fits all known data, reduction of nitrogen to ammonium (Figure 6.4) requires eight operations of the Fe-protein cycle (Figure 6.3). Nitrogen binds in an end-on manner to a reduced form of the enzyme and displaces hydrogen, thereby explaining the evolution of hydrogen that always accompanies nitro-

gen fixation. Reduction of the two nitrogen atoms of nitrogen is not simultaneous, and intermediates of the type shown in Figure 6.2 are probably involved. Treatment of E_4—M=N—NH_2 with acid or alkali can generate hydrazine (N_2H_2), but hydrazine is not a natural intermediate of nitrogen fixation.

Mo-independent Nitrogenases

In 1980 it was discovered that certain mutants of *Azotobacter vinelandii* which showed no Mo-nitrogenase activity could nevertheless fix nitrogen when grown in Mo-free medium[9]. It was therefore postulated that these organisms produced a nitrogenase that did not contain molybdenum and whose synthesis was normally repressed in medium containing molybdenum. Although there was some initial scepticism, more detailed studies supported the initial interpretation of the data. Conclusive evidence came in 1986 with the isolation of the vanadium-containing nitrogenase from a strain of the related bacterium, *Azotobacter chroococcum*, from which the genes encoding Mo-nitrogenase had been completely deleted[10]. A third nitrogenase containing only traces of molybdenum and vanadium was isolated two years later from *A. vinelandii*[11]; this enzyme is not found in *A. chroococcum*. The structural proteins for the three enzymes are encoded on separate genes. Table 6.3 shows the properties of the three different nitrogenases. V-nitrogenase contains an FeVaco centre, analogous to FeMoco, and seems to function in the same way. FeVaco can, for example, activate inactive Mo-nitrogenase apoprotein (lacking FeMoco). However, while the hybrid V-containing Mo-nitrogenase produced in this way can reduce acetylene, it cannot reduce nitrogen.

While Mo-independent nitrogenases resemble Mo-nitrogenase in many respects, they do have some unusual properties. For example, they reduce acetylene not only to ethylene but also to ethane (which constitutes up to 2% of the ethylene formed). Furthermore, reduction of H^+ to hydrogen accompanies acetylene reduction by V-nitrogenase (Eady, 1990). Mo-independent nitrogenases also generate proportionally more hydrogen during nitrogen reduction than does Mo-nitrogenase (Table 6.3). A notable difference in the structure of Mo-independent nitrogenases is their possession of an extra subunit. The larger of the two component proteins (equivalent to the MoFe-protein of Mo-nitrogenase) has an $\alpha_2\beta_2\delta_2$ structure. The α and β subunits are similar to those in Mo-nitrogenase, while the δ-subunit is a small polypeptide of molecular weight 14–15 kDa.

There is circumstantial evidence that Mo-independent nitrogenases occur in a variety of diazotrophs, though these enzymes do not appear to be universal. However, it is not clear why there should be three different nitrogen-fixing enzymes at all. It may be related to the relative availability of molybdenum and vanadium in nature or to some other environmental factor. It has, for example, been shown that V-nitrogenase is more efficient than Mo-nitrogenase at reducing nitrogen at lower temperatures. Perhaps the different nitrogenases represent temperature adaptation in diazotrophs. Recent reviews concerning Mo-independent nitrogenases include Pau (1989) and Eady (1990).

Integration with Metabolism

In order to catalyse the reduction of nitrogen, nitrogenase needs a source of reductant and of ATP. In addition, carbon skeletons must be supplied for assimilation of the ammonium produced. Finally, there is a need to protect nitrogenase from inactivation by oxygen. All of these requirements must be met by the metabolism of the diazotroph.

Assimilation of ammonium is not exclusive to diazotrophs. All organisms that use inorganic nitrogen first convert it to ammonium for assimilation into organic material. The usual pathway for ammonium assimilation in diazotrophs is through the activities of glutamine synthetase (GS) and glutamate synthase (GOGAT). This pathway is described in detail in chapter 7.

In order to reduce nitrogenase, reductant with a low redox potential (about -0.43 V) is needed. Natural electron donors that meet this requirement are flavodoxins and ferredoxins, both of which occur in diazotrophs. Which of these proteins actually functions to reduce nitrogenase varies from organism to organism. Similarly, how these proteins are themselves reduced varies from organism to organism. For example, in *Clostridium pasteurianum* ferredoxin is reduced directly by a pyruvate:ferredoxin oxidoreductase. This enzyme forms part of the system that also generates ATP in this strict anaerobe (see Gallon and Chaplin, 1987). In *Klebsiella pneumoniae*, flavodoxin acts as the electron donor to nitrogenase and is itself reduced by the action of a pyruvate:flavodoxin oxidoreductase. Photosynthetic diazotrophs can generate reduced ferredoxin as a direct consequence of photosynthetic electron transport (see chapter 1). However, in many diazotrophs, including *Azotobacter* and *Rhizobium*, metabolically generated $NADPH+H^+$

(redox potential -0.32 V) is used to reduce, in turn, ferredoxin or flavodoxin. The thermodynamic unfavourability of this reduction is overcome by the involvement of an energized membrane. From a variety of studies it appears that the membrane potential component ($\Delta\psi$) of the proton motive force is involved in the supply of reductant for nitrogen fixation. The position is complicated by the fact that the proton motive force is also involved in the generation of ATP, but, in experiments where ATP was not limiting, nitrogen fixation ceased when $\Delta\psi$ had a value of less than 80 mV[12].

Generation of ATP for nitrogen fixation also varies from diazotroph to diazotroph. In obligate anaerobes the only possible sources of ATP are substrate-level phosphorylation, and, in some cases, anaerobic electron transport. Aerobes, however, can generate ATP through respiratory electron transport using oxygen as a terminal electron acceptor, while phototrophs can also exploit photosynthetic electron flow, cyclic or non-cyclic, for ATP generation. For more detail concerning how individual diazotrophs provide ATP and reductant for nitrogen fixation, the reader is referred to Gallon and Chaplin (1987).

How diazotrophs cope with oxygen is one of the most fascinating areas of research into nitrogen fixation. An understanding of the various systems that operate to protect nitrogenase from inactivation by oxygen is essential if we are to succeed in our aim of transferring the ability to fix nitrogen to new crop plants. Several strategies for minimizing oxygen inactivation of nitrogenase have been identified among diazotrophs and it is common for more than one strategy to be found in a single organism. However, there is no universal method by which diazotrophs protect nitrogenase from the damaging effects of oxygen. Figure 6.5 shows the various strategies that have so far been identified: how they are distributed among individual diazotrophs is shown in Table 6.5. Diazotrophic cyanobacteria have a particularly acute problem, because they actually evolve oxygen as a consequence of their photosynthetic activity. These organisms are dealt with later in this chapter. However all diazotrophs must deal with atmospheric oxygen.

Strategies fall into three main categories: behavioural, barriers and biochemical. Behavioural strategies include avoidance of oxygen, as practised by strict, and facultative, anaerobes. In addition, certain microaerobes, though able to tolerate oxygen, can fix nitrogen only when the concentration of oxygen is lower than that in air. When grown in soft agar, cultures of these microaerobes migrate to a region where the environmental concentration of oxygen is low, but not zero, since they need some oxygen in order to sustain respiratory production of ATP. A microaerobic environment can also be generated by the clumping together of diazotrophic cells such that they produce a colony with a microaerobic central region. Such aggregations may also involve non-diazotrophs. The behaviour of non-heterocystous cyanobacteria when grown under alternating light and darkness, which is described below, constitutes another behavioural strategy.

Several structures act as barriers to the ingress of oxygen, and allow a microaerobic or anaerobic internal cell environment. The heterocyst envelope of certain cyanobacteria is a case in point, as is the architecture of root nodules. In the case of legume nodules, there appears to be a distinct region of resistance to diffusion of oxygen to the nitrogen-fixing bacteroids. This region lies within the inner cortex of the nodule. Diffusion of oxygen into the nodule is through gas spaces and these are greatly decreased, though not absent, in the inner cortex. However, the diffusion barrier is not merely static. A variable resistance to ingress of oxygen can be created by alterations in osmotic potentials in response to changes in the prevailing concentration of oxygen. For example, exposure to an increased concentration of oxygen can result in replacement of gas spaces by water-filled spaces, with a consequent increase in resistance to diffusion. In addition, intercellular spaces in this region may be plugged with glycoprotein. The high flux of oxygen needed to sustain aerobic metabolism is delivered to the bacteroids in a combined form as oxyleghaemoglobin. For a review, see Layzell and Hunt (1990).

Biochemical strategies include oxygen-consuming reactions such as respiration. At least some diazotrophs possess electron transport

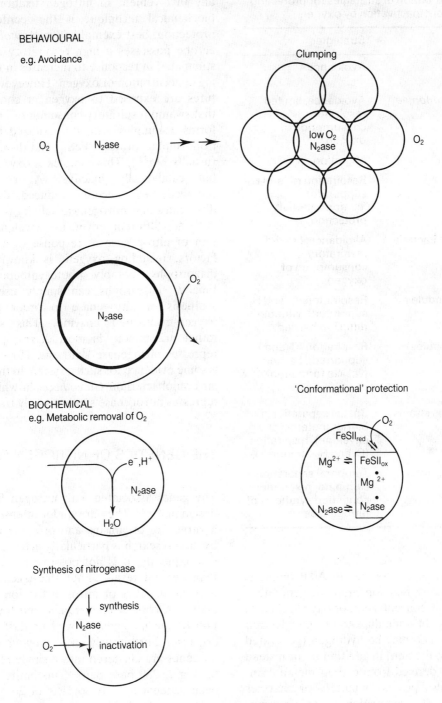

Figure 6.5. Some strategies by which diazotrophs cope with oxygen.

Table 6.5. Distribution of strategies for protecting nitrogenase from inactivation by oxygen.

Organism	Strategies
Clostridium pasteurianum	Avoidance
Klebsiella pneumoniae	Avoidance, limited respiratory consumption of oxygen
Azospirillum	Migration
Azobacter	Respiration (C- and H_2-supported), 'conformational' protection
Photosynthetic bacteria	Avoidance, limited respiratory consumption of oxygen
Legume root nodules	Respiration (C- and H_2-supported), variable diffusion barrier
Actinorhizal nodules	Respiration (C- and H_2-supported), barrier (vesicle formation)
Cyanobacteria:	
Heterocystous strains	Spatial separation, barrier (heterocyst envelope), respiration (C- and H_2-supported)
Non-heterocystous strains	Temporal separation, clumping, respiration, continued synthesis of nitrogenase

chains that are poorly coupled to ATP synthesis and that probably function more to generate a low intracellular concentration of oxygen than to provide energy. In some diazotrophs the electron donor to this chain may be hydrogen (generated during nitrogen fixation) in addition to, or instead of, NAD(P)H derived from carbon metabolism. Diazotrophs also possess a battery of enzymes associated with the destruction of toxic oxygen-derived radicals (Gallon, 1981), though these enzymes fulfil an important role independent of

any involvement in nitrogen fixation. Another biochemical technique is the 'conformational' protection, best exemplified in Azotobacter. This aerobe possesses a high respiratory rate that is stimulated in response to increases in the prevailing concentration of oxygen. However, when cultures are exposed to oxygen at concentrations that swamp respiratory consumption, nitrogenase forms a complex with the oxidized form of an iron–sulphur protein (FeS_{II}) and divalent cations (usually Mg^{2+}). This complex is oxygen-resistant but catalytically inactive. After oxygen is removed, FeS_{II} becomes reduced, the complex dissociates and nitrogenase activity is recovered. A rather different, reversible covalent modification of nitrogenase in response to a variety of factors, including oxygen, is known in other diazotrophs, notably photosynthetic bacteria. Finally, diazotrophs can simply use de novo synthesis of nitrogenase in order to replace oxygen-inactivated enzyme. This strategy is rather uncommon; in most diazotrophs oxygen represses nitrogenase synthesis. However, there is some evidence for such a system in the unicellular cyanobacterium Gloeothece, in which oxygen represses nitrogenase synthesis only transiently[13].

THE GENETICS OF NITROGEN FIXATION

The genes associated with nitrogen fixation are designated nif. They are under intensive study in a variety of organisms and progress in nitrogen fixation research is particularly active in this area. Consequently, it is not possible to talk in more than general terms about these genes. Studies into the genetics of nitrogen fixation were initiated in Klebsiella pneumoniae and the organization of the nif genes is well understood in this organism. Fortuitously, in K. pneumoniae, all the nif genes are clustered into a single regulon containing 23 280 base pairs. This entire region has been sequenced[14]. It consists of 20 genes organized most probably into seven operons (Figure 6.6). The characteristics of the protein products of these genes are shown in Table 6.6.

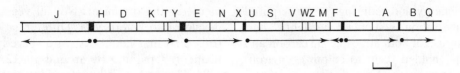

Figure 6.6. The *nif* gene cluster in *Klebsiella pneumoniae*. The arrows indicate the direction of transcription of individual operons. The bar represents 1000 base pairs.

The genes *nifH*, *nifD* and *nifK* encode the structural proteins of the Fe- and MoFe-proteins of nitrogenase. These genes are highly conserved among diazotrophs, emphasizing the similar nature of the enzyme in various nitrogen-fixing organisms. In addition, this has greatly facilitated the development of *nif* genetics, since these *nif* genes, isolated from one diazotroph, can be used to probe the DNA of other diazotrophs for homologous regions. Of the other *nif* genes whose function is known, all are involved in the processing of the three structural proteins except for *nifJ* and *nifF*, which are involved in the provision of reductant for nitrogen fixation in *K. pneumoniae*, and *nifL* and *nifA*, which are regulatory genes.

In *Azotobacter vinelandii* genes designated *vnfH*, *vnfD* and *vnfK* encode the equivalent proteins of V-nitrogenase. The additional δ subunit is encoded on a gene designated *vnfG*, which forms part of an operon *vnfDGK*. *VnfH* constitutes a separate operon. In contrast, the genes encoding 'alternative' nitrogenase, *anfHDGK*, are organized into a single operon. Some *nif* genes (*V*, *S*, *X*, *M* and *B*) are required for the expression of all three systems in *A. vinelandii*. Other genes, such as *E* and *N*, are reiterated—one set for processing of Mo-nitrogenase, another set apparently for processing of the Mo-independent enzymes (see Eady, 1990).

Rhizobia have a collection of genes, designated *nod*, that are involved with the formation of functional nodules (see Sprent and Sprent, 1990). In addition, host plant genes are involved in the establishment of a nitrogen-fixing symbiosis. Similarly in other diazotrophs, genes other than *nif* genes may be important in nitrogen fixation. As an example, in heterocystous cyanobacteria, genes involved in heterocyst differentiation clearly have a role in providing an environment suitable for active nitrogenase.

Table 6.6. Proposed functions of the products of the *nif* genes of *K. pneumoniae*.

Gene	Proposed function of gene product
J	Electron transport (pyruvate:flavodoxin oxidoreductase)
H	Fe-protein of nitrogenase
D	α-Subunit of MoFe-protein of nitrogenase
K	β-Subunit of MoFe-protein of nitrogenase
T	Unknown
Y	Processing of MoFe-protein
E	Synthesis of FeMoco
N	Synthesis of FeMoco
X	Unknown
U	Processing of MoFe-protein
S	Processing of MoFe-protein
V	Synthesis of FeMoco (homocitrate synthase)
W	Unknown
Z	Unknown
M	Processing of Fe-protein
F	Electron transport (flavodoxin)
L	Negative regulator (inactivator of *nifA* product)
A	Positive regulator (activator of *nif* transcription)
B	Synthesis of FeMoco
Q	Uptake and/or processing of Mo

PLANTS AND NITROGEN FIXATION

Nitrogen-fixing Plants: The Cyanobacteria

Among free-living diazotrophs, cyanobacteria have a distinct advantage in that, as photoautotrophs, the energy and reducing power required for nitrogen fixation is supplied by light. Their simultaneous use of both atmospheric carbon and nitrogen has enabled them to colonize environments that are unfavourable for other organisms. This photoautotrophic lifestyle enabled Precambrian cyanobacteria to spread far and wide across the surface of the earth, followed closely by heterotrophic organism that could exploit the cyanobacteria-enriched environments. However, diazotrophic cyanobacteria possess an oxygen-evolving photosystem II very similar to that found in higher plants. Consequently their photoautotrophic metabolism is based on two incompatible processes. The extreme oxygen lability of both nitrogenase proteins effectively prevents nitrogen fixation and photosynthesis from operating in the same cell at the same time. Being prokaryotes the cyanobacteria do not possess intracellular inclusions apart from the thylakoid membranes. They are unlikely therefore to effect an intracellular compartmentalization of the two processes.

Nevertheless, cyanobacteria have evolved answers to this paradox. The present-day unicellular forms separate the two processes in time, while the multicellular, filamentous forms effect a spatial separation of photosynthesis and nitrogen fixation. To achieve this, these latter organisms produce a specialized, differentiated cell called the heterocyst.

The Heterocyst

When diazotrophic cyanobacteria such as *Nostoc* and *Anabaena* species are deprived of an external source of combined nitrogen such as nitrate or ammonium, some of the vegetative cells of the filament develop into specialized, thick-walled cells called heterocysts[15]. This complex cytodifferentiation event involves the cessation of transcription of vegetative cell genes and the initiation of new transcripts from heterocyst-specific genes. As a result, both the structure and enzymic capabilities of the cell change over a period of 20–30 hours. The end-product is the terminally differentiated heterocyst, which can no longer participate in cell division and thus represents a heavy invest-

ment for the organism. Not all vegetative cells develop into heterocysts. Differentiation is controlled so that heterocysts are spaced along the filament, separated by around 10–12 vegetative cells. This pattern is maintained; new heterocysts form midway between two mature heterocysts[16].

The prime function of the heterocyst is to harbour nitrogenase in an anaerobic environment. The thick cell wall consists of three layers that prevent the entry of oxygen. In an unknown manner, the cell wall is more impermeable to oxygen than to nitrogen. In addition, the ends of the heterocyst, where it joins the adjacent vegetative cells, have 'plugs' that may function to limit the entry of oxygen. Microplasmodesmata connect adjacent vegetative cells with the heterocyst[15] and presumably permit entry of nitrogen and other necessary metabolites while preventing entry of oxygen.

Photosystem I remains active in the heterocyst, so cyclic photophosphorylation can provide ATP. However, photosystem II is inactivated and the heterocyst lacks Rubisco (chapter 1) so that carbon dioxide fixation is restricted to the undifferentiated vegetative cells. The primary source of reducing power must therefore be imported into the heterocyst from the vegetative cells. Current hypotheses suggest that a disaccharide, possibly maltose, is imported, hydrolysed and the resultant glucose used to provide NADPH by oxidation via the first two enzymes of the pentose phosphate pathway (Figure 6.7). In exchange, fixed nitrogen is exported. GS is responsible for the assimilation of ammonium produced by nitrogenase. Glutamine is exported to the vegetative cell and converted by GOGAT to glutamate. In this way, fixed nitrogen becomes available for the production of other metabolites, though some glutamate is used to complete the transport cycle by returning to the heterocyst[15,17]. A division of labour is therefore created in which two metabolically active cells exchange necessary products.

Non-heterocystous Cyanobacteria

Unlike their heterocystous relatives, the non-heterocystous cyanobacteria do not usually effect

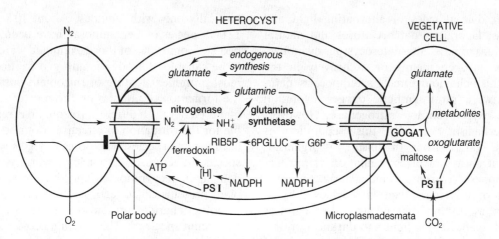

Figure 6.7. Diagrammatic representation of carbon and nitrogen exchange between the heterocyst and vegetative cells. The heterocyst of filamentous cyanobacteria is the site of nitrogen fixation. The heterocyst lacks photosystem II activity, which prevents the formation of oxygen in the vicinity of the oxygen-labile nitrogenase, but also prevents carbon fixation. Consequently carbon compounds, possibly in the form of disaccharides such as maltose, must be provided from the vegetative cell in order to supply the necessary reductant (NADPH) via the oxidative pentose phosphate pathway enzymes (glucose 6-phosphate and 6-phosphogluconate dehydrogenases). The ammonia produced by nitrogen fixation is used to form glutamine through condensation with glutamate by the enzyme glutamine synthetase (GS). The glutamate is generated either by endogenous synthesis or by transport from the vegetative cell. In return the glutamine is transported back to the vegetative cell, where it forms glutamate via glutamate synthase (GOGAT). The heterocyst and vegetative cell thus demonstrate an early evolved form of cellular specialization and a division of labour of the type normally associated with higher plants and animals.

a spatial separation between the oxygen-sensitive process of nitrogen fixation and photosynthetic oxygen evolution. There is one exception—a marine filamentous organism, *Trichodesmium*, which may confine nitrogenase to certain, undifferentiated, cells that have a reducing environment like that of heterocysts. However, recent studies have indicated that the situation in *Trichodesmium* may be more complex than previously thought[18]. This general lack of spatial separation between photosynthesis and nitrogen fixation simplifies things in one respect: photolysis of water can directly supply ATP and reductant for nitrogen fixation by non-heterocystous cyanobacteria. However, the two incompatible processes of photosynthesis and nitrogen fixation are now located within a single, undifferentiated cell.

Actually, there is no good evidence that photosynthesis directly provides reductant for nitrogen fixation in non-heterocystous cyanobacteria. With the notable exception of *Trichodesmium*, natural populations, which are exposed to alternating light and darkness, fix nitrogen mainly in the dark. Under these conditions, photosynthesis supports nitrogen fixation only indirectly through accumulation in the light of carbon reserves that are broken down to support nitrogen fixation in the dark. However, studies with batch cultures of the unicellular organism, *Gloeothece*, suggest that, even under continuous illumination in the laboratory, photosynthesis does not normally directly support nitrogen fixation[19]. On the other hand, recent studies, using continuous cultures, suggest that photosynthesis may, at least under some conditions, directly support nitrogen fixation in *Gloeothece*[20]. Perhaps nitrogenase activity is supported by a pool of reductant and ATP that can be maintained in a variety of ways, including photosynthesis. The relative contribution of individual metabolic processes that replenish this pool may vary according to environmental conditions.

Under diurnal cyles of alternating light and darkness, laboratory batch cultures and natural populations of most non-heterocystous cyanobacteria fix nitrogen mainly or exclusively in the dark. This behaviour effects a temporal separation between photosynthetic oxygen evolution and the oxygen-sensitive nitrogenase. However, the mechanism whereby this separation is achieved varies from organism to organism, though it always involves regulation of nitrogenase synthesis. In some cyanobacteria, the diurnal pattern of nitrogen fixation is achieved by an endogenous rhythm that persists after cultures exposed to alternating light and darkness are returned to constant illumination. In one species of unicellular cyanobacterium, cultures grown under alternating light and darkness exhibit synchronous cell division, with nitrogen fixation confined to a particular phase in the cell cycle[21]. In others, the diurnal pattern has a metabolic basis that is imposed by alternating light and darkness and does not persist under continuous illumination[22]. However, the situation has been complicated by the recent observation that, unlike batch cultures and natural populations, continuous cultures of *Gloeothece* fix nitrogen predominantly in the light, simultaneously with photosynthetic oxygen evolution[20].

Symbiotic Nitrogen Fixation

The symbiotic association between higher plants and the bacterial diazotrophs takes many forms, ranging from the complex, invasive nodulation of legumes by *Rhizobium* species and of woody plants by *Frankia* to the simpler colonization of leaf cavities in *Azolla* by *Anabaena*. While the manner of these associations varies they all show the basic exchange of fixed carbon and nitrogen compounds.

The *Rhizobium*–Legume Symbiosis

With the single known exception of *Parasponia*, a non-leguminous tree, rhizobia fix nitrogen symbiotically only with legumes. About 10% of the 12 000 species of Leguminosae have been examined and some 90% of those examined were found to be capable of symbiotic nitrogen fixation. The most prominent feature of rhizobial symbioses is the formation of nodules on the root of the host plant. Generally a different rhizobium is responsible for the formation of effective root nodules on each legume species. The process begins with the specific infection of a root hair, which is dependent upon recognition by the rhizobium of host plant proteins called lectins. These proteins bind to polysaccharides present on the cell surface of rhizobium species[23]. This binding provides a recognition system and explains why a particular host plant species is infected only by certain species of rhizobia. Infection proceeds via an infection thread through which the rhizobia penetrate to the cells of the inner cortex of the root. The infected cortex cells increase in size and divide to form a ball surrounded by uninfected cells and an outer fibrous layer. Within the infected cells the rhizobia divide and enlarge to form bacteroids which are separated from the cytoplasm of the cell by an envelope membrane (Figure 6.8).

The process of nodulation is controlled by both host and rhizobium genes. The rhizobial nodulation (*nod*) genes are encoded on plasmids known as symbiosis or sym plasmids. In the process of nodule development interaction takes place between the host and rhizobia to promote the expression, in both organisms, of genes that are required to form the nodule and bacteroids. Some *nod* genes, for example, are activated only in the presence of certain plant-derived flavonoids[24]. Different nodule structures are formed on infection of different plants, varying from cylindrical to spherical, and from annular to irregular (see Sprent and Sprent, 1990). These differences demonstrate the expression of different genes in different hosts, although the overall process of nodulation appears to be essentially the same. The analysis of nodulation and the isolation and characterization of *nod* genes is well advanced and several recent reviews are available (Sprent, 1989)[25,26].

One function of the nodule is to protect nitro-

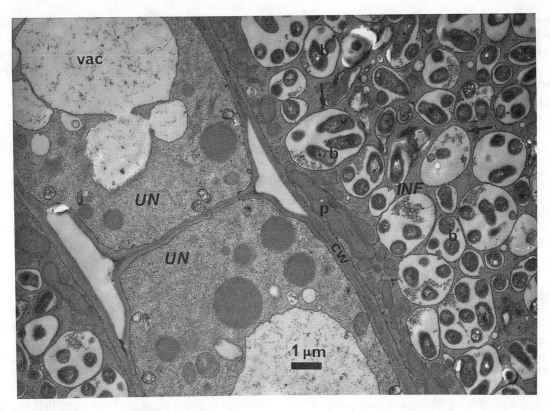

Figure 6.8. Electron micrograph of a soya bean root nodule showing infected cells (INF) and uninfected cells (UN). The bacteroids (b) are enclosed within the peribacteroid membrane (→). Note the absence of vacuoles (vac) in the infected cells, cw = cell wall; p = plastid. Photograph by courtesy of Dr F.W. Wagner, University of Nebraska; reproduced by permission from J.R. Gallon and A.E. Chaplin, 1987, *An Introduction to Nitrogen Fixation*, Cassell, London

genase within the bacteroids from oxygen. Shortly before the maturing nodule becomes capable of nitrogen fixation, a new protein, leghaemoglobin, is synthesized. This can accumulate to form 25% of the soluble protein in the infected cell and confers a characteristic pink/red colour on effective nodules. Leghaemoglobin is a nodulin, one of a class of proteins whose synthesis is restricted to the nodule. Several different forms of leghaemoglobin are found in each plant. Although some forms result from post-translational modifications, the rest are the product of multiple host-plant genes. Leghaemoglobin has a high affinity for oxygen and, like its animal counterpart, myoglobin, acts as an oxygen carrier. It controls the concentration of free ox-

ygen in the nodule and facilitates the controlled diffusion of oxygen to the bacteroids. Since leghaemoglobin has a high affinity for oxygen, the concentration of free oxygen in the vicinity of the bacteroids is kept low, usually to 10 nM, though it may be as high as 20–30 nM at the periphery of the infected plant cell. The outer fibrous layer of the nodule also acts as an effective and variable barrier to diffusion. Channels within the fibrous layer permit the diffusion of oxygen and nitrogen into the nodule, but limit the rate of entry and so may prevent the saturation of leghaemoglobin.

The energy and reducing power for nitrogen fixation in the nodule are thought to be supplied entirely by the plant in the form of photosynthate, mainly sucrose, transported from the leaves.

Growth of the legume at decreased incident light intensity for several days eventually limits nitrogen fixation. Under normal conditions, however, the supply of carbohydrate does not limit nitrogen fixation. It is more likely that the partial pressure of oxygen, required for oxidation of the carbohydrate, limits the production of ATP and reductant. For example, by artificially increasing the partial pressure of oxygen surrounding the nodulated roots of the legume, it has been shown that the increased flux of oxygen into the nodule stimulates the rate of nitrogen fixation. Sucrose that enters the infected plant cell is converted into usable sugars by a catabolic form of sucrose synthetase. This enzyme is also a nodulin. As the first enzyme in the catabolism of carbohydrate in the cell it may control carbon metabolism and therefore, indirectly, the supply of carbohydrate to the bacteroids. The glucose 6-phosphate generated from sucrose enters two relatively inert storage compounds: starch within the amyloplasts of the plant cell and poly-β-hydroxybutyrate (PHB) granules in the bacteroid. Presumably these compounds function as buffers against fluctuations in the supply of photosynthate and may account for the lag that precedes the inhibitory effect on nitrogen fixation of compounds that limit photosynthesis[27].

In infected cells, glucose is catabolized by the Embden–Meyeroff (glycolysis) pathway to form acetyl-CoA. The plant cells of the nodule are rich in PEP (phosphoenolpyruvate) carboxylase which, it is postulated, has an anaplerotic role in providing oxaloacetate that, by condensing with acetyl-CoA, produces the 2-oxoglutarate needed for assimilation of ammonium by GS/GOGAT. The study of carbon metabolism in the bacteroid is complicated not only by the surrounding host cell but also by the separation of the bacteroid from the cytoplasm of the host cell by two membranes: the peribacteroid membrane derived from the host cell and the bacteroid's own membrane. Furthermore species differences occur. Active uptake of disaccharides has been demonstrated for fast-growing *Rhizobium* bacteroids, but slow-growing bacteroids, such as those derived from *Bradyrhizobium*, accumulate disaccharides by passive diffusion. Glucose appears to be poorly taken up by either, except in circumstances where exceptionally low oxygen tensions occur. *Rhizobium* species catabolize glucose using the Entner–Doudoroff pathway, while *Bradyrhizobium* species use the oxidative pentose phosphate pathway. However, the rates of uptake of di-saccharides appear to be insufficient to support the needs of nitrogen fixation in the bacteroid. Fur-thermore mutants unable to metabolize sugars, especially those lacking pyruvate dehydrogenase, still form nodules that are able to fix nitrogen (i.e. Fix$^+$).

The discovery that dicarboxylate transport mutants and mutants lacking succinate dehydrogenase are Fix$^-$ (unable to fix nitrogen) implicated dicarboxylic acids (succinate, and perhaps also malate and fumarate) in the fuelling of nitrogenase activity. The high levels in bacteroids of enzymes of the tricarboxylic acid cycle support this hypothesis, as does the observation that bacteroids assimilate carboxylic acids 50 times more rapidly than they assimilate carbohydrates. Additional evidence comes from the finding that the gene on the host plant choromosome that encodes the dicarboxylic acid transport protein is preceded by a *nifA*-type promoter. As a consequence, expression of this gene would be coupled to expression of genes for nitrogen fixation, suggesting that dicarboxylic acid transport and nitrogen fixation are intimately coupled. Since dicarboxylate transport mutants grow, divide and produce normal nodules, well stocked with bacteroids, it is currently accepted that carbohydrate metabolism supports the growth of bacteroids while metabolism of dicarboxylic acids through the TCA cycle fuels nitrogen fixation[28].

In free-living diazotrophs, ammonium produced by nitrogenase is assimilated by the GS/GOGAT uptake system. Synthesis of these enzymes is suppressed in the bacteroids and ammonia diffuses freely across the bacteroid and peribacteroid membranes into the cytoplasm of the host cell. Both glutamate dehydrogenase (GDH) and GS/GOGAT are present in the host cytoplasm. Research has shown that GDH has no effective assimilatory role. Removal of ammonium from the host cell maintains the concentration gradient for ammonium between the bacteroids and the host cytoplasm.

Two different pathways are used by legumes to transport fixed nitrogen out of the nodule. Aspartate, synthesized by the transamination of oxaloacetate, is converted to asparagine by glutamine-dependent asparagine synthetase. The asparagine produced is transported via surrounding uninfected cells to the xylem for distribution. In ureide-producing nodules, the metabolic conversions are more complicated. Fixed nitrogen is channelled from glutamine, via serine/glycine and asparate, into purine biosynthesis, which takes place in the plastids of the host cell. The purine is transported to the large peroxisomes of surrounding uninfected cells where, in the form of xanthine, it is converted into uric acid and then into allantoin and other ureides before export via the xylem[29]. Ureide transport is a feature of tropical legumes, whereas asparagine transport is more common in temperate zone legumes. However, the significance of the different pathways is not clear.

In some strains of rhizobia, hydrogen produced as a by-product of the nitrogenase-catalysed reduction of nitrogen is consumed through a hydrogen-oxidizing system. This system consists of an uptake hydrogenase and an electron transport chain that uses oxygen as its terminal acceptor. During passage of electrons derived from hydrogen, ATP can be generated, thereby recovering some of the energy input to the nitrogenase reaction. In addition, oxygen consumption may benefit nitrogen fixation by creating a relatively low intracellular concentration of oxygen. In at least some cases, legumes infected with strains of *Rhizobium* or *Bradyrhizobium* that contain an active uptake hydrogenase (Hup$^+$) produced a greater crop yield than did legumes infected with strains lacking uptake hydrogenase (Hup$^-$). However, this seems not always to be the case[30].

Other Nitrogen-fixing Associations

Although rhizobial symbioses are by far the most well studied of the nitrogen-fixing symbiotic associations with higher plants, many others exist. Symbiosis between woody dicotyledonous plants and the actinomycete, *Frankia* (a filamentous prokaryote), has been reported in 21 genera covering eight different families and is believed to have a role in nitrogen-fixation comparable to that of the rhizobia symbioses.

Many similarities exist between rhizobial and *Frankia* symbioses. Both lead to the formation of nodules on the roots of the host plant in which nitrogen fixation occurs. In both cases, there is complementarity with some of the *nif* gene sequences of *K. pneumoniae* and a number of associated proteins are related. However, *Frankia* does not invade the host cell and is not therefore surrounded by a peribacteroid membrane. Instead it is encapsulated in carbohydrate material derived from the host plant. This may not be distinctive since recent studies of rhizobium-induced nodules of a woody legume (*Andria*) have shown that a similar encapsulation occurs[31]. Other evidence suggests that the *Frankia* symbiosis involves a haemoglobin similar to that found in legumes and that the mechanisms regulating expression of nitrogenase in the presence of oxygen and ammonium are very similar in both *Rhizobium* and *Frankia*. Although these indications imply that actinomycete symbioses respond to the requirements for nitrogen fixation in ways similar to that shown for *Rhizobium*, more research is needed before clear parallels and distinctions can be drawn.

Nitrogen-fixing cyanobacteria occur in symbiosis with a variety of other living organisms. Some of these symbiotic systems have been extensively studied (see Rai, 1990) and some general characteristics have emerged. For example, in symbiosis, the cyanobacterium, if heterocystous, often exhibits increased heterocyst frequency. As many as 65% of the cyanobacterial cells may be heterocysts, in marked contrast to the 5–10% heterocyst frequency typical of free-living cyanobacteria. Decreased ammonium uptake is also often seen, accompanied by a transfer of fixed nitrogen, usually as ammonium, to the host. This appears to be a consequence of an inhibition of cyanobacterial GS. However, cycads are a notable exception to this rule, since the symbiotic cyanobacteria found in coralloid roots of these plants possess normal levels of GS[32]. Nitrogen is nevertheless exported from these cyanobacteria to the host plant, probably as glutamine or citrulline. In symbiotic cyanobacteria whose partner is also photosynthetic, there is usually an inhibition of cyanobacterial carbon dioxide fixation though, again, this is not an invariable rule. For example, in certain symbioses of cyanobacteria with diatoms, and also in the *Azolla* symbiosis (Table 6.2), both partners continue to fix carbon dioxide.

Another association is found with C_4 plants, particularly tropical grasses such as *Paspalum notatum*. Photosynthate is released into the area around the roots (the rhizosphere), and this attracts *Azospirillum* species (Table 6.1). The bacteria fix nitrogen and release organic nitrogen, which is taken up by the plant. Although there is no apparent invasion of the plant by *Azospirillum*, nor any formation of specialized structures such as nodules, the association appears to be a symbiosis involving mutual benefit to the two participating species.

REGULATION

Regulation of nitrogen fixation can be at the level of nitrogenase activity or of nitrogenase synthesis, the latter exerted predominantly by regulation of *nif* gene transcription.

Regulation of Nitrogenase Activity

For activity, nitrogenase requires ATP and reductant. Clearly, therefore, nitrogenase activity can be regulated by the availability of these requirements. Because ADP is an inhibitor of nitrogenase, the ADP/ATP ratio may have a special role in regulating activity. Indeed, in experimental systems, there is often a close relationship between the value of the ADP/ATP ratio and the rate of nitrogen fixation. The supply of reductant to nitrogenase often involves an energized membrane, so nitrogen fixation is susceptible to agents

that affect membrane energization, not only because of an effect on ATP synthesis but also through disruption of reductant supply. Notable among such agents is ammonium. Nitrogenase is not inhibited by ammonium *in vitro*, but addition of ammonium to some diazotrophs results in a rapid, but reversible, inhibition of nitrogenase activity. This inhibition is caused by a decrease in $\Delta\psi$ that interrupts reductant supply to nitrogenase.

There is now a large body of evidence that, in certain diazotrophs, nitrogenase activity can be regulated by covalent modification of the Fe-protein of nitrogenase. This phenomenon has been most extensively studied in the photosynthetic bacterium *Rhodospirillum rubrum*[33]. In this diazotroph, inactivation of the Fe-protein is achieved by covalent attachment of an ADP-ribose moiety, derived from NAD^+, to a specific arginine residue (Arg-100) of one of the two subunits of this protein (Figure 6.9). ADP ribosylation is catalysed by dinitrogenase reductase ADP-ribosyl transferase (DRAT) in a reaction that requires a divalent cation (usually, but not exclusively Mn^{2+}) and ADP. When ADP ribosylated, the Fe-protein cannot complex with the MoFe-protein, so nitrogen fixation is inhibited. Removal of ADP-ribose, with consequent activa-

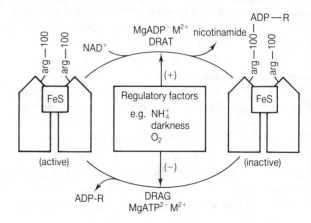

Figure 6.9. Covalent modification of the Fe-protein of nitrogenase in *Rhodospirillum rubrum*. DRAT = dinitrogenase reductase ADP-ribosyl transferase; DRAG = dinitrogenase reductase activating glycohydrolase; ADP-R= ADP-ribose.

tion of nitrogenase, is catalysed by dinitrogenase reductase-activating glycohydrolase (DRAG). Again, a divalent cation is required, but activation also requires ATP. The genes encoding DRAT and DRAG (*draT* and *draG*) have been identified (they are contiguous with *nifHDk* and have been transferred to *K. pneumoniae*. This diazotroph does not normally exhibit covalent modification of nitrogenase but, after introduction of *draT* and *draG*, behaves like *Rhodospirillum rubrum*. A variety of factors regulate nitrogenase activity through DRAT-and DRAG-mediated covalent modification. Among these are addition of ammonium, the presence of oxygen and the absence of light, all of which stimulate ADP ribosylation. However, the metabolic signal regulating the activities of DRAT and DRAG remains to be elucidated, though a promising contender is the reductant status of cells since, in *R. rubrum*, NAD^+ effects inactivation while NADH does not.

Other diazotrophs exhibit what appears to be covalent modification of nitrogenase. In some of these systems there is evidence consistent with ADP ribosylation (e.g. the presence of DRAG, incorporation of ^{32}P into the Fe-protein of nitrogenase), but in others the nature of the modification may be different. For example, it has recently been shown that, in certain cyanobacteria, the electrophoretic mobility of the Fe-protein of nitrogenase is altered in response to environmental changes such as addition of ammonium or exposure to oxygen[34,35]. This may be due to covalent modification of the protein but there is, as yet, no evidence for ADP ribosylation. However, in cyanobacteria, both subunits of the Fe-protein appear to be modified so, even if the mechanism is the same as in photosynthetic bacteria, a significant difference remains.

Genetic Regulation of Nitrogen Fixation

The regulation of *nif* gene transcription in *K. pneumoniae* responds to ammonium, to oxygen and to a general nitrogen regulation system (*ntr* system). The *ntr* system is common to several enteric bacteria and consists of the products of the *ntrA*, *ntrB* and *ntrC* genes. The product of the *ntrA* gene (also known as the *rpoN* gene) is a sigma factor (σ^{54}) that is required by RNA polymerase for recognition of the *nif* and *ntr* type of promoter. These promoters differ substantially from the characteristic hexanucleotide -10 (Pribnow box) and -35 sequences that require the vegetative sigma factor encoded by the gene *rpoD* in order to be recognized by RNA polymerase. Transcription from *ntr* promoters requires, in addition to σ^{54}, positive activation either by the

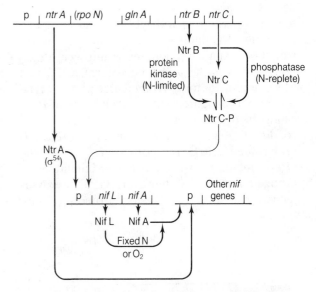

Figure 6.10. Regulation of nitrogen fixation (*nif*) gene expressions in *K. pneumoniae*. The *ntrA* gene product (NtrA; σ^{54}) is required for recognition of *ntr*-activated promoters by RNA polymerase. These promoters include those of the *nif* genes. For transcription of the *nifLA* operon, the phosphorylated form of the *ntrC* gene product (NtrC-P) is also needed. NtrB, the product of the *ntrB* gene, acts as a protein kinase/phosphatase and controls the phosphorylation state of NtrC. Whether NtrB acts as a protein kinase or as a phosphatase depends upon the nitrogen status of the cell (see Figure 6.11). NtrC-P predominates under conditions of nitrogen limitation. The products of *nifLA* act as a positive effector of transcription of the other *nif* genes (NifA) and as an inhibitor of this positive effect (NifL). However, NifL inhibition of NifA-induced transcription of the other *nif* genes occurs only in the presence of fixed nitrogen or oxygen.

general nitrogen regulatory protein NtrC or by the *nif*-specific activator NifA (Figure 6.10). In order to avoid confusion it should be noted that genes are always expressed in lower-case italics (e.g. *nifA*), but the protein encoded by a gene is expressed as standard characters with a capital first letter, e.g. NifA.

Under nitrogen-limiting conditions, NtrC is phosphorylated by NtrB. In its phosphorylated form NtrC acts in conjunction with σ^{54} as a positive activator of *nifL* and *nifA* expression (Figure 6.10). However, in addition to NtrB, phosphorylation of NtrC requires the product of a gene designated *glnB*[36]. The *glnB* gene product, a protein known as P_{II}, is uridylylated in response to a low ratio of glutamine to 2-oxoglutarate (which would occur under conditions of nitrogen limitation), and it is the uridylylated form of P_{II} that acts as a cofactor for NtrB-mediated phosphorylation of NtrC (Figure 6.11). Uridylylation of P_{II} is catalysed by a protein known as GlnD. Under conditions of nitrogen-limitation, therefore, P_{II} is uridylylated, NtrC is phosphorylated and the *nifLA* operon is expressed. Under conditions of nitrogen repleteness, however, GlnD catalyses the removal of the UMP residue from P_{II} (Figure 6.11). This, in turn, favours NtrB-mediated dephosphorylation of NtrC, which would, by removal of phosphorylated NtrC, interrupt transcription of *nifLA*. In this way, expression of *nifLA* responds to the nitrogen status of the cell.

Along with σ^{54}, NifA protein acts as a positive regulator of the expression of the other *nif* genes, though NifA interacts at a different site from σ^{54}[36]. This function of NifA is controlled by NifL in response to nitrogen status and also to oxygen (Figure 6.10). Thus, while the effect of fixed nitrogen on *nif* gene expression may involve 'external' (NtrC, NtrB) and 'internal' (NifA, NifL) regulation, inhibition of nitrogenase synthesis by oxygen is mediated only internally. Nevertheless, by the operation of these regulatory systems, the cell is guarded against unnecessary *nif* gene transcription either if organic nitrogen or ammonium is available, or if the prevailing concentration of oxygen would be sufficient to inactivate nitrogenase.

CONCLUDING REMARKS

Nitrogen fixation is an essential part of the nitrogen cycle, balancing as it does the loss of nitrogen

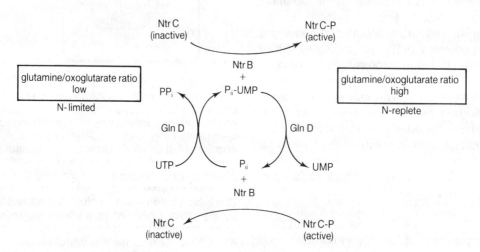

Figure 6.11. Control of NtrC phosphorylation by nitrogen status. Phosphorylation of NtrC requires NtrB (Figure 6.10) and also the uridylylated form of P_{II}. Uridylylation of P_{II} is regulated by GlnD, with nitrogen limitation (when the intracellular ratio glutamine (gluNH₂)/2-oxoglutarate (2-OG) is low) favouring uridylylation and nitrogen repleteness (when gluNH₂/2-OG is high) favouring de-uridylylation. Non-uridylylated P_{II} supports NtrB-catalysed dephosphorylation of NtrC.

to the atmosphere through the action of denitrifying bacteria. The ability to fix nitrogen probably evolved in response to the depletion of available nitrogen sources in the Precambrian ocean but, despite its apparent antiquity, nitrogen fixation remains confined to a few representatives of the bacteria. This could be taken to suggest that insurmountable difficulties may lie in the path of those who wish to generate new nitrogen-fixing crop plants, but hope is offered on two fronts. Firstly, many diazotrophs form beneficial associations of varying degrees of complexity with higher plants, implying that it may be possible to establish novel associations between nitrogen-fixing bacteria and agronomically important plants that currently need to be provided with nitrogen-based fertilizer. Secondly, the ability of many cyanobacteria to fix nitrogen testifies that nitrogen fixation and oxygenic photosynthesis can coexist and that it may be possible to use the techniques of genetic engineering to construct a nitrogen-fixing chloroplast. After all, according to the endosymbiont theory of the origin of subcellular organelles, present-day chloroplasts have their origin in the engulfment by a primitive anaerobe of an ancestral free-living cyanobacterium.

The possibility of constructing novel nitrogen-fixing crop plants has spurred recent research into nitrogen fixation. However, the fact remains that our knowledge of this process, though considerable, remains far from complete. Thus, while the structure, function and genetics of the nitrogenase complex itself are relatively well described, rather less is known about the integrated regulation of energy and reductant supply, and of auxiliary functions such as the protection of nitrogenase from inactivation of oxygen and the need for efficient assimilation of newly fixed nitrogen. Agricultural and ecological considerations have focused research on symbiotic nitrogen fixation, particularly those involving rhizobia, but the additional complexity of systems that involve two distinct partners emphasizes the need to adopt a wide approach to the problem. It is not sufficient to study only the nitrogen-fixing apparatus itself—we need to know how nitrogen fixation integrates with the metabolism of both partners in the symbiosis, what bacterial and plant genes are needed for successful establishment of a nitrogen-fixing symbiosis, where these genes are located on the appropriate chromosome and whether they are amenable to genetic manipulation. There is plenty of room for more work in this fascinating area.

REFERENCES

1. Dreyfus, B., Garcia, J.L. and Gillis, M. (1988) Characterization of *Azorhizobium caulinodans* gen, nov., sp. nov., a stem nodulating nitrogen-fixing bacterium isolated from *Sesbiana rostra*. *International Journal of Systematic Bacteriology*, **38**, 89–98.
2. Eaglesham, A.R.J., Ellis, J.M., Evans, W.R., Fleischman, D.E., Hungria, M. and Hardy, R.W.F. (1990) The first photosynthetic N_2 fixing *Rhizobium*: characteristics. In *Nitrogen Fixation: Achievements and Objectives* (eds P.M. Gresshoff, L.E. Roth, G. Stacey and W.E. Newton), pp. 805–811. Chapman and Hall, New York.
3. Carnahan, J.E., Mortenson, L.E., Mower, H. and Castle, J.C. (1960) Nitrogen fixation in cell-free extracts of *Clostridium pasteurianum*. *Biochimica et Biophysica Acta*, **44**, 520–535.
4. Eady, R.R., (1980) Methods for studying nitrogenase. In *Methods for Evaluating Biological Nitrogen Fixation* (ed. F.J. Bergersen), pp. 213–264. Wiley, Chichester.
5. Koch, B. and Evans, H.J. (1966) Reduction of acetylene to ethylene by soybean root nodules. *Plant Physiology*, **41**, 1748–1750.
6. Burris, R.H. (1975) The acetylene reduction technique. In *Nitrogen Fixation by Free-living Microorganisms* (ed. W.D.P. Stewart), pp 249–257. Cambridge University Press, Cambridge.
7. Smith, B.E., Dixon, R.A., Hawkes, T.R., Liang, Y.-C., McClean, P.A. and Postgate, J.R. (1984) Nitrogenase from *nifV* mutants of *Klebsiella pneumoniae*. In *Advances in Nitrogen Fixation Research* (eds C. Veeger and W.E. Newton), pp. 139–142. Nijhoff/Junk/Pudoc, The Hague.
8. Lowe, D.J. and Thorneley, R.N.F. (1984) The mechanism of *Klebsiella pneumoniae* nitrogenase action: pre-steady state kinetics of H_2 formation. *Biochemical Journal*, **224**, 877–886.
9. Bishop, P.E., Jarlenski, D.M.L. and Hetherington, D.R. (1980) Evidence for an alternative nitrogen fixation system of *Azotobacter vinelandii*.

Proceedings of the National Academy of Sciences of the USA, **77**, 7342–7346.

10. Robson, R.L., Eady, R.R., Richardson, T.J., Miller, R.W., Hawkins, M. and Postgate, J.R. (1986) The alternative nitrogenase of *Azotobacter chroococcum* is a vanadium enzyme. *Nature (London)*, **322**, 388–390.

11. Chisnell, J.R., Premakumar, R. and Bishop, P.E. (1988) Purification of a second alternative nitrogenase from a *nifHDK* deletion strain of *Azotobacter vinelandii*. *Journal of Bacteriology*, **170**, 27–33.

12. Haaker, H., Laane, C. and Veeger, C. (1980) Dinitrogen fixation and the proton motive force. In *Nitrogen Fixation* (eds W.D.P. Stewart and J.R. Gallon), pp. 113–138. Academic Press, London.

13. Maryan, P.S., Eady, R.R., Chaplin, A.E. and Gallon, J.R. (1986a) Nitrogen fixation by the unicellular cyanobacterium *Gloeothece*: nitrogenase synthesis is only transiently repressed by oxygen. *FEMS Microbiology Letters*, **34**, 251–255.

14. Arnold, W., Rump, A., Klipp, W., Priefer, V.B. and Pühler, A. (1988) Nucleotide sequence of a 24, 206-base-pair DNA fragment carrying the entire nitrogen fixation gene cluster of *Klebsiella pneumoniae*. *Journal of Molecular Biology*, **203**, 715–738.

15. Haselkorn, R. (1978) Heterocysts. *Annual Review of Plant Physiology*, **29**, 319–344.

16. Wilcox, M., McIchinson, G.J. and Smith R.J. (1975) Mutants of *Anabaena cylindrica* altered in heterocyst spacing. *Archives of Microbiology*, **103**, 219–223.

17. Wolk, C.P. (1982) Heterocysts. In *The Biology of Cyanobacteria* (eds N.G. Carr and B.A. Whitton), pp. 359–386. Blackwell Scientific Publications, Oxford.

18. Carpenter, E.J., Cheng, J., Cottrell, M., Schubauer, J., Paerl, H.W., Bebout, B.M. and Capone, D.G. (1990) Re-evaluation of nitrogenase oxygen protective mechanism in the planktonic marine cyanobacterium *Trichodesmium*. *Marine Ecology Progress Services*, **65**, 151–158.

19. Maryan, P.S., Eady R.R., Chaplin, A.E. and Gallon, J.R. (1986b) Nitrogen fixation by *Gloeothece* sp. (PCC 6909): respiration and not photosynthesis supports nitrogenase activity in the light. *Journal of General Microbiology*, **132**, 789–796.

20. Stal, L.J., Myint, K.S. and Ortega-Calvo, J.J. (1991) Photosynthesis and nitrogen fixation in the unicellular cyanobacterium *Gloeothece* PCC 6909. In *Proceedings of the Fifth International Symposium on Nitrogen Fixation with Non-legumes* (ed. M. Polsinelli), pp. 437–442. Kluwer, Dordrecht.

21. Mitsui, A., Kumazawa, S., Takahashi, A., Ikemoto, H., Cao, S. and Arai T. (1986). Strategy by which nitrogen-fixing unicellular cyanobacteria grow photoautotrophically. *Nature (London)*, **323**, 720–722.

22. Gallon J.R., Perry, S.M., Rajab, T.M.A., Flayeh, K.A.M., Yunes, J.S. and Chaplin, A.E. (1988) Metabolic changes associated with the diurnal pattern of N_2 fixation in *Gloeothece*. *Journal of General Microbiology*, **134**, 3079–3087.

23. Lerouge, P., Roche, P., Faucher, C., Maillet, F., Truchet, G., Promé, J.C. and Dénairié, J.N. (1990) Symbiotic host-specificity of *Rhizobium meliloti* is determined by a sulphated and acylated glucosamine oligosaccharide signal. *Nature (London)*, **344**, 781–784.

24. Rossen, L., Davies, E.O. and Johnston, A.W.B. (1987) Plant-induced expression of *Rhizobium* genes involved in host specificity and early stages of nodulation. *Trends in Biochemical Sciences*, **12**, 430–433.

25. Stacey, G. (1990) Compilation of the *nod, fix* and *nif* genes of Rhizobia and information concerning their function. In *Nitrogen Fixation: Achievements and Objectives* (eds P.M. Gresshoff, L.E. Roth, G. Stacey and W.E. Newton), pp. 239-244. Chapman and Hall, New York.

26. Vance, C.P. (1990) Symbiotic nitrogen fixation: recent genetic advances. In *The Biochemistry of Plants*, Vol. 16 (eds B.J. Miflin and P.J. Lea), pp. 43–88. Academic Press, San Diego.

27. Streeter, J.G. and Salminen, S.O. (1985) Carbon metabolism in legume nodules. In *Nitrogen Fixation Research Progress* (eds H.J. Evans, P.J. Bottomley and W.E. Newton), pp. 277–283. Nijhoff, Dordrecht.

28. Mellor R.B. and Werner, D. (1990) Legume nodule biochemistry and function. In *Molecular Biology of Symbiotic Nitrogen Fixation* (ed. P.M. Gresshoff), pp. 111–129. CRC Press, Boca Raton.

29. Schubert, K.R. and Boland, M.J. (1990) The ureides. In *The Biochemistry of Plants*, Vol. 16 (eds B.J. Miflin and P.J. Lea), pp. 197–282. Academic Press, San Diego.

30. Evans, H.J., Russell, S.A., Hanus, F.J., Papen, H., Sayavedra Soto, L., Zuber, M. and Boursier, P. (1988) Hydrogenase and nitrogenase relationships in *Rhizobium*: some recent developments. In *Nitrogen Fixation: Hundred Years After*. (eds H. Bothes, F.J. de Bruijn and W.E. Newton), pp. 577–582. Fischer, Stuttgart.

31. DeFaria, S.M., Sutherland, J.M. and Sprent, J.I. (1986) A new type of infected cell in root nodules of *Andria* spp. *Plant Science*, **45**, 143–148.

32. Lindblad, P. (1990) Nitrogen and carbon metabolism in coralloid roots of cycads. *Memoirs of the New York Botanical Garden*, **57**, 104–113.

33. Ludden, P.W., Roberts, G.P., Lowery, R.G., Fitzmaurice, W.P., Saari, L.L., Lehman, L., Lies, D., Woehle, D., Wirt, H., Murrell, S.A., Pope, M.R. and Kanemoto, R.H. (1988) Regulation of nitrogenase activity by reversible ADP-

ribosylation of dinitrogenase reductase. In *Nitrogen Fixation: Hundred Years After* (eds H. Bothe, F.J. de Bruijn and W.E. Newton), pp. 157–162. Fischer, Stuttgart.

34. Reich, S. and Böger, P. (1989) Regulation of nitrogenase activity in *Anabaena variabilis* by modification of the Fe protein. *FEMS Microbiology Letters*, **58**, 81–86.

35. Ernst, A., Reich, S. and Böger, P. (1990) Modification of dinitrogenase reductase in the cyanobacterium *Anabaena variabilis* due to C starvation and ammonia. *Journal of Bacteriology*, **172**, 748–755.

36. Buck, M. (1990) Transcriptional activation of nitrogen fixation genes in *Klebsiella pneumoniae*. In *Nitrogen Fixation: Achievements and Objectives* (eds P.M. Gresshoff, L.E. Roth, G. Stacey and W.E. Newton), pp. 451-457. Chapman and Hall, New York.

37. Magasanik, B. (1988) Reversible phosphorylation of an enhancer binding protein regulates the transcription of bacterial nitrogen utilization genes. *Trends in Biochemical Sciences*, **13**, 475–479.

FURTHER READING

Dixon, R.O.D. and Wheeler, C.T. (1986) *Nitrogen Fixation in Plants*. Blackie, Glasgow.

Eady, R.R. (1990) Vanadium nitrogenases. In *Vanadium in Biological Systems* (ed. N.D. Chasteen), pp. 99–127. Kluwer, Dordrecht.

Gallon, J.R. (1981) The oxygen sensitivity of nitrogenase: a problem for biochemists and microorganisms. *Trends in Biochemical Sciences*, **6**, 19–23.

Gallon J.R. and Chaplin, A.E. (1987) *An Introduction to Nitrogen Fixation*. Cassell, London.

Layzell, D.B. and Hunt, S. (1990) Oxygen and the regulation of nitrogen fixation in legume nodules. *Physiologia Plantarum*, **80**, 322–327.

Pau, R.N. (1989) Nitrogenases without molybdenum. *Trends in Biochemical Sciences*, **14**, 183–186.

Postgate, J. (1987) *Nitrogen Fixation*, 2nd edn. Arnold, London.

Rai, A.N. (ed.) (1990) *A Handbook of Symbiotic Cyanobacteria*. CRC Press, Boca Raton.

Richards, R.L. (1988) Biological nitrogen fixation. *Chemistry in Britain*, **24**, 133–136.

Sprent, J.I. (1989) Tansley Review No. 15. Which steps are essential for the formation of functional legume nodules? *New Phytologist*, **111**, 129–153.

Sprent, J.I. and Sprent, P. (1990) *Nitrogen Fixing Organisms: Pure and Applied Aspects*. Chapman and Hall, London.

7

NITROGEN METABOLISM

Peter J. Lea

Division of Biological Sciences, Lancaster University, UK

INTRODUCTION

Nitrogen is a constituent of a large number of important compounds found in all living cells. Particular notable examples are amino acids, proteins (enzymes) and nucleic acids (RNA and DNA), while others, e.g. polyamines and chlorophyll, may play a major role in some organisms. Most animals do not have the capacity to assimilate inorganic nitrogen, or to synthesize half the amino acids found in proteins, unless assisted by bacteria (e.g. in the rumen of sheep and cattle).

Plant Biochemistry and Molecular Biology. Edited by P.J. Lea and R.C. Leegood
© 1993 John Wiley & Sons Ltd

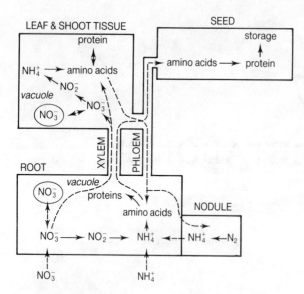

Figure 7.1. A simplified outline of the path taken by nitrogen following uptake in the root to the final deposition as protein in the seed. Based on an original idea by B.J. Miflin.

Although nitrogen is readily available in the air, only certain bacteria (see chapter 6) have the ability to carry out nitrogen fixation and synthesize ammonia. In the root nodule of legumes this is a particularly important process. However, in the majority of plants, nitrate is the sole source of nitrogen and is taken up from the soil by the roots. Ammonia is present as NH_4^+ in certain acidic anaerobic soils and may be taken up directly, but normally it is rapidly oxidized to nitrate by nitrifying bacteria.

Following the uptake of nitrate, the nitrogen is transported to the growing parts of the plant. Ultimately nitrogen is stored in the seed, and in agronomically important crops, e.g. cereals and legumes, is of considerable commercial value. Figure 7.1 shows a brief outline of the major routes of metabolism of nitrogen within a plant. This chapter will attempt to describe the metabolic process in greater detail. During the last five years there have been major advances in our understanding of the regulation of the genes coding for a number of the enzymes involved in nitrogen metabolism. For some reason this work

has tended to concentrate on the early enzymes of nitrate and ammonia assimilation; there are still some pathways, e.g. the synthesis of lysine, proline and histidine, which are very poorly understood.

THE UPTAKE OF NITRATE

Studies on nitrate uptake have been hampered by the lack of a suitable radioactive tracer. Attempts have been made using $^{36}ClO_3^-$, and more recently studies with the short-lived isotope $^{13}NO_3^-$ have been successful. The data may be complicated by the presence of an efflux system which transports NO_3^- out of the root.

In barley plants that have been grown on nitrate, $^{13}NO_3^-$ influx displays typical Michaelis–Menten kinetics (Figure 7.2). The rate of uptake is, however, dependent upon the length of the previous treatment with nitrate, suggesting that a nitrate transport system is induced by nitrate. In plants that have not previously been exposed to nitrate, the rate of uptake is low at low concentrations of nitrate, but above 0.2 mM the rate of uptake increases linearly (Figure 7.3)[1]. The data suggest that nitrate uptake is carried out by a low-

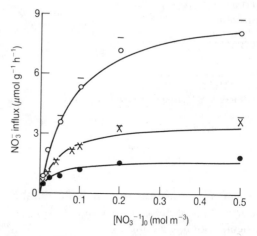

Figure 7.2. $^{13}NO_3^-$ influx into barley roots of plants pretreated with 0.1 mM NO_3^- for 1 day (●), 4 days (X) or with 10 mM NO_3^- for 4 days (○). From Siddiqui et al., 1990[1], by permission of the American Society of Plant Physiologists.

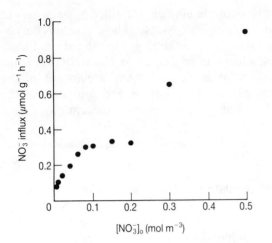

Figure 7.3. $^{13}NO_3^-$ influx into barley roots that have not been pretreated with NO_3^-. From Siddiqui *et al.*, 1990[1], by permission of the American Society of Plant Physiologists.

concentration saturable system and a high-concentration non-saturable system. Only the low-concentration system is stimulated by nitrate. Utilizing ion-specific electrodes, it has been suggested that there is a cotransport of NO_3^-/H^+ across the plasmalemma of maize root cells[2], and that the cytoplasmic level of nitrate is kept relatively constant by storage of nitrate in the vacuole[3].

NITRATE REDUCTION

Nitrate is reduced to ammonia by a two-step process catalysed by the enzymes nitrate reductase and nitrite reductase[4]:

$$NO_3^- + 2H^+ + 2e^- \rightarrow NO_2^- + H_2O$$
$$NO_2^- + 8H^+ + 6e^- \rightarrow NH_4^+ + 2H_2O$$

The reducing power for nitrate reduction is supplied by NADH, while for the six-electron transfer of nitrite reduction, ferredoxin is required. Nitrate reduction can take place in the root or the shoot, depending upon the plant species, developmental age or nitrate supply. In general, as the external nitrate concentration increases,

the proportion that is transported to the shoot for reduction increases.

Nitrite reductase is totally located in the chloroplast or plastid. Despite numerous investigations, a chloroplast location for nitrate reductase has never been confirmed. The enzyme is localized in the cytoplasm, but may be loosely attached to the chloroplast membrane during times of nitrate reduction. Such a system would allow a rapid transport of nitrite into the chloroplast and prevent the build-up of the potentially toxic metabolite.

NITRATE REDUCTASE

Structure

NADH-dependent nitrate reductase (NR) comprises two identical subunits of 110–115 kDa, each containing the prosthetic groups FAD, haem (cytochrome b_{557}) and a molybdenum cofactor (Figure 7.4)[5]. Electrons are passed from NADH to nitrate as shown below:

$$NADH \rightarrow FAD \rightarrow Cytochrome\ b_{557} \rightarrow$$
$$\rightarrow Moco \rightarrow Nitrate$$

The enzyme can also carry out a number of partial reactions (e.g. the reduction of cytochrome c), the physiological role of which is not clear. In addition NR can also convert chlorate to the toxic metabolite chlorite, a reaction which forms the basis of the action of the herbicide chlorate.

Genetics and Molecular Biology

Following initial work in bacteria and fungi, mutants lacking NR enzyme activity have been extensively studied in higher plants. In barley, one structural gene (*nar1*) has been identified. In mutants deficient in NADH-dependent NR, a second NAD(P)H-dependent enzyme has been shown to be present. In other plants there is evidence that there are at least two genes coding for NADH-dependent NR.

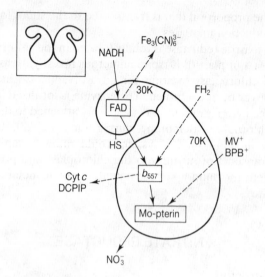

Figure 7.4 Structure–function model of higher plant nitrate reductase. Features include two identical subunits, each composed of domains connected by protease-sensitive regions. Solid arrows denote physiological electron transfer. Dashed arrows denote artificial electron transfer. Reproduced with permission from L. P. Solomonson and M. J. Barber, 1989, in *Molecular and Genetic Aspects of Nitrate Assimilation*, J.L. Wray and J.R. Kinghorn (eds), Oxford University Press, pp. 88–100.

In addition to structural gene mutants, NR activity is also lost due to mutations in genes controlling the molybdenum cofactor. These mutations are readily identified as they also lack the enzyme xanthine dehydrogenase. Detailed analysis of mutants of *Nicotiana* and *Hordeum* have indicated that there are at least six different genes involved in the synthesis and function of the molybdenum cofactor. Although mutants of genes involved in the regulation of nitrate assimilation have been identified in fungi, the existence of such regulatory genes have not yet been confirmed in higher plants[4].

NR has now been cloned from a range of higher plants and fungi. A comparison of the complete sequence of the *Arabidopsis* cDNA with other known protein sequences has shown that NR is composed of three functional domains. These are (a) FAD binding, (b) haem binding and (c)

molybdopterin binding, identified by homology to rat liver sulphite oxidase. These domains correspond to the prosthetic groups shown in Figure 7.4. Genomic sequences of plant NR have just become available and there is evidence of the presence of three introns and four exons—the sizes of the introns are very different.

Regulation

The regulation of NR activity has been studied in a wide variety of plants. As would be expected there are species differences, but a clear general pattern has emerged. When a plant is exposed to nitrate there is a dramatic increase in the extractable NR activity and protein in both the leaves and roots. NR was in fact one of the first enzymes for which clear evidence of *de novo* synthesis was obtained in higher plants. The response appears to be dependent on an increase of a flux into the nitrate metabolic pool, rather than the storage of nitrate in the vacuole.

The literature has, however, been confused with difficulties over the type of assay used and the inherent instability of the protein, coupled with the presence of active protease enzymes.

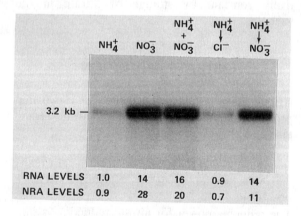

Figure 7.5. Analysis of mRNA coding for nitrate reductase from *Arabidopsis thaliana* treated with various nitrogen sources as shown. Nitrate reductase activities (NRA) were measured in the same tissues. Reproduced with permission from N.M. Crawford *et al.*, 1988, *Proc. Natl Acad. Sci. USA*, **85**, 5006–5010.

The situation has become clearer with the availability of cDNA probes, allowing the detection of NR mRNA levels. Using these procedures nitrate has been shown to control NR activity at the transcription level (Figure 7.5)[6]. Light, probably through the action of phytochrome, also plays a major role in the stimulation of NR activity. Recent evidence suggests that there is a strong diurnal rhythm in the level of NR mRNA, which anticipates the onset of the light period and ensures maximum NR activity during the day[7]. There is also a substantial volume of evidence to show that NR activity may be regulated by carbohydrate supply and a variety of growth regulators and is subject to inhibition following the application of a range of environmental stresses[4].

NITRITE REDUCTASE

Structure

Ferredoxin-dependent nitrite reductase (NiR) was initially considered to comprise one subunit of molecular weight 60–64 kDa. The polypeptide contains a sirohaem prosthetic group and (4Fe–4S) cluster at its active site. There is some evidence to suggest that in order to use reduced ferredoxin as a substrate, the complete enzyme requires a second subunit of molecular weight 24 kDa.

Genetics and Molecular Biology

In contrast to the vast assay of plant mutants lacking various activities of NR, mutants lacking NiR are conspicuous by their absence. This may be due in part to the toxic nature of nitrite, which would have a serious deleterious effect on metabolism. A recent report has established the isolation of the mutant of barley lacking NiR. The plant has to be maintained on low concentrations of ammonia or glutamine as nitrogen source.

NiR cDNA clones have been characterized from spinach and maize. The gene is nuclear coded and a 32-amino-acid leader sequence is present, which is probably the transit peptide that facilitates entry into the chloroplast. There is considerable homology (86%) between the amino acid sequence of the maize and spinach enzymes. However, the homology is less apparent in the cDNA sequence (66%), owing to the consistent use in the maize gene of G/C in the third codon position.

Regulation

NiR is also regulated by nitrate and light, but the effects are less dramatic than observed with NR. The increase in NiR activity has been shown to be due to *de novo* protein synthesis, but enzyme activity and protein are present even in the total absence of nitrate. Transcription of NiR mRNA is rapidly induced by nitrate reaching a peak within 5 hours and then declining to a lower level even in the presence of continued nitrate. The NiR mRNA appears to be short lived with a half-life of 30 minutes. Although nitrite has also been shown to induce NiR mRNA synthesis, it is not clear at the present time whether nitrate or nitrite is the molecule active in regulating transcription[4].

GLUTAMINE SYNTHETASE

We have seen in the previous section that the product of nitrate assimilation is ammonia. It was originally thought that ammonia was incorporated into the organic form by the reductive amination of 2-oxoglutarate catalysed by glutamate dehydrogenase. It is now considered that the enzyme functions in the direction of glutamate breakdown, yielding 2-oxoglutarate for metabolism in the tricarboxylic acid cycle. A substantial body of evidence has been built up to show that glutamine synthetase (GS) is the sole port of entry of ammonia into amino acids in higher plants[8]. GS catalyses the ATP-dependent conversion of glutamate to glutamine (Figure 7.6). The enzyme has a very high affinity for ammonia (K_m 3–5 μM) and is present in all plant tissues.

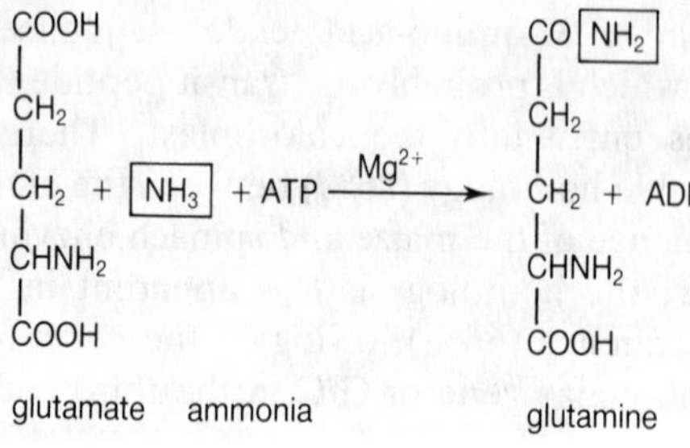

$$\begin{array}{c} \text{COOH} \\ | \\ \text{CH}_2 \\ | \\ \text{CH}_2 + \boxed{\text{NH}_3} + \text{ATP} \xrightarrow{\text{Mg}^{2+}} \\ | \\ \text{CHNH}_2 \\ | \\ \text{COOH} \end{array} \qquad \begin{array}{c} \text{CO}\,\boxed{\text{NH}_2} \\ | \\ \text{CH}_2 \\ | \\ \text{CH}_2 + \text{ADP} + \text{P}_i \\ | \\ \text{CHNH}_2 \\ | \\ \text{COOH} \end{array}$$

glutamate   ammonia   glutamine

Figure 7.6.

Structure

Higher plant GS is an octameric protein with a native molecular weight of 350–400 kDa. In size and quaternary structure it strongly resembles the enzyme isolated from mammals, but is distinct from the bacterial enzyme, which consists of 12 subunits. Extremely high levels of GS activity have been isolated from nitrogen-fixing root nodules. In the bean *Phaseolus vulgaris*, nodule GS can be separated into isoenzymes designated GS_{n1} and GS_{n2}, the latter being very similar to the non-nodulated root enzyme. Analysis of the individual GS subunits has shown that there are three distinct isoelectric forms in the nodule (designated α, β and γ) with a molecular weight of 43 kDa. GS_{n1} was shown to be composed of β and γ subunits, while GS_{n2} contains mainly β subunits. In leaves, two major isoenzymes have been isolated and shown to be localized in the cytoplasm (GS_1) and chloroplast (GS_2) respectively. The proportion of the two isoenzymic forms may vary with the plant studied and is dependent upon the developmental age of the tissue. The chloroplast comprises one larger subunit of 43–45 kDa termed δ. The polypeptide is synthesized on cytoplasmic ribosomes and transported into the chloroplast following the removal of a 4–5 kDa peptide[9].

Molecular Biology and Regulation

It is now well established in *P. vulgaris* that GS is coded for by a small multigene family, and much

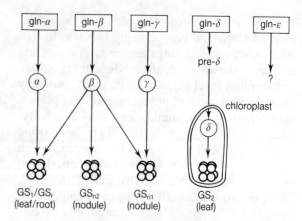

Figure 7.7. Genetic control of glutamine synthetase (GS) isoenzymes in *Phaseolus vulgaris*. Reproduced with permission from B. G. Forde and J.V. Cullimore, 1989, in *Oxford Surveys of Plant Molecular and Cell Biology*, B.J. Miflin (ed.), Oxford University Press, pp. 246–296.

of the previously described subunit structure has been explained at the gene level. A simplified model of the mechanism of the genetic regulation of GS synthesis in *P. vulgaris* is shown in Figure 7.7. The role of the fifth gene gln-ε has not yet been ascertained.

There is a relatively high degree of amino acid similarity (74%) between the chloroplast and cytoplasmic sequences. It has been argued that the two genes arose from duplication of a pre-existing nuclear gene, followed by modification. Since the two dicotyledonous GS_2 cDNAs are more closely related to barley GS_2 than to dicotyledonous GS_1 sequences, it appears that the duplication and divergence of the ancestral GS gene preceded the divergence of monocotyledons and dicotyledons[9].

The appearance of the chloroplastic GS in leaves is apparently regulated by light. In wheat leaves the enzyme activity increases with leaf age (Figure 7.8) and there is a clear correlation with photosynthetic and photorespiratory capacity. In peas, GS_2 mRNA increases about 20-fold in abundance within 72 hours of illumination of etiolated shoots, while the GS_1 mRNA is unaffected[10].

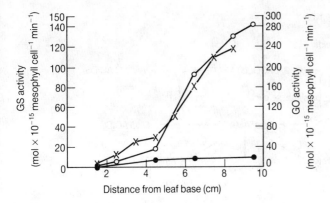

Figure 7.8. Changes in enzyme activity during the development of a wheat leaf grown in the light. Cytoplasmic glutamine synthetase (●), chloroplastic glutamine synthetase (GS) (○), glycollate oxidase (GO) (X). Reproduced by courtesy of Dr A. Tobin.

The activity of plant GS has been shown to increase severalfold during the nodulation of many legume species. This increase occurs at about the same time that nitrogenase is expressed in the *Rhizobium* bacteroid and leghaemoglobin is produced in the plant cell cytosol. The increase in GS mRNA during the nodulation of *P. vulgaris* may be accounted for by a strong induction of the *gln-γ* gene (Figure 7.9)[9].

GLUTAMATE SYNTHASE

This enzyme is responsible for the transfer of the amide group of glutamine to 2-oxoglutarate to yield two molecules of glutamate (Figure 7.10). The combined operation of GS and glutamate synthase in the assimilation of ammonia is shown in Figure 7.11.

Two different forms of glutamate synthase are present in higher plants: one utilizes reduced ferredoxin as a source of reductant and the other utilizes NADH[8].

Structure

Ferredoxin-dependent glutamate synthase, which is present in high concentrations in leaves, is

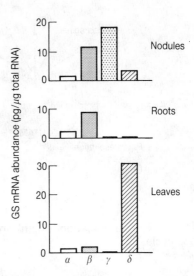

Figure 7.9. Abundance of mRNA related to four glutamine synthetase (GS) genes (*gln-α, gln-β, gln-γ and gln-σ*) of *Phaseolus vulgaris* in the nodules, roots and leaves. Reproduced with permission from B.G. Forde and J.V. Cullimore, 1989, in *Oxford Surveys of Plant Molecular and Cell Biology*, B.J. Miflin (ed.), Oxford University Press, pp. 246–296.

located solely in the chloroplast. The enzyme is an iron–sulphur flavoprotein with a single polypeptide of molecular weight 140–160 kDa. NADH-dependent glutamate synthase is present in low levels in leaves but appears to play a major role in nitrogen-fixing root nodules. The enzyme from root nodules is a monomer of approximately 200 kDa and is probably one of the highest-molecular-weight enzyme subunits known. Two

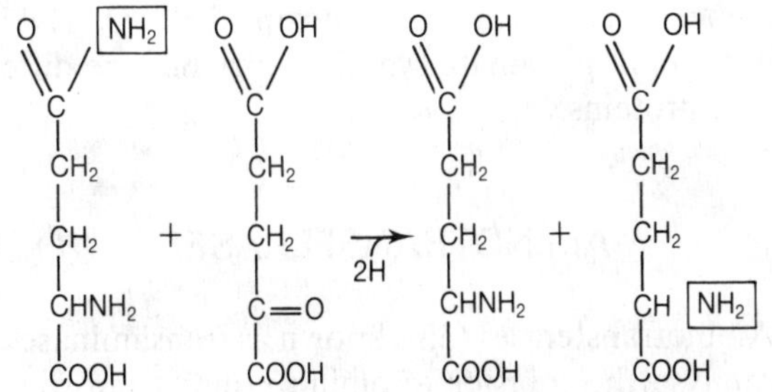

Figure 7.10.

162   P.J. LEA

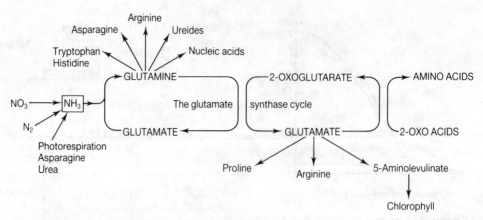

Figure 7.11. The assimilation of ammonia in higher plants via the glutamine synthetase/glutamate synthase cycle.

plastid localized isoenzymes have been isolated in *Phaseolus* nodules, one form of which increases 20-fold during the development period.

Molecular Biology and Regulation

The gene for ferredoxin-dependent glutamate synthase has recently been cloned from maize[11]. The nucleotide sequence shows a 40% similarity to the sequence of the *E. coli* NADPH-dependent enzyme. The amount of mRNA coding for the enzyme was shown to increase 6 hours after the illumination of etiolated leaves, confirming previous data showing a light-dependent increase in enzyme activity. A transit peptide of 97 amino acids required for transport into the chloroplast was also detected. As yet the gene coding for the NADH-dependent enzyme has not been isolated. However, immunological studies suggest that in higher plants the ferredoxin and NADH-dependent glutamate synthase enzymes are different proteins.

AMINOTRANSFERASES

Aminotransferases (also known as transaminases) catalyse the transfer of amino group from the 2 position of an amino acid to a 2-oxo acid to yield a

new amino and a new oxo acid. Glutamate is a major amino donor and takes part in two important reactions (Figure 7.12). Both reactions liberate 2-oxoglutarate, which may then return to the GS/glutamate synthase cycle. Aminotransferase enzymes have been detected in plants that can synthesize all the protein amino acids (except proline) provided the correct oxo acid precursor is available. Thus nitrogen can be readily distributed from glutamate, via aspartate and alanine,

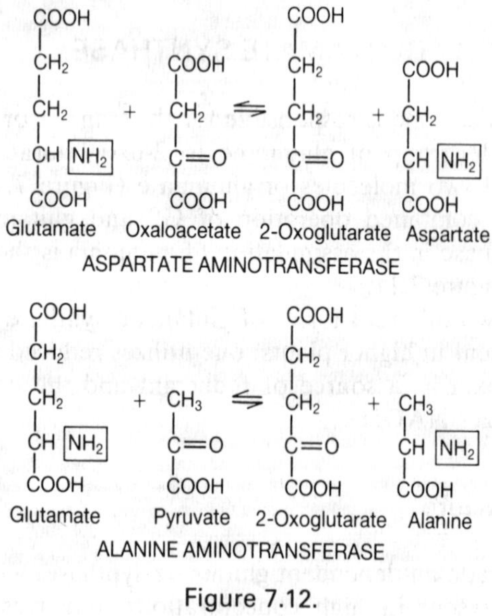

Figure 7.12.

to all amino acids. The reversibility of the reactions allows for sudden changes in demand for key amino acids, although in specific pathways, e.g. photorespiration (chapter 2), certain enzymes tend only to operate in one direction.

There is good evidence that aminotransferases are 'promiscuous' and use a wide range of both amino and oxo acid substrates. A full description of all the aminotransferases is beyond the scope of this chapter, but some mention must be made of aspartate aminotransferase. Numerous isoenzymic forms have been detected in higher plants, and there is good evidence that the cytoplasm, mitochondria, chloroplast and peroxisome each contain their own isoenzyme. The molecular weight of aspartate aminotransferase varies between 95 and 110 kDa and contains two subunits.

Aspartate aminotransferase may be involved in the transport of reducing power from the mitochondria and chloroplast into the cytoplasm as shown in Figure 7.13. The cycle allows for the production of NADH in the cytoplasm for important metabolic reactions, e.g. nitrate reduction, or for the reduction of hydroxypyruvate as in the peroxisome. Frequently oxaloacetate is transported directly, but the conversion to aspartate is extremely rapid. A similar procedure operates in C_4 photosynthesis when aspartate is transported

between the mesophyll and bundle-sheath cells following the initial synthesis of oxaloacetate. The pathway is described in chapter 3.

NITROGEN STORAGE AND TRANSPORT

There are occasions when the plant needs to transport nitrogen from one organ to another. Typical examples are:

1. Nitrogen-fixing root nodule to the leaves and fruit.
2. Senescing leaves to the young leaves and fruit.
3. Cotyledons of germinating seed to expanding shoot and root tip.

On these occasions the availability of carbon is usually limited and the nitrogen is transported in a compound with a high N : C ratio.

Asparagine is almost universally used by higher plants as a storage and transport compound. The amide nitrogen is derived directly from the amide group of glutamine, synthesized by the enzyme asparagine synthetase (Figure 7.14)[12]. Thus ammonia can be incorporated into the amide group of asparagine in two ATP-driven reactions. The glutamate molecule recycles in order to accept ammonia in the glutamine synthetase reaction. Note that the N : C ratio of asparagine is 2 : 2 as opposed to 1 : 5 in glutamate.

The cotyledons of germinating seeds have proved to be a major source of asparagine synthetase. The enzyme requires Cl^- for activity and is inhibited in the presence of Ca^{2+}. Attempts to isolate the enzyme in the leaves have been unsuccessful owing to the presence of an inhibitor. The gene for the plant enzyme has been cloned utilizing a heterologous DNA probe encoding the human enzyme. Two clones of asparagine synthetase cDNA (AS1 and AS2) that encode homologous but distinct proteins were obtained from pea cDNA libraries. Northern blot analysis revealed that dark treatment induced the accumulation of high levels of AS1 mRNA in pea leaves, while light treatment repressed this effect by as much as 30-fold. Both AS1 and AS2 mRNA accumulated

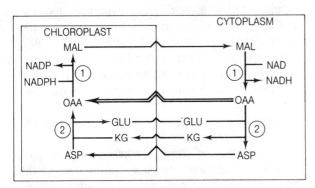

Figure 7.13. The generation of NADH in the cytoplasm via the operation of malate dehydrogenase (1) and aspartate aminotransferase (2) oxaloacetate may either be transported directly into the chloroplast ($\Rightarrow$) or in the form of aspartate following transamination ($\rightarrow$).

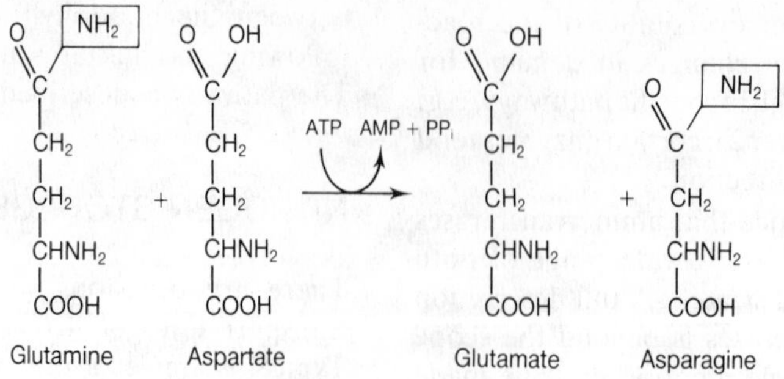

Figure 7.14.

to high levels in the cotyledons of germinating seedlings and in nitrogen-fixing root nodules[10]. The work is a dramatic demonstration of the power of molecular biology in examining the regulation of an enzyme which has proven extremely difficult to study by standard enzymological techniques.

Asparagine Metabolism

The simplest route of asparagine metabolism is via the enzyme asparaginase, which catalyses hydrolysis to yield aspartate and ammonia (Figure 7.15). The enzyme is present in young expanding leaves and plays a major role in the development of the maturing seeds of legumes. The ammonia liberated is reassimilated via glutamine synthetase and glutamate synthase. In the maturing legume seed the enzyme may be potassium dependent or independent, the molecular basis of this difference has not yet been established.

^{15}N-labelling data have shown that asparagine is metabolized in green leaves during the process of photorespiration and is involved in the synthesis of glycine. The enzyme responsible is seri-

Figure 7.15.

ne:glyoxylate aminotransferase, which has a wide substrate specificity and is able to use serine, alanine or asparagine as an amino donor. The product of asparagine transamination is 2-oxosuccinamate, which may be either deaminated directly or reduced to 2-hydroxysuccinamate[8,12].

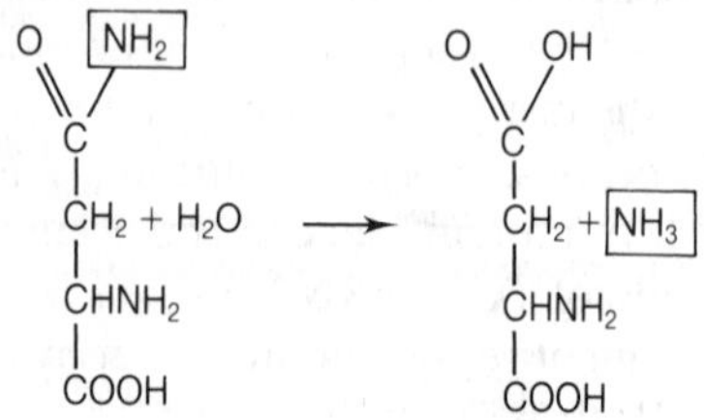

UREIDES

Three major compounds based on the structure of urea are known to occur in plants, i.e. allantoin, allantoic acid and citrulline (Figure 7.16). Allantoin and allantoic acid account for 50–90% of the organic nitrogen in the xylem of many tropical legumes (e.g. soya bean, cowpea and *Phaseolus* beans). There is now good evidence that the ureides are synthesized as products of recently fixed nitrogen in tropical legume root nodules. After 10 minutes' exposure of soya bean nodules to ^{13}N$_2$, 40% of the radioactivity was found in allantoic acid. In similar experiments with cowpea, ureides accounted for over 90% of the ^{15}N in the xylem exudate 2 hours after exposing the nodues to ^{15}N$_2$.

The pathway of ureide synthesis in legume root nodules is long and complex and far beyond the scope of this chapter. Allantoin and allantoic acid are derived from the purine inosine monophosphate, which is synthesized from ribose 5-phosphate and requires an input of nitrogen from glutamine, aspartate and glycine. An excellent article by Schubert and Boland (1990) discusses the pathway in detail[13].

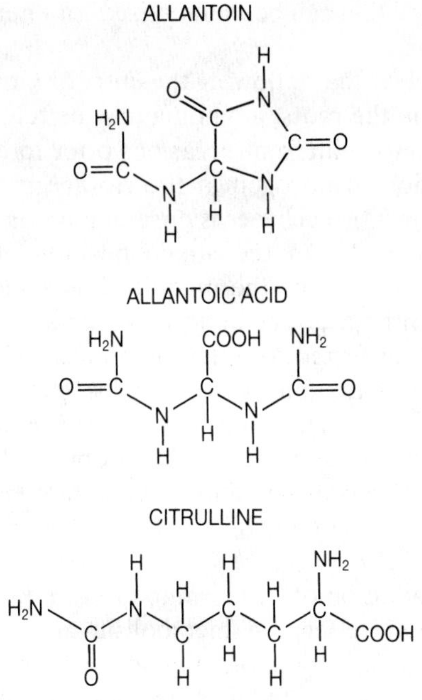

ALLANTOIN

ALLANTOIC ACID

CITRULLINE

Figure 7.16.

It is essential that once allantoin and allantoic acid reach the developing seed or shoot, they are metabolized to a usable nitrogen source. It was originally considered that following the hydrolysis of allantoin to allantoic acid by allantoinase, the ureide was cleaved to yield two molecules of urea and glyoxylate. However, the role of urea and its subsequent metabolism by urease to carbon dioxide and ammonia have been questioned and a second pathway has been proposed. In this pathway, the four nitrogen atoms in allantoic acid are released as ammonia directly, without urea as an intermediate (Figure 7.17).

THE ROLE OF AMMONIA IN PLANT METABOLISM

An interesting point that has consistently emerged in studies on plant metabolism is that ammonia is continuously formed during various metabolic processes (Figure 7.18)[12]:

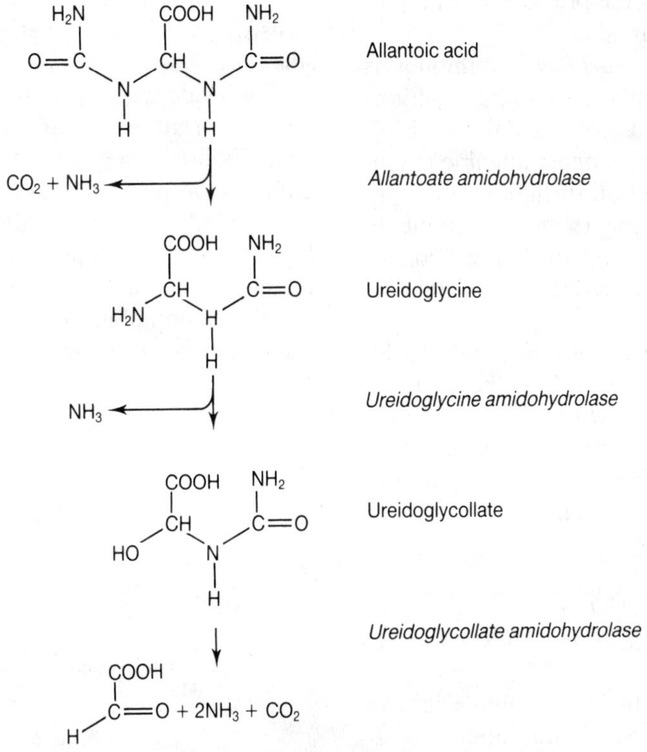

Allantoic acid

$CO_2 + NH_3$ ← Allantoate amidohydrolase

Ureidoglycine

NH_3 ← Ureidoglycine amidohydrolase

Ureidoglycollate

Ureidoglycollate amidohydrolase

$C=O + 2NH_3 + CO_2$

Figure 7.17.

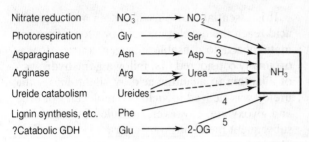

Nitrate reduction	NO_3^- ⟶ NO_2^-	1
Photorespiration	Gly ⟶ Ser	2
Asparaginase	Asn ⟶ Asp	3
Arginase	Arg ⟶ Urea ⟶ NH_3	
Ureide catabolism	Ureides	
Lignin synthesis, etc.	Phe	4
?Catabolic GDH	Glu ⟶ 2-OG	5

Figure 7.18. Metabolic reactions that produce ammonia in plants. Pathways 1–5 are described in the text. GDH = glutamate dehydrogenase, 2-OG = 2-oxoglutarate. Based on a diagram from K.W. Joy, 1988, *Can. J. Bot.*, **66**, 2103–2109.

1. *Initial assimilation*: Ammonia is the product of nitrate reduction and nitrogen fixation.
2. *Photorespiration*: Ammonia is liberated at a very high rate in the conversion of glycine to serine during photorespiration the major portion is reassimilated in the chloroplast (chapter 2).
3. *Metabolism of transport compounds*: Ammonia is liberated during the breakdown of asparagine, arginine and ureides.
4. *Specific amino acid reactions*: Ammonia is liberated during specific metabolic reactions involving amino acids. Particular reactions include the conversion of phenylalanine to cinnamate in the production of lignins, the conversion of cystathionine to homocysteine in methionine synthesis, and the conversion of threonine to 2-oxobutyrate in isoleucine biosynthesis.
5. *Protein catabolism*: As mentioned previously, proteins may be hydrolysed during the germination of seedlings or following leaf senescence. There is now good evidence that prior to the synthesis of the transport compounds, ammonia is released probably from the operation of glutamate dehydrogenase:

$$\text{Glutamate} + NAD^+ + H_2O \rightleftharpoons$$
$$\text{2-oxoglutarate} + NH_3 + NADH + H^+$$

The amino groups of all the other amino acids can be channelled through glutamate, liberating 2-oxo acids that can be metabolized for energy production.

It is clear that following the entry of nitrogen as nitrate in the roots, ammonia may be released on a number of different occasions prior to the final deposition of nitrogen in the storage protein of the seed. On each occasion ammonia is rapidly reassimilated into the amide position of glutamine, catalysed by the ubiquitous enzyme glutamine synthetase. The majority of the amide nitrogen is transferred to other amino acids via glutamate synthase, but some may be used directly (see Figure 7.11). Confirmation of the importance of ammonia recycling has been obtained by the use of the specific glutamine synthetase inhibitors methionine sulphoximine and phosphinothricin (Figure 7.19).

The addition of these compounds to plants has a dramatic effect on metabolism and causes a rapid accumulation of ammonia within the tissues. In C_3 plants photosynthesis is inhibited owing to a shortage of amino acids required for the conversion of glyoxylate to glycine in the photorespiratory cycle. Phosphinothricin is the basis of an important herbicide, named gluphosinate or 'Basta'[14].

Key evidence concerning the importance of photorespiration in ammonia production has been obtained using mutants of C_3 plants that lack either chloroplastic glutamine synthetase (barley) or ferredoxin-dependent glutamate synthase (barley, *Arabidopsis* and pea). Such mutants are unable to assimilate the ammonia released during the conversion of glycine to serine in photorespiration. The plants show severe symptoms of stress

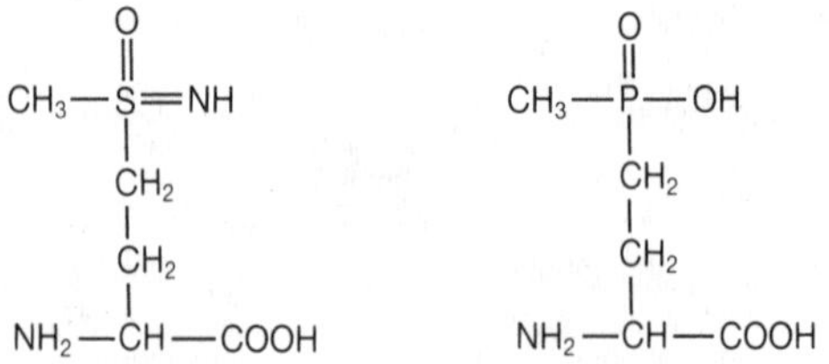

$CH_3-S{=}NH$ (with =O above S)	CH_3-P-OH (with =O above P)
CH_2	CH_2
CH_2	CH_2
$NH_2-CH-COOH$	$NH_2-CH-COOH$
Methionine sulphoximine (MSO)	Phosphinothricin (PPT)

Figure 7.19.

when exposed to air at 0.034% carbon dioxide but grow normally in air at 0.7% carbon dioxide, when photorespiration is suppressed. The metabolism of phosphinothricin-treated and glutamine synthetase-deficient plants is remarkably similar and gives clear support to the suggestion that the photorespiration pathway (chapter 2) is quantitatively by far the most important nitrogen-requiring metabolic pathway in the leaves of C_3 plants[8,14].

AMINO ACID FAMILIES

It is possible to divide amino acids into families, each with a 'head' or precursor, based on their biosynthetic pathway[15]. The divisions are artificial, and there are obviously occasions when an arbitrary decision has to be taken as to the location of an individual amino acid:

1. *Aspartate*: Asparagine, lysine, threonine, methionine, isoleucine. Although leucine and valine do not derive carbon from aspartate, they share a common enzyme pathway with isoleucine.
2. *Glutamate*: Glutamine, arginine and proline.
3. *Pyruvate*: The amino acids derived from three-carbon precursors are the most heterogeneous and include alanine, serine, cysteine and glycine. Pyruvate donates carbon to lysine, isoleucine and valine, and phosphoenolpyruvate is involved in the synthesis of aromatic amino acids.
4. *Erythrose 4-phosphate*: The aromatic amino acids phenylalanine, tyrosine and tryptophan.
5. *Ribose 5-phosphate*: Histidine.

Aspartate Pathway

The biosynthesis of lysine, threonine, methionine and isoleucine is shown in Figure 7.20. Experiments with ^{14}C-labelled aspartate have shown that chloroplasts are capable of synthesizing lysine, threonine, isoleucine and homocysteine in light-driven reactions which can therefore be termed photosynthetic (see later section). Subsequent subcellular localization studies have shown that all the enzymes required are present in the chloroplast. The final methylation step involved in the conversion of homocysteine to methionine has been shown to take place in the cytoplasm. It is no coincidence that all the essential amino acids required by animals in their diet are synthesized by plants in the chloroplast. Animals do, however, have the capacity to convert homocysteine to methionine.

Regulation

A number of the enzymes in the aspartate pathways are subject to end-product feedback inhibition. The process allows for the individual amino acids to be synthesized as they are required, but prevents their build-up, which would be expensive in terms of carbon and nitrogen[16]. The enzyme aspartate kinase catalyses the ATP-dependent phosphorylation of the β-carboxyl group of aspartate (enzyme 1, Figure 7.20; see Figure 7.21).

Three isoenzymes of aspartate kinase have been isolated from higher plant tissues and their presence confirmed by genetic analysis[17]. The first isoenzyme (AKI) is subject to feedback inhibition by threonine alone and is only present in low levels in rapidly growing tissue. The other two isoenzymes (AKII and AKIII) are similar to each other and are both inhibited by lysine. Methionine alone has no direct effect on any of the aspartate kinase isoenzymes, but is able to regulate the flow of carbon through the activated form of S-adenosylmethionine. AKII and AKIII are both subject to synergistic inhibition in the presence of S-adenosylmethionine and lysine[18]. S-Adenosylmethionine has very little inhibitory action on its own, but may greatly reduce the concentration at which lysine exerts an effect; this can be easily seen by scrutiny of Figure 7.22 (see also Figure 7.35).

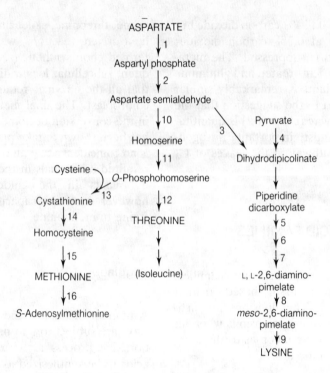

Figure 7.20 The metabolic pathway required for the synthesis of the aspartate-derived amino acids lysine, threonine, methionine and isoleucine in higher plants. The enzymes 1–16 are described in the text.

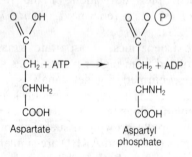

Figure 7.21.

Lysine Synthesis

Following the reduction of aspartyl phosphate to aspartate semialdehyde (2), the first branch in the pathway leads to the synthesis of lysine. Dihydrodipicolinate is formed from the combination of pyruvate and aspartate semialdehyde catalysed by dihydrodipicolinate synthase (3) (Figure 7.24). The enzyme is extremely sensitive to lysine, which is a competitive inhibitor with respect to aspartate semialdehyde and non-competitive with respect to pyruvate. It is likely that dihydrodipicolinate synthase is the major site of regulation of lysine synthesis in higher plants.

Although the pathway from dihydrodipicolinate to meso-2,6-diaminopimelate is somewhat obscure in higher plants, circumstantial evidence still indicates that the route shown in Figure 7.20 is correct[15]. Although one enzyme capable of carrying out reactions (5–8) has been proposed, the result has not been confirmed. There is no evidence that in higher plants lysine can be synthesized via α-aminoadipic acid, a pathway that operates in the fungi.

Threonine Synthesis

Aspartate semialdehyde is converted to homoserine by homoserine dehydrogenase (10) (Figure 7.25). The enzyme exists in two isoenzymic

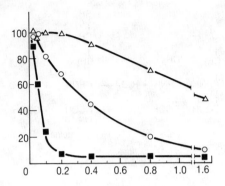

Figure 7.22. Synergistic inhibition of barley aspartate kinase. *S*-Adenosylmethionine alone (△); lysine alone (○); lysine plus *S*-adenosylmethionine, equimolar concentrations (■). From Rognes *et al.*[18], reprinted with permission from *Nature*, **287**, 357–9, © 1980, Macmillan Magazines Ltd.

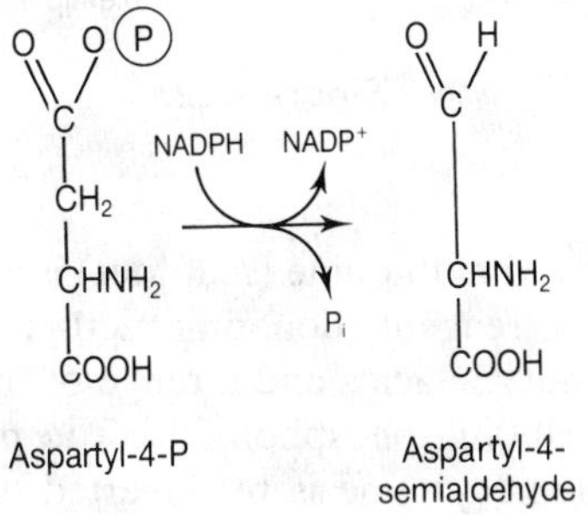

Figure 7.23.

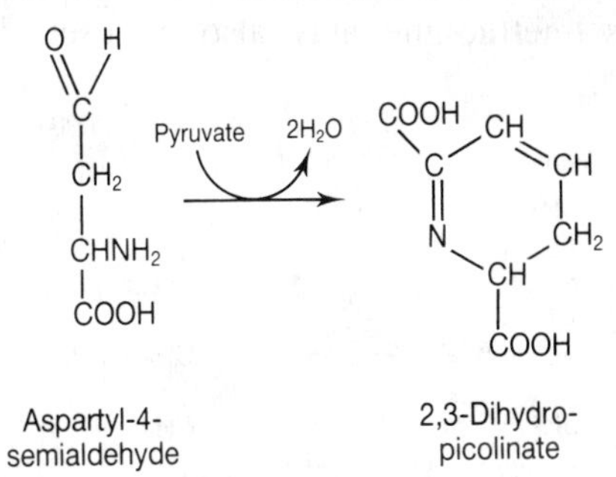

Figure 7.24.

forms, one which is sensitive to inhibition by threonine and is localized in the chloroplast and one which is insensitive to threonine and is situated in the cytoplasm.

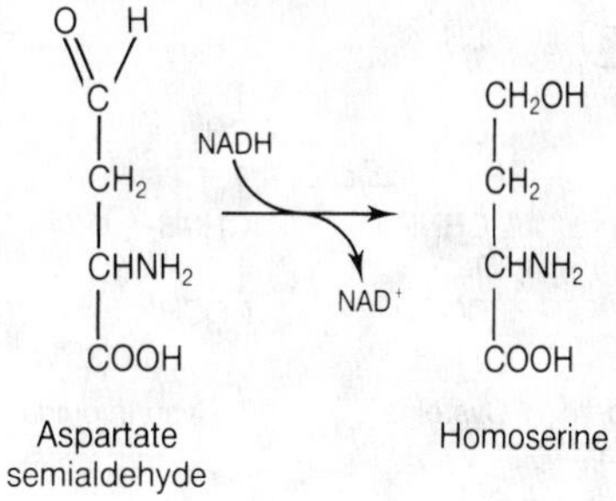

Figure 7.25.

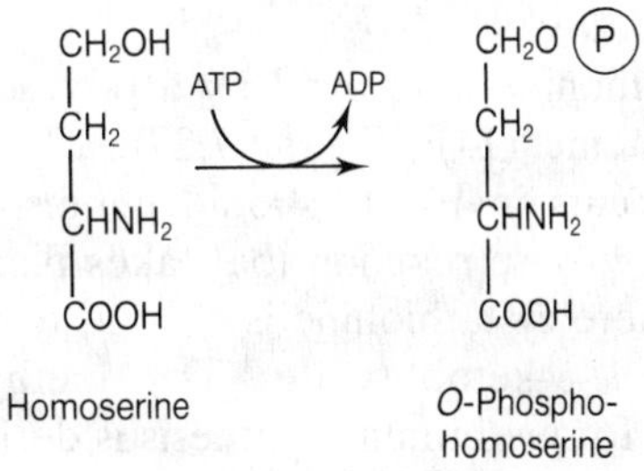

Figure 7.26.

Homoserine is phosphorylated by homoserine kinase (11) in an ATP-dependent reaction to yield phosphohomoserine (Figure 7.26), which may then be converted into threonine by threonine synthase (12) (Figure 7.26). The enxyme is unusual in that it requires *S*-adenosylmethionine for the reaction to take place. Thus once again the activated form of methionine (*S*-adenosylmethionine) is able to exert a regulation effect by directing carbon towards threonine[15]. In certain plants (e.g. pea), homoserine may accumulate in high concentrations and is used as a nitrogen storage and transport compound[12].

Methionine Synthesis

Phosphohomoserine is also the precursor of methionine and reacts with cysteine (see later section) to yield cystathionine, catalysed by cystathionine-γ-synthase (13) (Figure 7.27)[16]. The level of enzyme activity has been shown to be repressed by the presence of methionine and

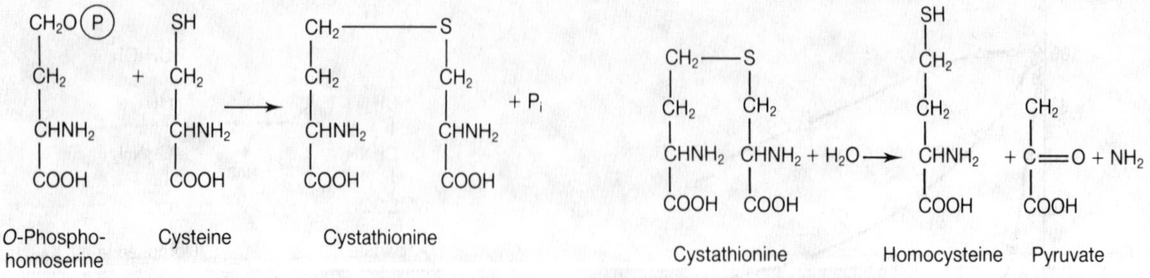

Figure 7.27.

O-Phosphohomoserine + Cysteine → Cystathionine + P$_i$

Cystathionine + H$_2$O → Homocysteine + Pyruvate (C=O + NH$_2$)

increased at times of methionine starvation (Figure 7.28)[16,19].

Cystathionine is cleaved by a β-lyase to yield homocysteine (14) (Figure 7.29). The reaction which occurs in plants should be distinguished from the γ-lyase reaction that takes place in animals, where cystathionine is cleared to yield cysteine and α-ketobutyrate. The methyl group required for methionine synthesis is derived from $5N$-methyltetrahydrofolate[15]. *S*-Adenosylmethionine is synthesized following the reaction of methionine with ATP (16) (Figure 7.30).

Homocysteine + N^5-methyltetrahydrofolate → Methionine + tetrahydrofolate

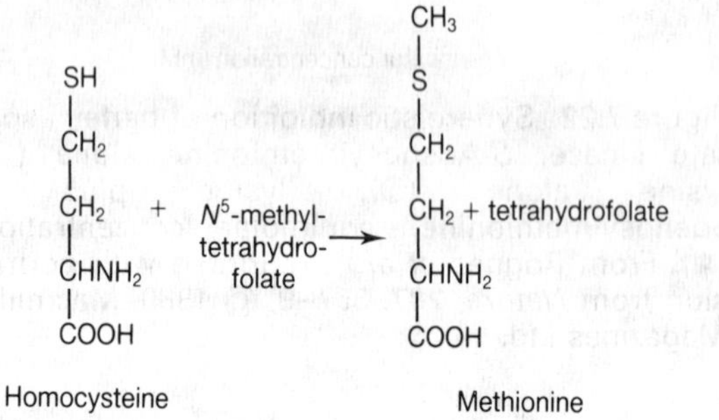

Figure 7.29.

S-Adenosylmethionine (AdoMet) is a universal methylating reagent, donating methyl groups to proteins, nucleic acids and a range of other compounds including phospholipids. The product *S*-adenosylhomocysteine is reconverted to methionine.

In higher plants the central carbon atoms of *S*-adenosylmethionine may also be used for the

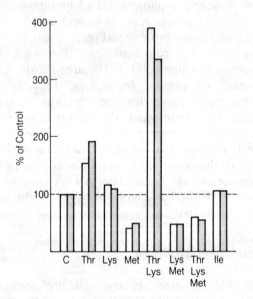

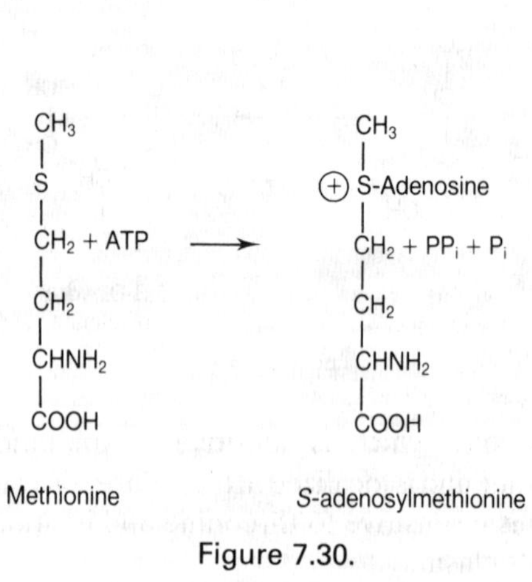

Figure 7.28. Effect of various amino acids on the level of extractable cystathionine synthase, isolated from barley seedlings grown under sterile conditions. Reproduced with permission from S.E. Rognes *et al.*, 1986, *Plant Sci.*, **43**, 45–50.

Methionine + ATP → *S*-adenosylmethionine + PP$_i$ + P$_i$

Figure 7.30.

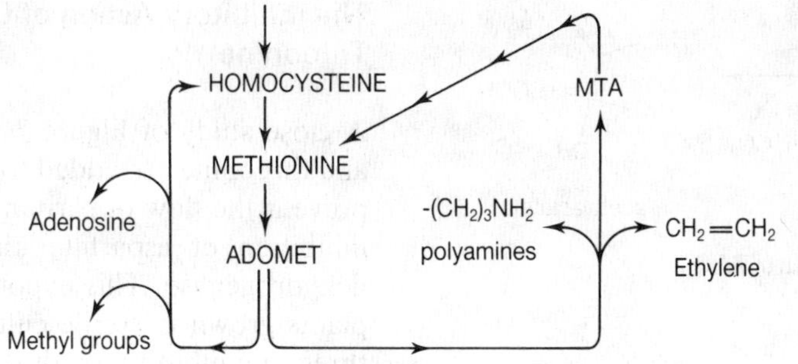

Figure 7.31.

synthesis of ethylene and polyamines (Figure 7.31)[20]. Methylthioadenosine (MTA), the product of both reactions, may be recycled to synthesize methionine without homocysteine as an intermediate. The C_4 skeleton of methionine is derived from the ribosyl group of ATP[15,16].

Isoleucine Synthesis

Threonine is deaminated by threonine dehydratase to form 2-oxobutyrate and ammonia. Isoenzymic forms have been detected but both are sensitive to feedback inhibition by isoleucine. A key step in the conversion of threonine to isoleucine is the combination of 2-oxobutyrate with pyruvate to yield acetohydroxybutyrate (Figure 7.32). The reactions are closely paralleled in the synthesis of valine and leucine, in which two molecules of pyruvate react to form acetolactate (Figure 7.33). The combination of 2-oxobutyrate and pyruvate and of two molecules of pyruvate appear to be catalysed by the same enzyme, termed acetohydroxy acid synthase. The enzyme is subject to feedback inhibition by leucine and valine. In addition, leucine is able to regulate the first enzyme unique to its own synthesis, α-isopropylmalate synthase.

The enzyme acetohydroxy acid synthase has been subject to a considerable amount of interest, following the discovery that it was the site of action of two major classes of herbicides—the sulphonylureas and imidazolinones (Figure 7.34)[21]. The enzyme requires thiamine pyrophosphate and FAD for activity, and 50% inhibition has been detected in the presence of only 12 nM sulphometuron methyl. Such high affinities for the enzyme have ensured that the herbicides need only be applied to plants in extremely low doses. Mutant plants resistant to the herbicides have been isolated; acetohydroxy acid synthase extracted from the resistant plants have been shown to exhibit a reduced sensitivity to the inhibitors *in vitro*[21].

Overall Regulation of Aspartate Pathway

A simple scheme of the overall regulation is shown in Figure 7.35; the process can be summa-

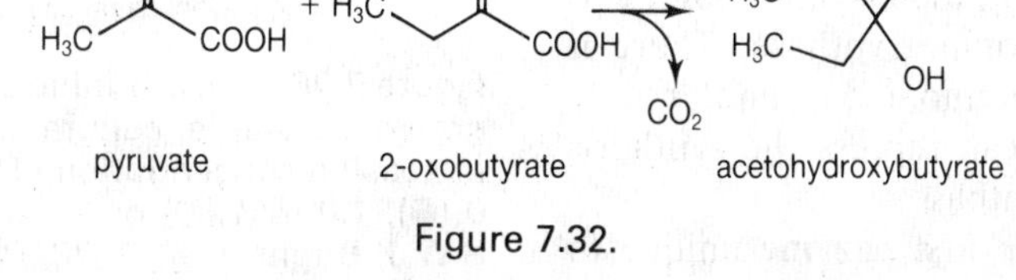

| pyruvate | 2-oxobutyrate | acetohydroxybutyrate |

Figure 7.32.

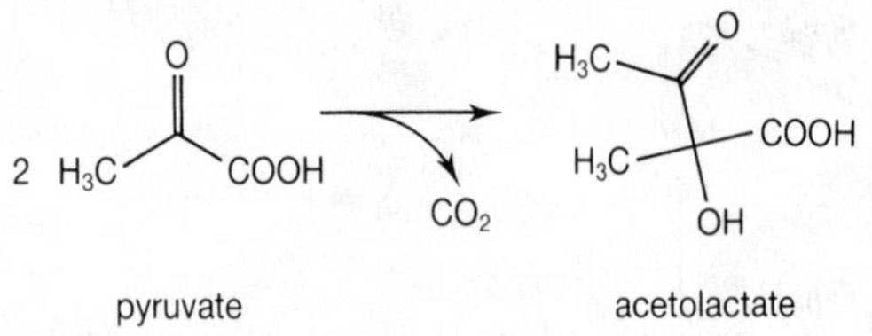

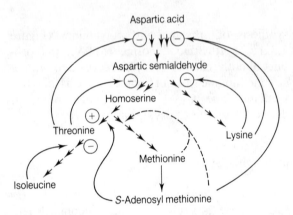

Figure 7.33.

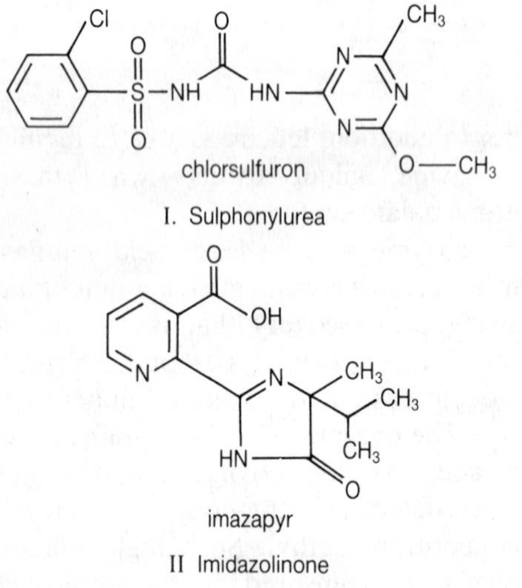

chlorsulfuron

I. Sulphonylurea

imazapyr

II Imidazolinone

Figure 7.34.

The Inhibitory Action of Lysine Plus Threonine

A close study of Figure 7.35 shows that if lysine and threonine are added together they will totally prevent the flow of carbon to methionine, by the inhibition of aspartate kinase and homoserine dehydrogenase. This hypothesis can be tested on plants grown in sterile culture, where lysine and threonine inhibit growth (Figure 7.36)[22]. The ad-

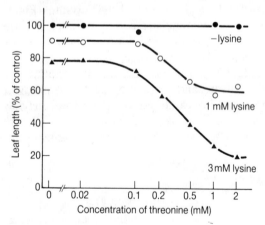

Figure 7.35. Regulation of biosynthesis of the aspartate-derived amino acids. ⊖, enzyme inhibition; ⊕, enzyme activation; ----enzyme repression.

rized as follows:

1. *Lysine* inhibits two isoenzymes of aspartate kinase and the first enzyme unique to its own synthesis, dihydrodipicolinate synthase.
2. *Theonine* inhibits one isoenzyme of aspartate kinase and one isoenzyme of homoserine dehydrogenase.
3. *Methionine* appears to operate through the activated form S-adenosylmethionine. Two isoenzymes of aspartate kinase are synergistically inhibited by S-adenosylmethionine in the presence of lysine. S-Adenosylmethionine is able to activate threonine synthase. There is evidence that methionine and/or S-adenosylmethionine can repress the synthesis of cystathionine-γ-synthase.
4. *Isoleucine* inhibits the first enzyme unique to its own synthesis, threonine dehydratase.

Figure 7.36. Growth inhibition of barley seedlings grown in sterile culture in the presence of an increasing concentration of threonine and lysine at 0 (●), 1.0 mM (○) or 3.0 mM (▲). Redrawn from S.W.J. Bright *et al.*, 1978, *Planta*, **139**, 113–117, by permission of Springer-Verlag.

dition of methionine totally relieves this growth inhibition. Mutant lines of barley, tobacco and maize have been isolated that are resistant to the toxic action of lysine plus threonine. Individual lines of barley have been shown to have mutated forms of AKII or AKIII that are no longer subject to inhibition by lysine (Figure 7.37)[23]. Such mutant plants have been shown to have elevated levels of threonine in the leaves and the seed. Increases in the content of lysine have been less pronounced, presumably due to the tight regulation of dihydrodipicolinate synthase. The potential to increase the concentration of the essential amino acids in animal feedstuffs may have an important future in plant breeding[17]*.

PROLINE AND ARGININE

Proline

Glutamate is converted to glutamate semi-aldehyde in two reactions similar to that already seen for aspartate. The aldehyde is cyclized non-enzymically to form Δ'-pyrroline-5-carboxylate, which is reduced to yield proline (Figure 7.38). The enzymology of the reactions is not well established in higher plants. Only the last enzyme in the pathway, Δ'-pyrroline-5-carboxylate reductase, has been characterized and is shown to be localized in the chloroplast[15].

There is evidence that ornithine (see below) may also act as a precursor of proline via either Δ'-pyrroline-5-carboxylate or Δ'-pyrroline-2-

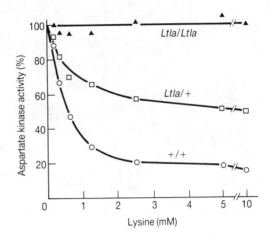

Figure 7.37. Feedback inhibition by lysine of aspartase kinase isoenzyme II extracted from wild-type plants (+/+) and lysine plus threonine-resistant progeny derived from mutant R2501 (*Ltla/Ltla*) and a heterozygous plant (*Ltla/*+). Redrawn from Arruda et al.[23].

carboxylate. The route under normal conditions has not been confirmed.

As well as being a major constituent of proteins, in particular in cereal seeds, proline is also thought to act as an osmotic protectant in plants that have been subject to either drought or high salt stress. An example of the accumulation of proline in barley subject to drought stress is shown in Figure 7.39. Several mechanisms to explain such dramatic increases in the concentration of proline have been proposed. It would appear that there is an increase in the rate of synthesis, possibly through the induction of Δ'-pyrroline-5-carboxylate reductase, and also a decrease in the rate of mitochondrial oxidation of proline.

Arginine

The synthesis of arginine proceeds through ornithine. which is derived from glutamate in an acetylated pathway (Figure 7.40). Note that the formation of the semialdehyde is similar to the initial stages of proline synthesis. However, the presence of the N-acetyl group prevents the non-

---

*Since the completion of this chapter, there have been two major advances in the over production of essential amino acids; (1) transgenic tobacco plants have been obtained that contain the E. coli gene for dihydrodipicolinate synthase and aspartate kinase in the chloroplast. These genetically manipulated plants were shown to contain elevated levels of lysine and threonine in the leaves[28,29]; (2) also in tobacco, mutant lines have been selected that are resistant to lysine plus threonine and to the lysine analogue S-aminoethylcysteine. Such lines have been shown to contain altered aspartate kinase and dihydrodipicolinate synthase respectively. Heterozygous crosses between the two mutant lines have been shown to accumulate lysine in the leaves[30].

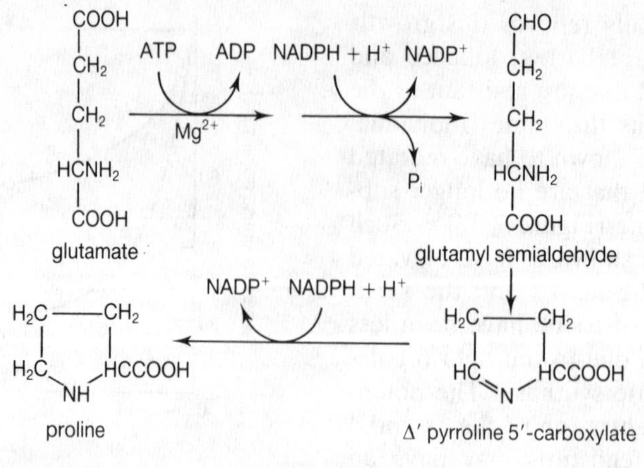

Figure 7.38.

enzymic cyclization reaction and allows the trans-amination reaction to proceed. The acetyl groups rapidly recycle from ornithine to glutamate.

Ornithine is converted into arginine, following the input of two additional amino groups which are derived from carbamoyl phosphate and aspartate. The nitrogen of carbamoyl phosphate is derived from the amide nitrogen of glutamine, catalysed by carbamoyl phosphate synthetase (Figure 7.41). Ornithine carbamoyltransferase catalyses the synthesis of citrulline from ornithine and carbamoyl phosphate; the remainder of the pathway of arginine synthesis is as shown in Figure 7.42[15].

Under certain conditions (e.g. anaerobiosis), glutamate may be decarboxylated to yield γ-aminobutyrate: $NH_2CH_2CH_2CH_2COOH$.

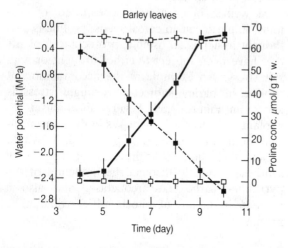

Figure 7.39. Relationship between water potential and proline content of barley leaves subject to drought stress by withholding of water. Note the massive accumulation of proline in the leaves of the stressed plants. Water potential, watered plants (---□); water potential, stressed plants (---■); proline concentration, watered plants (—□); proline concentration, stressed plants (—■). Reproduced by courtesy of A. Al-Sulaiti.

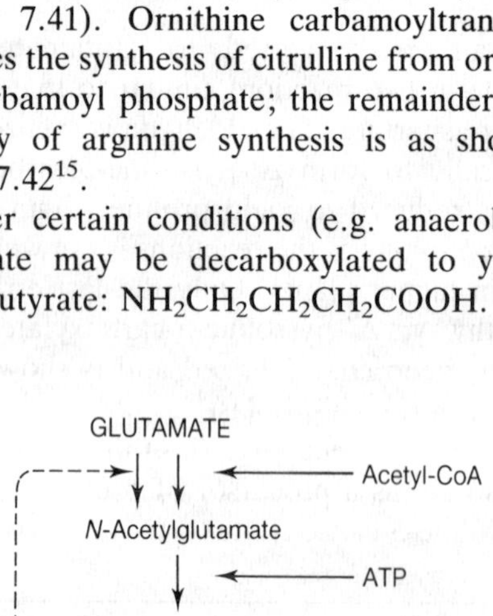

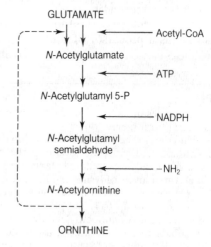

Figure 7.40.

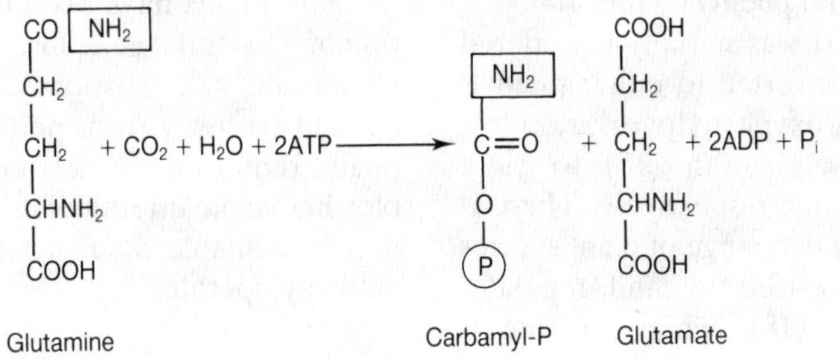

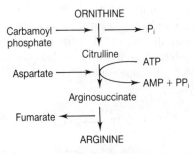

Figure 7.41.

(structures shown)

CO—NH₂ | CH₂ | CH₂ + CO₂ + H₂O + 2ATP → NH₂—C=O—O—(P) + COOH | CH₂ | CH₂ | CHNH₂ | COOH + 2ADP + Pᵢ

Glutamine   Carbamyl-P   Glutamate

ORNITHINE

Carbamoyl phosphate ⟶ ↓ ⟶ Pᵢ

Citrulline
⟶ ATP
Aspartate ⟶ ⟶ AMP + PPᵢ

Arginosuccinate

Fumarate ⟵ ↓

ARGININE

Figure 7.42.

THE AROMATIC AMINO ACIDS

The biosynthetic pathway of the three aromatic amino acids phenylalanine, tyrosine and tryptophan is long and complex. The structures and names of the majority of the intermediates are beyond the scope of this chapter. However, there are a number of interesting points of regulation that deserve a mention. A brief outline of the pathway is shown in Figure 7.43. By analogy with the aspartate pathway described previously, it would be expected that the first enzyme (1), 3-deoxy-arabinoheptulosonate-7-P (DAHP) synthase, would be subject to feedback inhibition by phenylalanine, tyrosine and tryptophan. However, two isoenzymic forms have been detected in plants, one requiring Mn^{2+} and one Co^{2+} for activity. Arogenate and prephenate have been shown to inhibit the Mn^{2+}-dependent form and in some circumstances tryptophan has been shown to activate the enzyme. Tryptophan

does, however, inhibit the first enzyme unique to its own synthesis, anthranillate synthase (8).

Chorismate mutase (13) catalyses the conversion of chorismate to prephenate and has also been shown to exist in two isoenzymic forms. Chorismate mutase-1 is subject to feedback

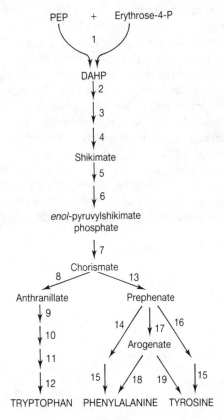

Figure 7.43.

inhibition by tyrosine and phenylalanine and activation by tryptophan. It was initially considered that prephenate was converted to either phenylpyruvate (14) or hydroxyphenylpyruvate (16), with subsequent transamination (15) to yield phenylalanine and tyrosine respectively. There is now good evidence that in a range of plant species phenylalanine is synthesized by and regulates arogenate dehydratase (18), while tyrosine is synthesized by and regulates arogenate dehydrogenase (19). A diagram similar to that in Figure 7.35 may be constructed to show the regulation of the aromatic amino acid synthetic pathway[15].

One apparently inconspicuous enzyme involved in the synthesis of chorismate, enolpyruvylshikimate phosphate (EPSP) synthase (6), has been the subject of intense study over the past few years owing to the fact that it is the target enzyme of the herbicide glyphosate[24]: ℗—CH_2NHCH_2COOH. The enzyme catalyses the reaction between shikimate 3-phosphate and phosphenolpyruvate (PEP) to yield 5-enolpyruvylshikimate 3-phosphate (Figure 7.44).

Glyphosate is a potent inhibitor of the enzyme and apparently binds at the active site in place of PEP. The herbicide causes a predictable build-up in shikimate and shikimate 3-phosphate, and the toxic action can be reversed by the addition of phenylalanine, tyrosine and tryptophan. The gene for EPSP synthase has been cloned from various plant sources. The enzyme is initially synthesized as a 55 kDa precursor on cytoplasmic ribosomes and is transported into the chloroplast with the removal of a 7 kDa peptide. Herbicide-

resistant plants have been obtained by the insertion of a bacterial gene insensitive to inhibition by glyphosate[15,24]. Histidine is synthesized in a straight pathway from phosphoribosyl pyrophosphate, requiring ten enzyme steps involving complex imidazole intermediates. Very little information is available in higher plants as to how the pathway operates[15].

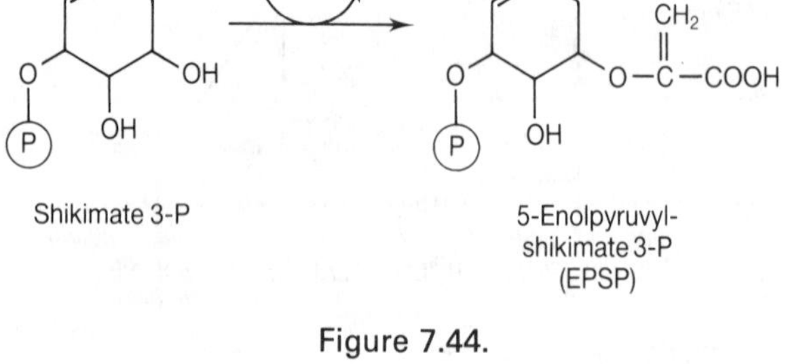

SULPHATE REDUCTION

As sulphur is an important constituent of cysteine, methionine and glutathione, it is essential to consider its metabolism in this chapter[25]. Sulphate is initially activated in the presence of ATP to form adenosine 5'-sulphatophosphate or adenosine phosphosulphate (APS) catalysed by ATP sulphurylase:

$$ATP + SO_4^{2-} \rightarrow APS + PP_i$$

The electrons required for sulphate reduction are derived from reduced ferredoxin, which may be formed in the chloroplast directly from photosystem I. However, prior to reduction, APS is bound to a carrier molecule, which is probably glutathione, and contains a free thiol group. The enzyme APS sulphotransferase catalyses the following reaction:

APS + carrier–SH $\rightarrow$ AMP + carrier–S — SO_3^-
carrier–S — SO_3^- + 6 reduced ferredoxin $\rightarrow$
carrier–S — S^- + 6 oxidized ferredoxin + 3H_2O

Free sulphide may react with O-acetylserine to form cysteine catalysed by the enzyme cysteine synthase (Figure 7.45). O-Acetylserine is synthesized by the acetylation of serine, utilizing acetyl-CoA as a substrate and the enzyme serine transacetylase:

serine + acetyl-CoA $\rightarrow$ O-acetylserine + CoA

Although the cysteine synthase reaction described above procedes rapidly in the test tube, it is likely that the bound carrier–S — S^- acts as the

Shikimate 3-P

5-Enolpyruvyl-
shikimate 3-P
(EPSP)

Figure 7.44.

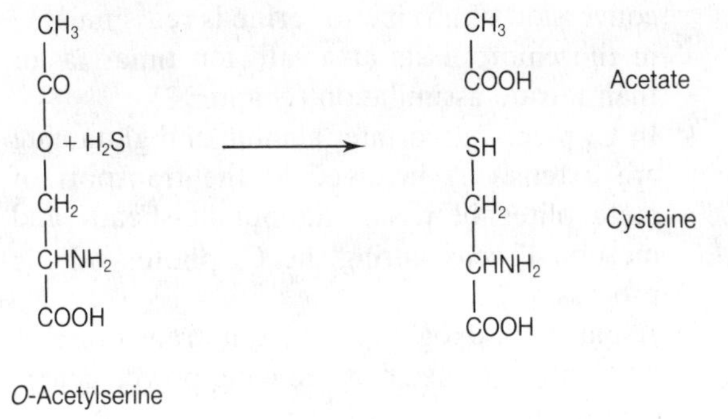

CH₃
|
CO
|
O + H₂S ⟶
|
CH₂
|
CHNH₂
|
COOH

O-Acetylserine

CH₃
|
COOH   Acetate
|
SH
|
CH₂   Cysteine
|
CHNH₂
|
COOH

Figure 7.45.

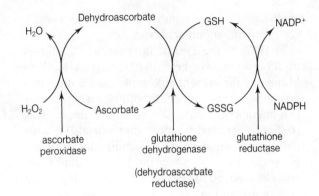

Figure 7.46. The glutathione/ascorbate cycle showing the detoxification of hydrogen peroxide in the chloroplast. NADPH is regenerated from PSI. After C.H. Foyer and B. Halliwell, 1976, *Planta*, **133**, 1–25, by permission of Springer-Verlag.

substrate *in vivo*[25]. Concentrations of free cysteine are very low, owing to the reactivity of the free thiol group. Cysteine either reacts with phosphomoserine in the biosynthesis of methionine or is rapidly converted to the tripeptide glutathione[26]: glutamate–cysteine–glycine.

Glutathione may exist in the reduced form as GSH or in the oxidized state as a dimer, GSSG, where two molecules are linked by a disulphide bond, similar to that found in the tertiary structure of many proteins. Glutathione can act as a store of sulphur, but it has also been proposed that the tripeptide may play a key role in the defence of a plant against various environmental stresses, e.g. cold, heat, drought, high light, fungal attack and herbicides (atrazine). Phytochelatins are polymers of γ-glutamylcysteine, which have been shown to be involved in the chelation of certain toxic metals, e.g. cadmium[26].

Glutathione and ascorbate play a particularly important role in the chloroplast in the detoxification of activated oxygen species which may be formed at times when the rates of electron transport through PSI exceed the rates of carbon dioxide reduction, e.g. high light and low temperature.

Superoxide (O_2^-) formed by the reduction of oxygen is rapidly converted to peroxide through the operation of superoxide dismutase:

$$2O_2^- + 2H^+ \rightarrow H_2O_2 + O_2$$

Hydrogen peroxide is detoxified through a series of reactions termed the glutathione/ascorbate cycle which relies upon NADPH generated from the PSI light reaction as a source of reductant (Figure 7.46)[27].

The regulation of sulphate reduction appears to operate at two major sites. Cysteine has a strong inhibitory effect on the extractable level of APS sulphotransferase activity and also inhibits serine transacetylase, and thus prevents the formation of *O*-acetylserine.

A marked coordination between the assimilation of nitrate and sulphate has been demonstrated. Reduced forms of nitrate stimulate the assimilation of sulphate and, conversely, reduced forms of sulphate stimulate the assimilation of nitrate[25].

PHOTOSYNTHESIS AND AMINO ACID METABOLISM

If the average biology student were asked the question 'What is photosynthesis?' the immediate reply would be 'the assimilation of carbon dioxide into carbohydrates carried out by chloroplasts in the light'. Unfortunately the answer does not tell the whole story. NADPH and ATP generated by the two photosystems may be utilized for the synthesis of a large number of other compounds, ranging from amino acids to lipids.

In 1977, at the Fourth International Congress on Photosynthesis held in Reading, UK, Ben Miflin and the present author placed a notice on a poster board stating 'Photosynthesis is also nitrogen metabolism'. At the time the notice was studiously ignored by most (but not all!) of our colleagues involved in carbon metabolism. However, by 1980, when the International Congress had moved to Greece, I was invited to give a lecture presenting evidence for our original proposal: this can be summarized as follows:

1. Nitrite (but not nitrate!) can be reduced by chloroplasts in a light-dependent reaction to form glutamine and glutamate.
2. Nitrite reductase, the major proportion of glutamine synthetase and glutamate synthase are localized inside chloroplasts.
3. The pathway of photorespiration involves the synthesis of glycine by the transamination of glyoxylate. Ammonia liberated following the

conversion of glycine to serine is reassimilated in the chloroplasts at a rate ten times faster than nitrate assimilation (chapter 2).

4. In C_4 plants, aspartate, alanine and glutamate are extensively involved in the transport of metabolites between the bundle-sheath and mesophyll cells during the C_4 photosynthetic process.
5. Aspartate:2-oxoglutarate aminotransferase is involved in transfer of reducing power generated in the chloroplast.
6. Chloroplasts have the capacity to convert aspartate to lysine, threonine, isoleucine and homocysteine (but not methionine) in light-dependent reactions. Chloroplasts also have the capacity to convert pyruvate to leucine, isoleucine and valine to synthesize the aromatic amino acids.
7. The major proportion (if not all) of the enzymes required for the synthesis of the essential amino acids lysine, threonine, isoleucine, homocysteine, leucine, valine, tryptophan, phenylalanine and tyrosine have been

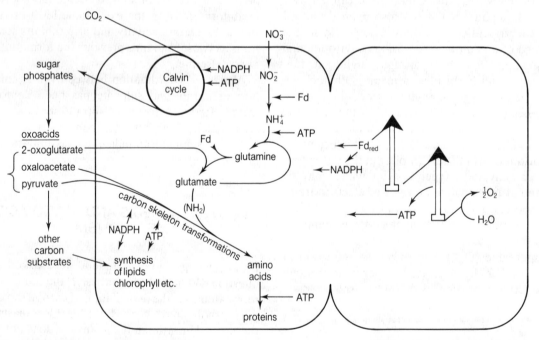

Figure 7.47. Nitrogen assimilation in the chloroplast, showing the interactions of ATP and NADPH generation and photosynthetic nitrogen metabolism.

shown to be located in the chloroplast. In addition some of the enzymes involved in the synthesis of arginine and proline also have a predominantly chloroplast location.

8. Chloroplasts have the capacity to reduce sulphate to cysteine, and the enzymes involved in the pathway have all been localized in the chloroplast. The reduction of glutathione as a superoxide detoxification mechanism is also a light-dependent chloroplast reaction.

9. A number of proteins (e.g. large subunit of Rubisco, subunits of ATP synthetase and certain cytochromes) are coded for by chloroplast DNA. Chloroplasts have the capacity to synthesize these proteins when fed with radioactive amino acids in light-dependent reactions.

In addition, chloroplasts have the capacity to synthesize a range of saturated fatty acids, nucleic acids and chlorophyll utilizing NADPH and ATP generated directly by the two photosystems. A summary of the biosynthetic capacity of chloroplasts is shown in Figure 7.47.

In non-green tissues, in particular roots, a number of the processes described above still occur, although often at a reduced level. Plastids are still an important site of nitrogen metabolism, although the energy requirements have to be met by the catabolism of sugars transported from the leaf.

REFERENCES

1. Siddiqi, M.Y., Glass, A.D.M., Ruth, T.J. and Rufty, T.W. (1990) Studies on the uptake of nitrate in barley. *Plant Physiology*, **93**, 1426–1432.
2. McClure, P.R., Kochian, L.V., Spanswick, R.M. and Shaff, J.E. (1990) Evidence for cotransport of nitrate and protons in maize roots. *Plant Physiology* **93**, 290–294.
3. Zhen, R.G., Koyro, H.W., Leigh, R.A., Tomos, A.D. and Miller, A.J. (1991) Compartmental nitrate concentrations in barley root cells measured with nitrate-selective microelectrodes. *Planta*, **185**, 356–361.
4. Kleinhofs, A. and Warner, R.L. (1990) Advances in nitrate assimilation. In *The Biochemistry of Plants* , Vol. 16 (eds B.J. Miflin and P.J. Lea), pp. 89–120. Academic Press, San Diego.

5. Solomonson, L.P. and Barber, M.J. (1989) Structure–function relationships of algal nitrate reductases. In *Molecular and Genetic Aspects of Nitrate Assimilation* (eds J.L. Wray and J.R. Kinghorn), pp. 88–100. Oxford University Press, Oxford.
6. Crawford, N.M., Smith, M., Bellissimo, D. and Davis, R.W. (1988) Sequence and nitrate regulation of the *Arabidopsis thaliana* mRNA encoding nitrate reductase, a metalloflavoprotein with three functional domains. *Proceedings of the National Academy of Sciences of the USA*, **85**, 5006–5010.
7. Deng, M.D., Moureaux, T., Leydecker, M.-T. and Caboche, M. (1990) Nitrate reductase expression is under the control of a diurnal rhythm and is light inducible in *Nicotiana tabacum* leaves. *Planta*, **180**, 257–261.
8. Lea, P.J., Robinson, S.A. and Stewart, G.R. (1990) The enzymology and metabolism of glutamine, glutamate and asparagine. In *The Biochemistry of Plants*, Vol. 16 (eds B.J. Miflin and P.J. Lea), pp. 121–159. Academic Press, San Diego.
9. Forde, B.G. and Cullimore, J.V. (1989) The molecular biology of glutamine synthetase in higher plants. In *Oxford Surveys of Plant Molecular and Cell Biology*, Vol. 5 (ed. B.J. Miflin), pp. 246–296. Oxford University Press, Oxford.
10. Coruzzi, G.M. (1991) A gene network controlling glutamine and asparagine biosynthesis in plants. *Plant Journal*, **1**, 275–280.
11. Sakakibara, H., Watanabe, M., Hase, T. and Sugiyama, T. (1991) Molecular cloning and characterisation of complementary DNA encoding for ferredoxin-dependent glutamate synthase in maize. *Journal of Biological Chemistry*, **266**, 2028–2034.
12. Joy, K.W. (1988) Ammonia, glutamine and asparagine: a carbon–nitrogen interface. *Canadian Journal of Botany*, **66**, 2103–2109.
13. Schubert, K.R. and Boland, M.J. (1990) The ureides. In *The Biochemistry of Plants*, Vol. 16 (eds B.J. Miflin and P.J Lea), pp. 197–282. Academic Press, San Diego.
14. Lea, P.J. (1991) The inhibition of ammonia assimilation: a mechanism of herbicide action. In *Herbicides* (eds N.R. Baker and M.P. Percival), pp. 267–298. Elsevier, Amsterdam.
15. Bryan, J.K. (1990) Advances in the biochemistry of amino acid biosynthesis. In *The Biochemistry of Plants*, Vol. 16 (eds B.J. Miflin and P.J. Lea), pp. 161–195. Academic Press, San Diego.
16. Giovanelli, J., Mudd, S.H. and Datko, A.H. (1989) Regulatory structure of the biosynthetic pathway of the aspartate family of amino acids in *Lemna pavcicostata*. *Plant Physiology*, **90**, 1584–1599.
17. Lea, P.J., Blackwell, R.D. and Azevedo, R.A.

(1992) Analysis of barley metabolism using mutant genes. In *Barley: Genetics, Biochemistry, Molecular Biology and Biotechnology* (ed. P.R. Shewry), pp. 180–207. CAB International, Wallingford.

18. Rognes, S.E., Lea, P.J. and Miflin, B.J. (1980) *S*-Adenosylmethionine: a novel regulator of aspartate kinase. *Nature*, **287**, 357–359.

19. Rognes, S.E., Wallsgrove, R.M., Kueh, J.S.H. and Bright, S.W.J. (1986) Effects of exogenous amino acids on growth and activity of four aspartate pathway enzymes in barley. *Plant Science*, **43**, 45–50.

20. Tiburcio, A.F., Kaur-Sanrey, R. and Galston, A.W. (1990) Polyamine metabolism. In *The Biochemistry of Plants*, Vol. 16 (eds B.J. Miflin and P.J. Lea), pp. 283–325. Academic Press, San Diego.

21. Stidham, M.A. (1991) Herbicidal inhibitors of branched chain amino acid biosynthesis. In *Herbicides* (eds N.R. Baker and M.P. Percival), pp. 247–266. Elsevier, Amsterdam.

22. Bright, S.W.J., Wood, E.A. and Miflin, B.J. (1978) The effect of aspartate-derived amino acids (lysine, threonine and methionine) on the growth of excised embryos of wheat and barley. *Planta*, **139**, 113–117.

23. Arruda, P., Bright, S.W.J., Kueh, J.S.H., Lea, P.J. and Rognes, S.E. (1984) Regulation of aspartate kinase isoenzymes in mutants of barley resistant to lysine plus threonine. *Plant Physiology*, **76**, 442–446.

24. Coggins, J.R. (1989) The shikimate pathway as a target for herbicides. In *Herbicides and Plant Metabolism* (ed. A.D. Dodge), pp. 97–112. Cambridge University Press, Cambridge.

25. Anderson, J.W. (1990) Sulphur metabolism in plants. In *The Biochemistry of Plants*, Vol. 16 (eds B.J. Miflin and P.J. Lea), pp. 327–381. Academic Press, San Diego.

26. Rennenberg, H. and Lamoureux, G.L. (1990) Physiological processes that modulate the concentration of glutathione in plant cells. In *Sulphur Nutrition and Sulphur Assimilation in Higher Plants* (eds H. Rennenberg, C. Brunold, L.J. DeKok and I. Stulen), pp. 53–65. SBP Academic Publishing, The Hague.

27. Foyer, C.H. and Halliwell, B. (1976) The presence of glutathione and glutathione reductase in chloroplasts: a proposed role in ascorbic acid metabolism. *Planta*, **133**, 1–25.

28. Shaul, O. and Galili, G. (1992) Increased lysine synthesis in tobacco plants that express high levels of bacterial dihydrodipicolinate synthase in their chloroplasts. *Plant Journal*, **2**, 203–209.

29. Shaul, O. and Galili, G. (1992) Threonine overproduction in transgenic tobacco plants expressing

a mutant desensitised aspartate kinase of *Escherichia coli*. *Plant Physiology*, **100**, 1157–1163.

30. Frankard, V., Ghislain, M. and Jacobs, M. (1992) Two feedback-insensitive enzymes of the aspartate pathway in *Nicotiana sylvestris*. *Plant Physiology*, **99**, 1285–1293.

FURTHER READING

The series *The Biochemistry of Plants* is a major work now comprising 16 volumes, published by Academic Press, San Diego, under the general guidance of P.K. Stumpf and E.E. Conn. The volumes contain a vast amount of information and are essential reading for the serious student of plant biochemistry. Volume 16, *Intermediary Nitrogen Metabolism* (eds B.J. Miflin and P.J. Lea), is particularly relevant to this chapter.

A companion series of volumes entitled *Methods in Plant Biochemistry*, published by Academic Press, London, under the guidance of P.M. Dey and J.B. Harborne, contain all the practical details required to carry out laboratory experiments in plant biochemistry. Other important books published recently on plant nitrogen metabolism include the following.

Mengel, K. and Pilbeam, D.J. (eds) (1991) *Nitrogen Metabolism in Plants*. Oxford University Press, Oxford.

Poulton, J.E., Romeo, J.T. and Conn, E.E. (eds) (1989) *Plant Nitrogen Metabolism*. Plenum Press, New York.

Wray, J.L. and Kinghorn, J.R. (eds) (1989) *Molecular and Genetic Aspects of Nitrate Assimilation*. Oxford University Press, Oxford.

For those with more of an interest in the molecular aspects of plant biochemistry, the series *Oxford Surveys of Plant Molecular and Cell Biology* edited by B.J. Miflin is an important one. The following two books discuss the action of herbicides and their effects on plant metabolism.

Baker, N.R. and Percival, M.P. (eds) (1991) *Herbicides: Topics in Photosynthesis*, Vol 10. Elsevier, Amsterdam.

Dodge, A.D. (ed) (1989) *Herbicides and Plant Metabolism*. Cambridge University Press, Cambridge.

Finally a good book on sulphur metabolism:

Rennenberg, H., Brunold, C.H., DeKok, L.J. and Stulen, I. (eds) (1990) *Sulphur Assimilation in Higher Plants*. SBP Academic Publishing, The Hague.

8

PLANT PIGMENTS

George A.F. Hendry

NERC Unit of Comparative Plant Ecology, University of Sheffield, UK

INTRODUCTION

Viewed from orbiting satellites, earth is a green and blue planet. Almost all of the pigmentation, outside polar and desert areas, is due to plants. Regional events such as drought, the onset of the rainy season, or autumn bring about marked changes in the coloration of vegetation. At this global level much the greater contribution comes from one group of pigments—the chlorophylls. It has been estimated that each year the total chlorophyll production globally exceeds 10^9 tonnes[1]— 75% from terrestrial plants, the remainder from aquatic organisms, largely marine phytoplankton. Production of carotenoids, the second major plant pigment, is probably of the order of about 2×10^8 tonnes annually. All other natural pigments—plants, fungal or animal are relatively minor in abundance. However, against the dominant green background of vegetation, these minor pigments of ripening fruits, of flower petals, of animals and particularly insects and birds, make for an extraordinarily rich display of biological colours. The following account covers almost all of the plant pigments but ends with a perspective of pigments in other biological systems.

COLOUR

The term colour is one based on human perception of light and refers to one narrow band of the

---

Plant Biochemistry and Molecular Biology. Edited by P.J. Lea and R.C. Leegood
© 1993 John Wiley & Sons Ltd

electromagnetic spectrum, called the visible spectrum. This band coincides with much, but not all, of the light radiation from the sun and extends from the boundaries of ultraviolet through to the infrared. More precisely, pigments are described by the wavelength (λ) of the maximum absorption (A_{max}) of the visible spectrum, expressed in nanometers (nm) (10^{-9} m). White light (wavelengths between 380 and 730 nm approximately) is seen as the simultaneous incidence of the full range of the visible spectrum at the *same relative intensity*. Any substance which lessens the intensity of one part of the spectrum of white light by absorption (as in a monochromatic filter) will bring about the perception of colour to the substance and to the residual transmitted (non-absorbed) light. This latter transmitted light will also show a reduction in intensity (measured in watts per square metre, einsteins or moles of photons) or more simply a darkening, which can be quantified. As more of the spectrum is filtered out the perception of colour will be reduced until ultimately the substance appears to be black.

Pigments are defined as having a particular maximum absorbance (A_{max}). A compound which is perceived as, say, deep blue will probably have an absorption maximum at around 600 nm, i.e. in the yellow part of the spectrum. A pigment perceived as, say, yellow will probably absorb light in the blue-green part of the spectrum, around 450–470 nm. At this point students can be forgiven for being thoroughly confused! The confusion is readily explained—light or more specifically visible radiation, passed through a prism, will be refracted to emerge as a series of hues familiar in a rainbow. A solution of an orange-coloured pigment, placed in front of the beam of white light, will *absorb* much of the blue and green part of the spectrum, leaving only violet, yellow and red light to pass through the solution. Mixing these three *transmitted* colours will give a colour which most human eyes will perceive as orange. A blue-coloured solution in white light absorbs light around the yellow part of the spectrum, leaving green, blue, violet and red to be transmitted, which we will probably perceive as a deep-blue hue. Confusion arises

because textbook physical descriptions of the pigment will refer to the *absorption* wavelength, while the human eye perceives the non-absorbed *transmitted* light. Table 8.1 provides a workable translation of absorption maxima and hues perceived.

As an example, a pigment such as β-carotene would be 'seen' as orange to yellow. Table 8.1 indicates that the region of the spectrum absorbed by β-carotene will be around 460 nm, i.e. it absorbs light in the blue to blue-green part of the spectrum. In contrast, a pigment of autumn leaves such as cyanidin, perceived as a red or mauve pigment, from Table 8.1 will be found to absorb green light around 530–540 nm. In reverse, chlorophyll familiar as a green pigment *absorbs* most of the light in the red and orange region (about 690–650 nm) and again in the blue to violet region (about 480 to below 400 nm). Table 8.1 shows that the perceived transmitted colours will be blue-green and yellow-green, which are indeed the dominant colours of the planet's vegetation.

The second and perhaps less cumbersome way of describing pigments is to use the chemical nomenclature which evolved largely at the hands of the great German chemists of the last century. These pioneers developed a scientific nomenclature based on classical languages. Once understood these names open up a world of colour, even if only on the laboratory shelf. Chlorophyll, for example, is derived from the Greek *chloros*, a yellow-green colour; riboflavin, a yellow pigment, comes from the Latin *flavus*, a yellow hue; indigo from the Greek word for deep-blue, *indikon*, and so on. Table 8.2 lists the more common chemical names (suffixes or prefixes) which imply a pigmented compound.

What determines whether or not a biological molecule is coloured or not is its structure, particularly the electronic structure, size of molecule, solubility and elemental composition. Almost all biological pigments have a number of features in common which immediately distinguish them from the larger number of colourless compounds found in biological material. These distinctive features can be summarized:

Table 8.1. The difference between colour as perceived by the average human eye and which provides the descriptive 'colour' of a pigment (top line) and the region of the spectrum (A_{max}) absorbed by the pigment (bottom line).

Colour perceived	Blue-green	Blue	Violet		Mauve	Red	Orange	Yellow	Yellow-green
Absorption wavelength (A_{max}) (nm)	675	600	585	570	540	525	495	460	410
Colour absorbed	Red	Orange	Yellow	Yellow-green	Green		Green-blue	Blue	Violet

Table 8.2. Chemical names implying colour.

Chemical prefix or suffix	Classical derivation or root	Implied colour	Chemical example
anil-	*al-nil* (arab)	Deep-blue	Aniline
arg-	*argentum* (l)	Silver	Arginine
aur-	*aurum* (l)	Gold	Aureomycin
-azur	*azul* (osp)	Sky-blue	Aplysioazurin
chloro-	*chloros* (gk)	Yellow-green	Chlorophyll
chrom- or -chrome	*chroma* (gk)	Coloured	Chromotropic acid
chryso-	*chrysos* (gk)	Gold	Chrysin
citr-	*citrus* (l)	Lemon-coloured	Citrulline
-cyanin	*kyanos* (gk)	Blue	Anthocyanin
erythro-	*erythros* (gk)	Red	Phycoerythrin
-flavin	*flavus* (l)	Pale yellow	Riboflavin
fulv-	*fulvus* (l)	Red-yellow	Fulvine
fusc-	*fuscus* (l)	Brown	Fuscin
haem- or heme-	*haima* (gk)	Blood-red	Haematin
indigo-	*indikon* (gk)	Deep-blue	Indigo
leuco-	*leukos* (l)	Colourless/white	Leucoptrein
lute-	*luteus* (gk)	Yellow	Lutein
mela-	*melas* (gk)	Black	Melanin
phaeo-	*phaios* (gk)	Dark-coloured	Phaeophytin
porph(o)-	*porphyros* (gk)	Purple	Porphyrin
purpur-	*purpura* (l)	Purple	Purpurin
-pyrrole	*pyrros* (gk)	Fiery-red	Pyrrole
rhodo-	*rhodon* (gk)	Rose-coloured	Rhodophyllin
rub-	*ruber* (l)	Red	Bilirubin
ruf-	*rufus* (l)	Red-brown	Anthrarufin
sepia-	*sepia* (l)	Brown-pink	Sepiapterin
stilb-	*stilbein* (gk)	Glitter	Stilbene
verd- or virid-	*viridis* (l)	Bright green	Biliverdin
viol-	*viola* (l)	Purple-blue	Aplysioviolin
xanth-	*xanthos* (gk)	Yellow	Xanthophyll

arab = Arabic; gk = Ancient Greek; l = Latin, osp = Old Spanish.

Table 8.3. Range of the electromagnetic spectrum absorbed by molecules, causing the excitation of electrons and the development of colour.

Molecular form	Light causing excitation	Electronic transition possible	Expected colour (transmitted)
Saturated hydrocarbons	Far UV	d → d*	Colourless
Unsaturated hydrocarbons	Near UV	p → p*	Probably colourless
Saturated hydrocarbon with N or O	Violet to blue	n → d*	Pale yellow
Unsaturated hydrocarbon with N or O	Visible	n → p*	Red, blue or green

1. Most biological pigments are relatively large molecules with, typically, molecular weights ranging from about 200 (quinones), 300 (flavonoids), 400 (betalains) and 500 (carotenoids) to 800 (chlorophylls).
2. Most are composed, at least in part, of *unsaturated* hydrocarbons. Many are ring (cyclic) structures.
3. Of the 17 or so elements found in biological material, just four predominate in pigments: H, C, N and O.

Given these limited variations it is possible to restrict an explanation of the electronic status of pigments to bare essentials. A fuller explanation of the physical events giving rise to colour are available (see Further Reading section). The electrons in the outer shells over the face of biological molecules, coloured or not, exist in distinct orbits, each shell holding a finite number of electrons maintained in the least energy-demanding orbit possible (referred to as the resting state). When irradiated with light of sufficient energy, the electrons in the outermost orbitals *absorb* the light energy, oscillate (wobble) and undergo a transition to a higher energy level before decaying (relaxing) back to the resting state. Simultaneously the exciting energy is liberated largely as heat or work (as in photosynthesis). Occasionally some of the absorbed light is emitted as fluorescence or phosphorescence, at longer, less energetic wavelengths, but this is less common.

Because of the limited number of elements in biological material and the narrow range of the *visible* light spectrum, only three orbitals are effectively involved in light-induced transitions and are designated d, p and n. On excitation the electrons are raised to orbitals designated d* and p*. The range of biological compounds undergoing these various electronic transitions is shown in Table 8.3 together with the type of light needed to bring about the transition.

The relatively low energy provided by visible radiation (as opposed to ultraviolet radiation) is only sufficient to excite the electrons in the non-bonding orbitals, n. Violet light (of say 400 nm) would be sufficiently energetic to bring about an n → d* transition in a limited range of molecules, essentially long and often branched saturated hydrocarbons, usually with N or O substitutions or additions. Because they absorb violet light they will appear, in white light, probably as pale-yellow pigments. The more interesting pigments come from the unsaturated hydrocarbons, again with N or O, and able to undergo n → p* transitions by absorbing blue, green, yellow or even the least energetic red light. This simplified picture has to be qualified, however. A number of structural modifications make it more possible for n → p* transitions to take place more readily and these are frequently found in biological pigments. Probably the most important are the modifications to form a ring structure (cyclization). Ring structures, particularly in relatively large molecules, permit sharing or delocalization of particular electrons over the face of the molecule. Such delocalized electrons are more readily excited by light of lower energy and in practice contribute to most of the non-yellow pigments in biology. The addition of certain metals may also help to lower the energy level. The prime example in biology of one class of compounds which answer to all of these modifications (large molecule, large number of double bonds, branched structure, ring arrangement with several rings linked, addition of N *and* O and a metal) are the porphyrins, which give a palette of colours unrivalled in biology.

THE PIGMENTS

Perspective

Almost all biological pigments can be reduced to no more than five or six major structural classes, but given the wealth of variation thrown up by 3000 million years of evolution there is often difficulty in fitting some of the more unusual pigments into this artificial classification. However, for much the greater number of pigments, particularly those of higher plants, the classification adopted in Table 8.4 is comprehensive. There are in addition many other pigments found in biological systems, particularly in fungi, lichens (fungal component) and invertebrates, which have not been recorded in plants. Therefore Table 8.4 is *not* a list of all pigments present in biological systems but only those present in plants, including algae.

Tetrapyrroles (examples: chlorophylls, haems and phycobilins).

This is a relatively small group of pigments but they contribute the greatest range of colours as well as being the most abundant globally. All are based on the same structure, the tetrapyrrole, either in its cyclic form (Figure 8.1) as in the chlorophylls and haems or in a linear form (Figure 8.2) as in phycobilins and phytochrome. In the cyclic form these tetrapyrroles are usually

Table 8.4. Naturally occurring pigments in plants, including eukaryotic algae.

Group	Alternative or familiar name	Major examples	Predominant colours
Tetrapyrrole	Porphyrins and porphyrin derivatives	Chlorophylls Haems Phycobilins	Green Red Blue and mauve
Tetraterpenoids	Carotenoids	Carotenes Xanthophylls	Yellow-orange to red
O-Heterocyclic	Flavonoids	Anthocyanins Flavonols Flavones	Pale blue to mauve Yellow to cream Cream-white
Phenol derivatives	Phenolics or quinoids	Naphthaquinones Anthraquinones Tannins, lignin	Red, blue (green) Red, purple Brown
N-Heterocyclic	Indigoids or indole derivatives	Betalains Phytomelanins Indigo	Yellow-red Black-brown Blue-pink

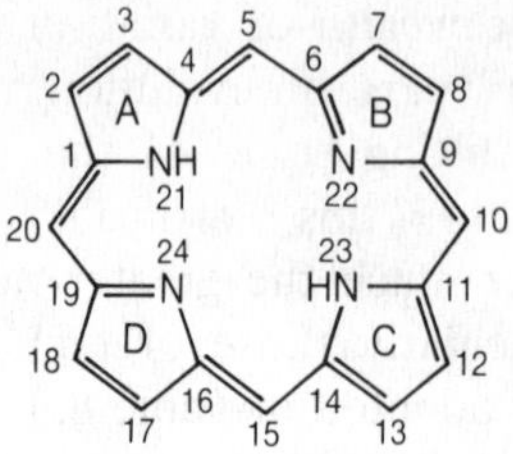

Figure 8.1. A cyclic tetrapyrrole with recommended numbering system.

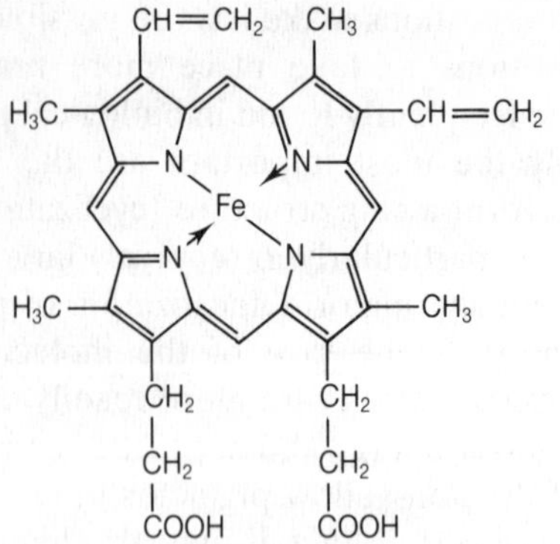

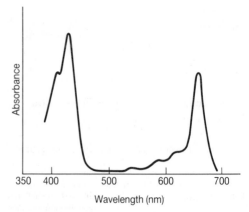

Figure 8.3. Haem b (protohaem).

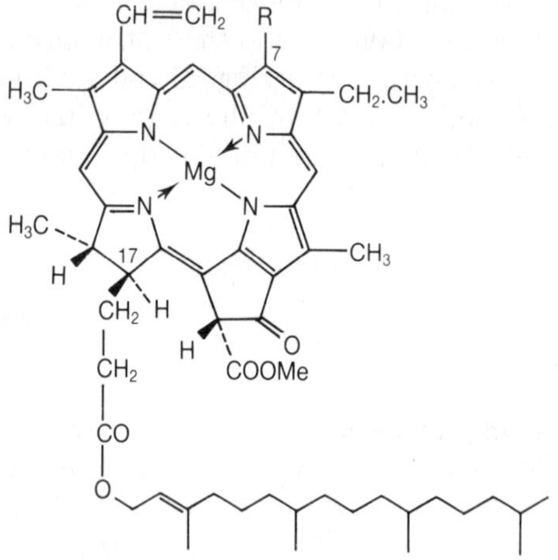

porphyrins or porphyrin derivatives. The linear forms are often referred to as bile pigments. In total the number of naturally occurring recorded tetrapyrroles in biology is about 36 of which just two, chlorophyll a and b, are dominant in terrestrial plants, with three more present in certain classes of algae[2]. They are generally green, green-blue or green-yellow. In addition two classes of bilins (phycobilins) are major pigments in a few algal groups and are blue-purple or red-mauve in colour. Many other tetrapyrroles, particularly the haems (Figure 8.3), are present in plants (as components of cytochromes, peroxidases, catalase and other proteins) but their concentration is unlikely to contribute significantly to the visible pigmentation of the organism. They are nevertheless pigments but more properly referred to as chromophores. Structurally the most abundant tetrapyrrole, chlorophyll a (Figure 8.4), is composed of four pyrrole rings (A, B, C and D in Figure 8.1), the central N atoms binding an Mg.

The chlorophylls have a characteristic light absorption pattern. That for chlorophyll a is shown in Figure 8.5 and shows the two major absorption bands: one in the red but with three

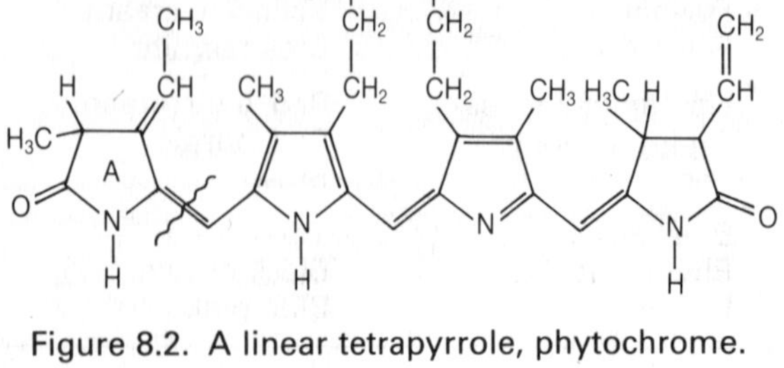

Figure 8.4. Chlorophyll a where R = CH_3 and chlorophyll b where R = CHO.

Figure 8.5. Absorption spectrum of chlorophyll a in diethyl ether.

Figure 8.2. A linear tetrapyrrole, phytochrome.

Table 8.5. Structures of some common tetrapyrroles. The numbering system is that shown in Figure 8.1.

Tetrapyrrole	Peripheral carbon position								Complexed metal
	2	3	7	8	12	13	17	18	
Protoporphyrin	Me	Vi	Me	Vi	Me	Pr	Pr	Me	–
Chlorophyll a	Me	Vi	Me	Et	Me	CP	Phy/H	Me/H	Mg
Chlorophyll b	Me	Vi	Fo	Et	Me	CP	Phy/H	Me/H	Mg
Chlorophyll c_1	Me	Vi	Me	Et	Me	CP	Acr	Me	Mg
Chlorophyll d	Me	Fo	Me	Et	Me	CP	Phy/H	Me/H	Mg
Bacterio-chlorophyll a	Me	Ri	Me/H	Et/H	Me	CP	Phy/H	Me/H	Mg
Haem a	Me	R2	Me	Vi	Me	Pr	Pr	Fo	Fe
Haem b	Me	Vi	Me	Vi	Me	Pr	Pr	Me	Fe
Haem c	Me	R3	Me	R3	Me	Pr	Pr	Me	Fe
Phycocyano-bilin	Me	Et	Me/H	R4	Me	Pr	Pr	Me	–
Phycoerythrin	Me	Vi	Me/H	R4	Me	Pr	Pr	Me	–
Phytochrome	Me	Vi	Me/H	R5	Me	Pr	R6	Me	–
Bilirubin	Me	Vi	Me	Vi	Me	Pr	Pr	Me	–

Abbreviations: Me = methyl; Vi = vinyl; Pr = propionic acid; Et = ethyl, Acr = acryl; Fo = formyl; Phy = phytol; CP = cyclopentanone; R2 = —CH_2OH—CH_2—(CH_2—CH=C.CH_3—$CH_2)_3$H; R3 = CH.S cysteine.CH_3; R4 = =CH—CH_3; R5 = =CH—CH_3.protein; R6 = —CH_2—CH_2—CO.protein.

minor peaks in the yellow part of the spectrum, together with at least one major absorption peak in the blue to purple end of the visible spectrum. This latter is known as the Soret band and is a characteristic of all cyclic tetrapyrroles, including the haems.

In the chlorophylls, but not the related haems, the side chains of ring C are further cyclized to form a fifth ring, E. The chlorophylls are also modified by the addition of a long-chain terpenoid phytol to ring D. The significant differences and, more importantly, the similarities between the various chlorophylls and haems are shown in Table 8.5.

The distribution of the various types of chlorophyll and bacteriochlorophylls is shown in Table 8.6. Chlorophyll a is found in all photosynthetic eukaryotes. The distribution of the second chlorophyll (b, c, d or e) may have an evolutionary significance.

Perhaps the most remarkable feature of Table 8.5 is the extraordinarily small amount of variation that has taken place among the porphyrins and their derivatives. Given that photosynthesis utilizing chlorophyll as a light receptor evolved some 2.6 million years ago and electron transfer processes using haem–proteins probably even earlier, the porphyrins demonstrate extreme conservation of biological molecules. Closer study of the table will show that the first tetrapyrrole protoporphyrin (in the older texts protoporphyrin IX) is the precursor of all tetrapyrroles—animal, plant, fungal and bacterial. All of the chlorophylls of plants and algae and bacteriochlorophylls of photosynthetic bacteria are structurally closely related to chlorophyll a. The haems are a little more varied but are derived from the common precursor haem b (Figure 8.3), the chromophore of all known peroxidases, catalase, many cytochromes as well as haemoglobin (and plant

Table 8.6. The distribution of chlorophylls and bacteriochlorophylls.

	Organism
Chlorophyll	
a	All oxygen-evolving photosynthetic organisms including higher plants, all algae and the photosynthetic bacteria Cyanophyta and Prochlorophyta
b	Higher plants, the algal Chlorophyta and Euglenophyta and bacterial Prochlorophyta
c	Algal Phaeophyta (brown algae), Pyrrophyta (dinoflagellates), Bacillariophyta (diatoms), Chrysophyta, Prasinophyta and Cryptophyta
d	Some Rhodophyta (red algae) and Chrysophyta
e	Reported from algal Xanthophyta
Bacterio-chlorophyll a and b	Purple bacteria including Chromatiaceae and Rhodospirillaceae
c, d and e	Green and brown sulphur bacteria including Chlorobiaceae and Chloroflexaceae

leghaemoglobin). The four bile pigments (including the animal bilirubin) are also derived from haem b.

All cyclic tetrapyrroles (chlorophylls and haems) are bound to proteins in the living cell. In the case of chlorophyll the attachments to proteins confer both an orientation for light harvesting as well as for electron flow. The attachment of haems confers specific properties appropriate to the enzymic function of the haem–proteins. Detached from their proteins it appears highly likely that the tetrapyrrole is promptly destroyed in living cells.

The function of the chlorophylls is in the capture of light energy as the first step in its transformation into chemical energy (adenosine triphosphate principally) and reducing power (NADPH) (see chapter 1) utilized in the reduction of carbon dioxide to carbohydrates (chapter 2) and nitrite to

ammonia and amino acids (chapter 7). To that extent the pigment chlorophyll is central to the sustained autotrophic existence of plant life on earth. The related phycobilins (phycoerythrins and phycocyanins) act as accessory pigments in the mainly marine red algae (Rhodophyta), the freshwater and marine Cryptophyta and in the prokaryotic blue-green Cyanophyta, present in water, soils and as lichen symbionts. A third bile pigment, phytochrome, a component of most if not all photosynthetic eukaryotes, is the photoreceptor for processes of growth and development which are regulated by light (photomorphogenesis). One interesting feature of the porphyrins is their remarkably limited function in plants. Despite the fact that the porphyrins come in more varied hues than any other biological pigment, their use or adaptation purely as a colorant is rare in plants. In contrast, in animals, porphyrins (or bile pigments) make a contribution to the colour of mollusc shells, and to the wings of some Lepidoptera and to the pigments in bird egg-shells and occasionally even the feathers of certain tropical birds. Instead, plants employ quite different pigments in pollinator attraction (petal colours) and seed dispersal (fruit colours).

The biosynthetic pathway of the tetrapyrroles is shown in Figure 8.6. All tetrapyrroles in biological systems are thought to originate from δ-aminolaevulinic acid. The assembly of the pyrroles and tetrapyrroles from this amino acid appears to follow identical routes in all organisms. The names, a delight to porphyrinologists but a nightmare to others, reflect in some cases the first source from which the intermediates were first isolated. Uroporphyrin was isolated from the urine of sufferers of a disease once known as the royal malady. The urine would turn a deep portred on standing. The pigment has little to do with King George III's favourite tipple but was due to the accumulation and excretion of a precursor of haem. Coproporphyrin was isolated from human faeces as an overproduction of another haem precursor. The branch point in the pathway, protoporphyrin, gives rise to either the iron porphyrins (the haems) or to the magnesium pathway (the chlorophylls). In angiosperms the

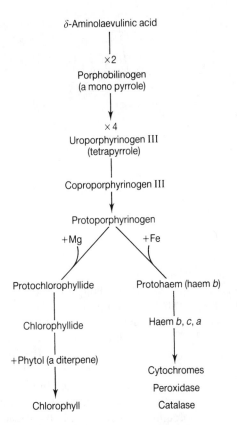

δ-Aminolaevulinic acid

×2

Porphobilinogen
(a mono pyrrole)

×4

Uroporphyrinogen III
(tetrapyrrole)

Coproporphyrinogen III

Protoporphyrinogen

+Mg   +Fe

Protochlorophyllide   Protohaem (haem *b*)

Chlorophyllide   Haem *b*, *c*, *a*

+Phytol (a diterpene)

Cytochromes

Peroxidase

Chlorophyll   Catalase

Figure 8.6. Biosynthetic pathway, in outline, of the tetrapyrroles.

penultimate step in the magnesium pathway—the reduction of protochlorophyllide to chlorophyllide—is dependent on light. Such plants if germinated and grown in the dark are yellow in colour. Only on exposure to light will they turn green as chlorophyll synthesis gets underway.

Despite its universality in biology, there is one important difference between animals and fungi on the one hand and plants on the other. The building block of δ-aminolaevulinic acid in animals, fungi and many bacteria is formed from condensation of succinic acid and glycine. In plants, including algae and a number of bacteria, including some but not all photosynthetic bacteria, δ-aminolaevulinic acid is formed directly from glutamate, a reaction catalysed most unusually by a specific glutamyl transfer RNA[3]. The significance of this relatively recent discovery

is not entirely clear but could indicate that tRNA in earliest times had a direct role in the catalysis of organic reactions, a role which was subsequently taken over by proteins. The retention of a tRNA-catalysed reaction in photosynthetic (and other) organisms may reflect extreme conservatism in evolution.

Tetraterpenoids   (example: carotenoids)

The second most abundant group of pigments on the planet are the carotenoids. They are present throughout the plant kingdom in both photosynthetic and non-photosynthetic tissue, in many bacteria, and are frequently found as colorants in insects, crustaceans (particularly), some echinoderms, fish and birds. In animals carotenoid derivatives also function as pigments of vision[4]. The carotenoids of plants are generally yellow, orange or occasionally red in colour. Animal carotenoids (more correctly carotenoid–protein complexes) extend into the purple, violet, blue, green and even black[5]. Again, as with tetrapyrroles, the use and variation of carotenoids in plants is comparatively limited. Carotenoids, with few exceptions, are tetraterpenoids and are based on or derived from a skeleton of 40 carbon atoms, linked symmetrically with alternating unsaturated bonds. The molecule is usually built of two identical parts, each of 20 carbon atoms as shown in lycopene (Figure 8.7), the red pigment of tomatoes.

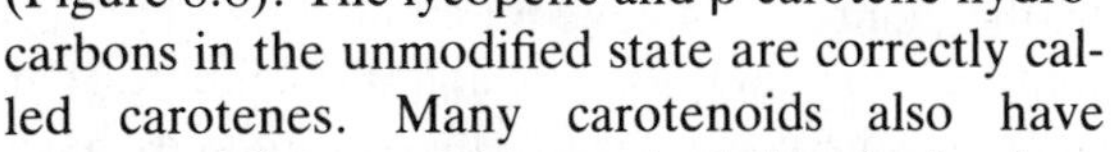

Figure 8.7. A carotene, lycopene.

Frequently the terminal carbons are arranged in a six-membered ring typified by β-carotene (Figure 8.8). The lycopene and β-carotene hydrocarbons in the unmodified state are correctly called carotenes. Many carotenoids also have oxygen additions to the terminal rings in the form of hydroxy, methoxy, aldehyde or carboxylic acid groups. The pigments are then known as xantho-

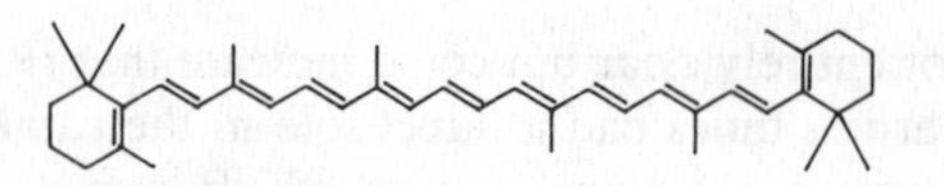

Figure 8.8. β-Carotene.

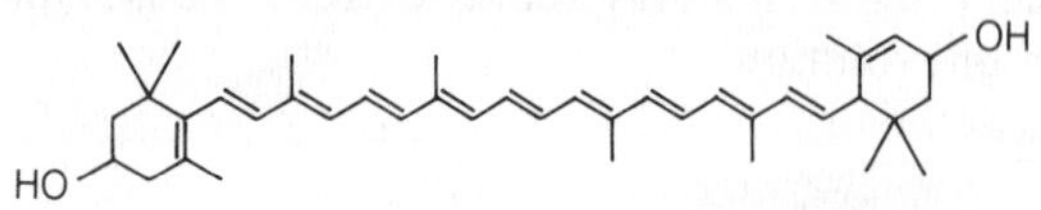

Figure 8.9. A xanthophyll, lutein.

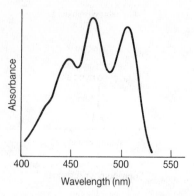

Figure 8.10. Absorption spectrum of a carotenoid, lycopene, in petroleum ether.

phylls, of which lutein (Figures 8.9) is a widespread example.

Of the 600 or so carotenoids described to date, much the greater number are animal in origin although *derived from* ingested carotenoids of plant origin. Little more than 10% of all known carotenoids are plant pigments. All green tissues of higher plants contain broadly similar carotenoids located in the chloroplast membranes. Perhaps because of the central role of photosynthesis in plant success, remarkably little variation exists in these pigments. In higher plants the most common carotenoids are β-carotene, and the xanthophylls lutein, violaxanthin and neoxanthin. Others, such as α-carotene and zeaxanthin, have limited distributions. Among the algae the carotenoids are more varied. Apart from the green algae (Chlorophyta) which contain carotenoids similar to those in higher plants, many algae have otherwise uncommon carotenoids. Almost all contain β-carotene but some, such as the dinoflagellates (Pyrophyta) and blue-green algae (Cyanophyta), have carotenoids unknown to other areas of biology. The presence of carotenoids of restricted distribution such as fucoxanthin in the Phaeophyta (brown algae), Bacillariophyta (diatoms) and the Chrysophyta has been taken to imply an evolutionary relationship between the groups.

The colour of plant carotenoids depends largely on the number of double bonds influenced by the form of the end groups and the addition of oxygen. Characteristically, the carotenoids have three distinct absorption maxima; that for lycopene is shown in Figure 8.10.

The carotenoids have a limited number of functions. The best-known perhaps are in photosynthesis. There the carotenoids, at least those in the thylakoid membranes, absorb light, typically beyond the blue absorbing peak of chlorophyll a (compare Figure 8.5 and Figure 8.10), and on relaxation (see above) the energy of light (at say 450 nm) is transferred to a longer but less energetic wavelength (at say 680 nm), when it may be passed to a neighbouring chlorophyll molecule. In addition the same carotenoids and others in the outer membrane of the chloroplast have the important function of protecting the chlorophyll molecule from destruction by light—a process known as photobleaching. The precise mechanism of protection is complex but can involve one or more forms of excited (activated) oxygen which have the potential to oxidize (bleach) the chlorophyll. The double bonds of carotenoids quench or absorb the appropriate activated oxygen molecule. A third and often overlooked function of carotenoids in plants is in the coloration of petals and fruits, where they serve to attract pollinators or seed dispersers. In most cases these carotenoids are located in modified (or degenerate) chloroplasts called chromoplasts. The yellow perianth of daffodils and the petals of many yellow Compositae contain violaxanthin in chromoplasts. Some of the more uncommon plant carotenoids are found not in chloroplasts but in pollinator attractant tissues. Fruit carotenoids are,

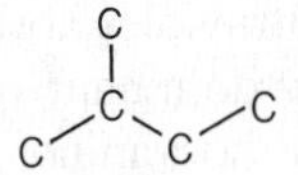

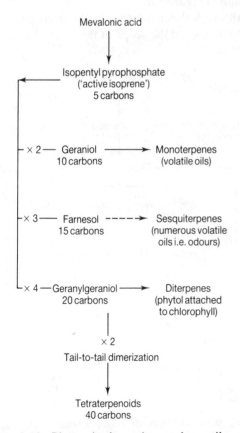

Figure 8.11. Isoprene, the functional unit of terpenoid synthesis.

however, often based directly on β-carotene.

The carotenoids, as with all terpenoids, are formed from a simple five-carbon isoprenoid unit (Figure 8.11), along with the mono- and sesquiterpenoids which contribute to the volatile scents of plants.

As tetraterpenoids of 40 carbons, the carotenoids are themselves assembled initially from a 20-carbon geranyl geraniol unit (itself assembled from a 10-carbon geraniol unit). The carotenoid pathway from the first 40-carbon carotene (phytoene) involves a series of desaturation reactions

Mevalonic acid

↓

Isopentyl pyrophosphate ('active isoprene') 5 carbons

× 2 — Geraniol ⟶ Monoterpenes
10 carbons (volatile oils)

× 3 — Farnesol ----→ Sesquiterpenes
15 carbons (numerous volatile oils i.e. odours)

× 4 — Geranylgeraniol ⟶ Diterpenes
20 carbons (phytol attached to chlorophyll)

↓

× 2

Tail-to-tail dimerization

↓

Tetraterpenoids 40 carbons

Figure 8.12. Biosynthetic pathway, in outline, of the terpenoids.

followed by cyclization of the terminal bonds and, in the case of the xanthophylls, by hydroxylation reactions. The pathway, though varied in the specific organism, follows the same overall order in all plants and microorganisms. This is summarized in Figure 8.12.

Flavonoids   (example: anthocyanins)

The flavonoids are a family of structurally related pigments largely confined to vascular plants. Many, but not all, are coloured and these include the anthocyanins, widespread as petal and fruit colorants. The group also contain a number of families of compounds with similar names, a minefield to the unwary!

All are based on the tricyclic skeleton flavone (Figure 8.13), consisting of two benzene rings (labelled A and B) joined with an oxygen-containing pyran group. The most colourful of the flavonoids are the anthocyanins (Figure 8.14), the pigments of many petals, of ripening fruits and of leaves in autumn. Some 250 have been described, giving rise to the widest range of hues from purple, blue, orange, red to mauve[6]. In living cells the anthocyanins are composed of the chromophore anthocyanidin (note spelling) and one or more sugars. Different substitutions on the rings A and B alter the colour and in addition associations between two or more anthocyanidin molecules give rise to numerous subtle shades.

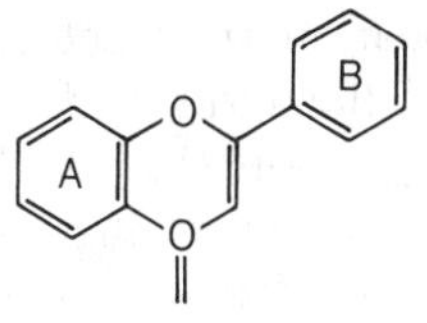

Figure 8.13. Basic structure of a flavone.

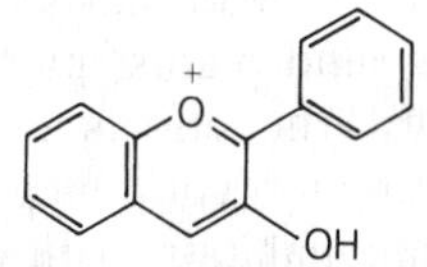

Figure 8.14. Basic structure of an anthocyanidin.

Table 8.7. The most widely occurring anthocyanidins.

Anthocyanidin	Colour *in vivo*
Cyanidin	Blue-red
Delphinidin	Purple-blue
Malvinidin	Purple
Pelargonidin	Scarlet-red
Peonidin	Blue-red
Petunidin	Purple-blue

Despite the extraordinary range of hues provided by the anthocyanins, only 17 anthocyanidins (the aglycone structure) are involved. Of these only seven are widespread (see Table 8.7). Their names—delphinidin, peonidin, petunidin and so on—reflect in most cases the species from which the pigment was first isolated and rather disguises the very widespread presence of the seven throughout the flowering plants. Exceptionally, perhaps, orange-red apigeninidin is associated with plants pollinated by hummingbirds.

Of the non-anthocyanin flavonoids, most are colourless or at best pale yellow, and nearly 4000 have been described to date, largely from higher plants and particularly the angiosperms[6]. Many, however, contribute to the white and cream colours of petals. Claims have occasionally been made of flavonoids in algae but most if not all are today considered to be unreliable. Appearance of these compounds begins with the mosses and liverworts (but not the Anthocerotales). Even so, rather less than half of the bryophyte species investigated appear to yield detectable amounts of flavonoids. In the ferns and related groups, flavonoids are much more the rule than the exception. This may be the clue to the original function of flavonoids. Flavonoids, as a group, have a strong absorption in the ultraviolet, particularly in the region 265–340 nm. It has been suggested that the early function of flavonoids was as a screen protecting land plants, particularly those of stature, against the effects of ultraviolet-B radiation. This function continues today. Plants of tropical montane habitats, with greatest exposure to ultraviolet-B radiation, are generally richly endowed with ultraviolet-absorbing flavonoids. Given the visually dominant role of one class of flavonoids, the anthocyanins, as fruit and petal colorants, it seems highly probable that the evolution of some flavonoids was closely connected to the evolution of insect pollination and animal-dependent seed dispersal. In fungi and animals flavonoids are rare and the role of ultraviolet-B protection appears to have been taken over by the melanins—indole-quinone-based brown to black polymers.

The biosynthesis of the flavonoids can be considered as the formation of a C_6–C_3–C_6 skeleton (see Figure 8.13). Ring A is derived from acetate. Ring B, together with the three carbons linking A with B, represent a phenyl unit with a propenoic acid side chain derived from phenylalanine (in common with numerous other phenolic compounds) (Figure 8.15).

Although the fine details of the flavonoid pathway are beyond the scope of this book, the essen-

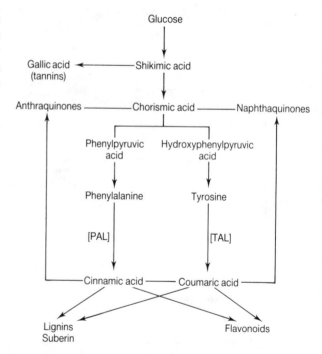

Figure 8.15. Outline of biosynthetic pathway of phenol derivatives, including quinoids, tannins, flavonoids, lignin and suberin.

tial and controlling point of flavonoid biosynthesis occurs with the deamination of the amino acid phenylalanine (tyrosine in some species), catalysed by the enzyme phenylalanine lyase (PAL). The activity of PAL is determined by light and although flavonoids are formed in roots (but usually only to low concentrations) the principal centres for flavonoid biosynthesis are the leaves, petals and ripening fruits.

Quinoid Compounds (examples: naphthaquinones, anthraquinones, tannins, lignins and suberin)

Of the several thousand phenolic compounds so far recorded from plants, a significant proportion are coloured but make little overall contribution to visible pigmentation of cells or tissues. Among these are the numerous quinones often present in trace concentrations, some functioning in electron transport in chloroplasts and mitochondria, others contributing to the toxic composition of some plants and described, sometimes rather loosely, as defence chemicals. The function of most of these toxic quinones is not known. They are thought to be bitter to taste and known to be highly toxic to most biological systems *in vitro*. Whether they actually perform the task of warding off would-be herbivores or fungal pathogens in the field is often not known. Another group of phenolic compounds with suspected defence properties are the dimers and trimers of simple phenols called naphthaquinones (Figure 8.16) and anthraquinones (Figure 8.17) respectively.

Many of these polymeric quinoids contribute to the coloration of the heartwood of trees and have been credited with anti-fungal properties. Others, such as hypericin (Figure 8.18), an anthraquinone

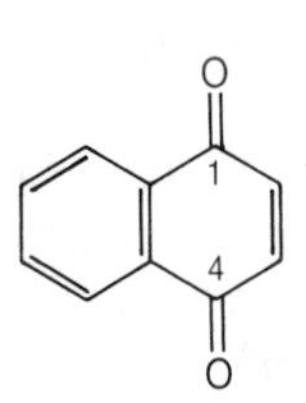

Figure 8.16. A naphthaquinone.

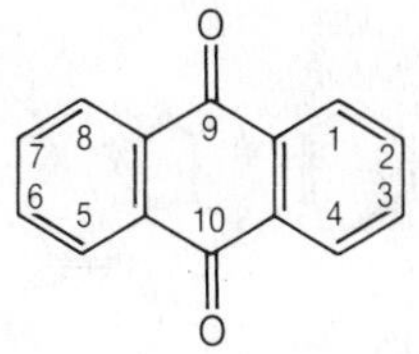

Figure 8.17. An anthraquinone.

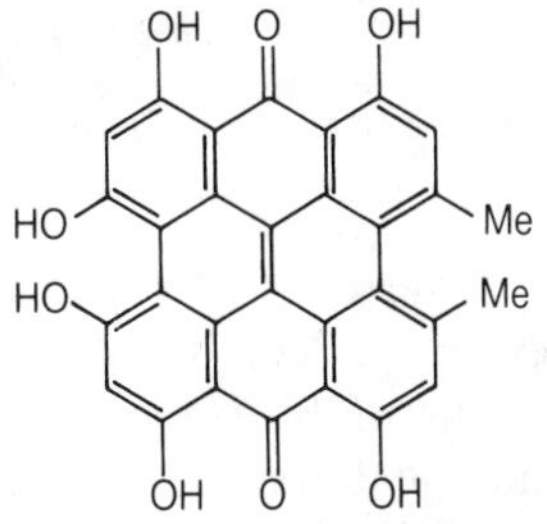

Figure 8.18. Hypericin.

of St John's wort (genus *Hypericum*), have proven toxicity in isolation. As the *Hypericum* genus has few natural enemies and appears to be avoided by herbivores, there is circumstantial evidence at least that anthraquinones function in plant defence chemistry.

Perhaps the most widespread phenolic pigments are the yellow, red-browns, browns and black that constitute the tannins, suberin and lignins. All have several features in common. They are formed in cells destined to die shortly afterwards. They function in various ways, including waterproofing, cell-wall strengthening and probably in defence chemistry. The tannins in particular, on hydrolysis, bind to almost any available protein, making this nitrogen source either indigestible or unpalatable to herbivores. Lignins, which are formed from the random assembly of three partly reduced forms of ferulic, sinapic and caffeic acids, defy breakdown by most organisms and are therefore ascribed with anti-fungal properties. The colour of dead plant remains as in humus is largely that of lignin. In addition, many plants contain a less well-defined polymer based on cross-links between quinones and proteins, often associated with dead structural tissue and called plant sclerotins, analogous to

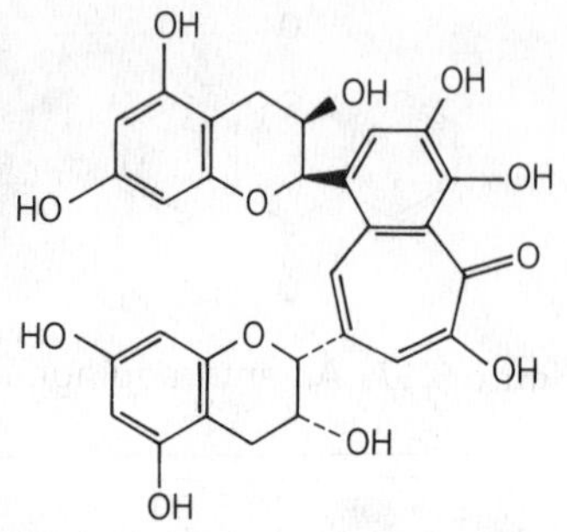

Figure 8.19. A theaflavin.

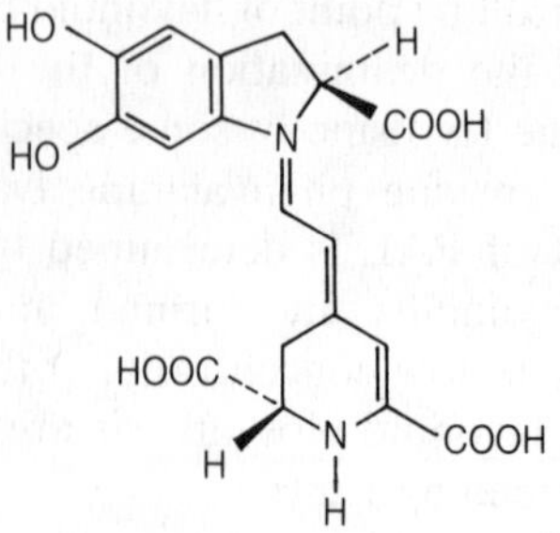

Figure 8.20. A betalain, betanidin.

the polymer in the exoskeletons of insects. On oxidation these form deep-brown pigments.

Another group of brown or orange-red pigments are the theaflavins, which contribute to the colour of dried tea leaves and, indeed, the rich red-brown colour of the beverage. These complex polymers are typically based on two or more benzopyran structures (Figure 8.19) and illustrate the point that many phenolic pigments are present in cells as oligomers (even polymers) whose structures cannot be determined with any precision.

Black pigments are not uncommon in plants, particularly in seed coats. Some of these pigments appear to be polymers of simple phenols and are known as allomelanins. The precise structures of these variable pigments are not known with certainty.

The biosynthesis of quinoids, as with many phenolic derivatives, is from phenylalanine (as in Figure 8.15). It will be apparent then that the 'phenol pathway', as it is often called, leads to the formation of several thousand compounds, many being formed in only one or a few related families. Phenolic derivatives are the basis for much of what is called 'secondary metabolism' and are of considerable medical and veterinary interest.

Indole Compounds   (example: betalains and indigo)

One order of dicotyledons, the Caryophyllales, is unique among flowering plants in not forming anthocyanins. Instead this order, which includes the Amaranthaceae, Cactaceae and Chenopodiaceae, forms a series of nitrogen-containing pigments derived from tyrosine—the betalains. These pigments function as petal colorants, of which betacyanin (Figure 8.20) is the dark-red pigment of beetroot (*Beta vulgaris*) and love-lies-bleeding (*Amaranthus* spp.), and betaxanthin is a yellow pigment.

The origin of the betalains is something of a mystery: the betalain-containing families appear to lack the genes for anthocyanin synthesis. The only other species with the capacity for betalain biosynthesis appear to be a few fungi, including the colourful fly agaric (*Amanita muscaria*) and one extinct group of primitive cycad-like plants, the Bennettitales.

The betalain biosynthetic pathway in plants closely resembles that found in mammals in the synthesis of melanins. The betalains originate from the amino acid tyrosine via dihydroxyphenylalanine (dopa) incorporated into both rings of the betanidin molecule (Figure 8.20). Unlike the anthocyanins, the betalains retain the nitrogen atoms from the amino acid precursor. But as with anthocyanins and other flavonoids, the betalain pathway in plants is substantially regulated by light. A dark biosynthetic pathway appears to exist in some species under certain nutritional conditions.

Occasionally certain plants have been found to contain unusual indole pigments. The blue textile dye indigo is derived from *Indigofera tinctoria* (and near relatives) and the woad plant *Isatis tinctoria*. The pigment is colourless in the intact living plant, where it is present as a glycoside of 3-

Table 8.8. Outline of the distribution and range of pigments in fungi, lichens, animals, protozoa and bacteria.

Organism	Pigment
Fungi	Flavins, betalains, melanins, sesquiterpenoids (carotenoids rare), malonate derivatives (ketides) including naphtha- and anthraquinones, terphenylquinones (largely unique to fungi)
Lichens (fungal component)	As for fungi plus depsides, xanthones, dibenzofuran derivatives (unique to group)
Vertebrate animals	Mainly melanins; carotenoids in fish, reptiles and birds. Tetrapyrroles in birds
Invertebrate animals	Carotenoids (numerous, particularly crustacea), polycyclic quinones (often unique to different classes), melanins (squid—ink), porphyrins (molluscs), pterins (Lepidoptera), flavins, unusual bilins, copper or vanadium-containing proteins
Protozoa	Chlorophylls and carotenoids (if photosynthetic), otherwise ± colourless
Bacteria	Unique carotenoids, phenazines (unique to group), dimers of indoles and pyridine (largely unique to bacteria)

hydroxyindole. On hydrolysis, oxidation and dimerization the blue pigment indigo is formed. Since the 1890s indigo has been produced from coal and later from oil.

COMMERCIAL APPLICATIONS

Many plant pigments have specific and known functions in the intact organism. Chlorophylls and carotenoids are integral parts of the photosynthetic apparatus: other pigments, including anthocyanins and betalain, function in pollinator attraction, flavonoids probably in protection from ultraviolet-B radiation, yet others, including naphthaquinones and anthraquinones, may act in the defence chemistry of the plant. From earliest times human economies have sought alternative uses for plant pigments. These uses have until recently included textile dyes. Similarly tannin extracts from bark and nutshells were the mainstay of the leather-tanning industry until comparatively recently. The one area where plant pigments are being increasingly exploited is as food colorants. The development of azo and other artificial dyes has dominated the food colorant market for many years, but with increasing concern over the safety of these additives there has been a renewed interest in the use of plant pigments as natural colorants. The four pigments most widely used in the colouring of foods, including drinks, are the carotenoids, mainly β-carotene from several sources including algae, bixin and norbixin, the carotenoids from a tropical bush (*Bixa orellana*) or annatto, anthocyanins from grape skins, betalains from *Beta vulgaris* and chlorophylls mainly from grasses[7]. Other pigments such as turmeric and saffron, though valuable as commodities, are of minor importance compared to the growing international trade, particularly in carotenoids, anthocyanins and betalains. Plant pigments are also used in a number of cosmetic preparations—an application of growing value.

PLANT PIGMENTS IN A BIOLOGICAL CONTEXT

Although the coloration of the planet is largely due to plants, as sources of pigments plants are in some ways remarkably restricted. The commonly occurring pigments in plants can be reduced to two chlorophylls, perhaps four or five carotenoids and three or four anthocyanidins. The rest are relatively uncommon. In contrast, the pigments from fungi, lichens (fungal component), invertebrates and bacteria are extraordinarily varied. Many, particularly among the lichens, are unique to the class. Table 8.8 gives an outline of the major pigments to be found in other organisms.

Quite why such a wide range of unusual, even unique, pigments accumulate in fungi is unclear. The wealth of pigments in invertebrates is partly explained by the evolution of camouflage and warning colours and sensory perception, particularly for mating. It will be apparent that plant pigments are remarkably restricted, the greatest scope for variation occurring in the quinoids and flavonoids. Apart from the anthocyanins, the natural function of these phenol derivatives is substantially a matter of speculation. Research into plant pigments is an Aladdin's cave, though in our ignorance a rather dark one!

REFERENCES

1. Brown, S.B., Houghton, J.D. and Hendry, G.A.F. (1991) Chlorophyll degradation. In *The Chlorophylls* (ed. H. Scheer), pp. 465–492, CRC Press, Boca Raton.
2. Hendry, G.A.F. and Jones, O.T.G. (1980) Haems and chlorophylls: comparison of function and formation. *Journal of Medical Genetics*, **17**, 1–14.
3. Schon, A., Krupp, G., Gough, S. Berry-Lowe, S., Kannangara, C.G. and Soll, D. (1986) The RNA required in the first step of chlorophyll synthesis is a chloroplast glutamate tRNA. *Nature*, **322**, 281–284.
4. Krinsky, N.I., Mathews-Roth, M.M. and Taylor, R.F. (1990) *Carotenoids: Chemistry and Biology*. Plenum Press, New York.
5. Needham, A.E. (1974) *The Significance of Zoochromes*. Springer-Verlag, Berlin.
6. Harborne, J.B. (1988) *The Flavonoids: Advances in Research since 1980*. Chapman and Hall, London.
7. Henry, B. (1991) Natural food colours. In *Natural Food Colorants* (eds G.A.F. Hendry and J.D. Houghton), pp. 39–78, Blackie, Glasgow.

FURTHER READING

Britton, G. (1983) *The Biochemistry of Natural Pigments*. Cambridge University Press, Cambridge.
Brown, S.B. (1980) *Introduction to Spectroscopy for Biochemists*. Academic Press, London.
Goodwin, T.W. (1988) *Plant Pigments*, Academic Press, London.

9

GENOME ORGANIZATION, PROTEIN SYNTHESIS AND PROCESSING IN PLANTS

M.D. Watson

Department of Biological Sciences, University of Durham, UK

D.J. Murphy

John Innes Centre for Plant Science Research, Norwich, UK

INTRODUCTION

All plant cells contain three genomes. These are localized respectively in the nucleus, the mitochondria and the plastids of the plant cells. The largest and most significant genome in plant cells is that of the nucleus. There is now increasing evidence that the two smaller plant genomes, i.e. those of the mitochondria and plastids, originated from endosymbiotic partners, which were

---

engulfed by a primitive eukaryotic cell. Following the original endosymbiotic event, there has been a tendency for the plastid and mitochondrial genomes to lose DNA to the nuclear genome. This has resulted in the present-day plastid and mitochondrial genomes containing relatively few genes. The expression of these organellar genomes is now under the control of the nuclear genome.

As a consequence of the vast majority of genes being located on the nuclear genome, almost all plant proteins are synthesized on cytoplasmic ribosomes. A mechanism must therefore exist for transporting and sorting these proteins to their correct subcellular destination. Each protein, other than those found in the cytoplasm, has a specific molecular 'flag' which indicates its final destination.

NUCLEAR GENOME ORGANIZATION

The nuclear genome of plants is similar to that of other advanced eukaryotes. The DNA in nuclei and chromosomes is highly organized on a number of different structural levels[1]. These range from localized secondary structures with which DNA-binding proteins can interact, through the winding of the DNA helix around histones, to the clustering of many hundreds of kilobase pairs (kbp) of DNA into the banding patterns, which are familiar attributes of metaphase chromosomes. These different levels of organization of plant chromosomes, ranging from the DNA double helix up to the metaphase chromosome, are shown schematically in Figure 9.1. It should be emphasized that mitotic chromosomes, such as the metaphase chromosome shown in Figure 9.1, are transcriptionally inert. All RNA synthesis ceases as the chromosomes condense, since the RNA polymerases are unable to gain access to the DNA in these condensed chromosomes. Transcriptionally active chromosomes are found during the interphase stage of the cell cycle. These chromosomes tend to have extended and unwound

structures relative to metaphase chromosomes. This is shown in the DAPI-stained chromosomes of rye in Figure 9.2[2]. Note that metaphase chromosomes are highly condensed regularly organized structures. In contrast, the interphase chromosomes are highly dispersed around the nucleus, with no clear organization at this level of magnification.

The detailed analysis of nuclear genome architecture is made more difficult by its dynamic nature, i.e. its organization changes during different stages of the cell cycle, its small size, typically 10 μm in diameter, and the similar morphology of many of the different chromosomes present in a given nucleus. Chromosome-spreading techniques have allowed for the examination of chromosome numbers and morphology. Chromosomes can easily be identified in banded two-dimensional spreads of metaphase nuclei by light microscopy. The three-dimensional position of chromosomes can be reconstructed from sections. The combination of these two techniques, however, is not at all simple to achieve. In more recent years, the newer techniques of confocal microscope reconstructions and computer processing of the images of chromosome sections have increasingly been used for chromosome analysis. The problem of chromosome identification at both interphase and metaphase is being tackled by *in situ* hybridization with various DNA probes.

Eukaryotic genes tend not to be clustered into the operons characteristic of most prokaryotic organisms. Neither are eukaryotic genes organized in a completely random fashion. Rather, they are organized into clusters which can be made visible as the transverse bands that can be visualized by various histochemical staining methods. The reason for the gene clustering is to facilitate replication of the genome and the co-ordinated transcription of certain sets of genes. The study of genome organization and nuclear architecture in plants is still in its infancy. It is likely, however, that the genomes of higher plants are organized in a broadly similar manner to that of other advanced eukaryotes, as discussed by Alberts *et al.* (1989).

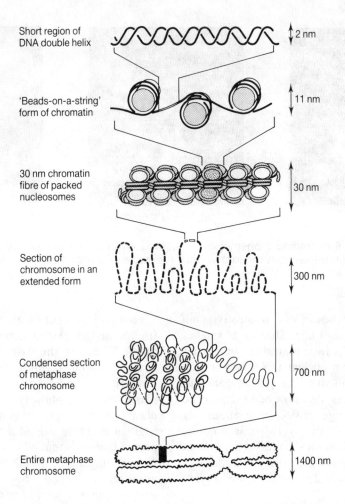

Short region of
DNA double helix

2 nm

'Beads-on-a-string'
form of chromatin

11 nm

30 nm chromatin
fibre of packed
nucleosomes

30 nm

Section of
chromosome in an
extended form

300 nm

Condensed section
of metaphase
chromosome

700 nm

Entire metaphase
chromosome

1400 nm

Figure 9.1. Schematic illustration of the hierarchy of chromatin packing which is believed to give rise to the highly condensed metaphase chromosome in plants. Courtesy of Professor K. Roberts and Garland Publishing Inc., reproduced by permission.

PLASTID GENOME ORGANIZATION

Plastids are a type of organelle which is unique to plant cells. All plant cells contain plastids and all the plastids within a given plant will contain identical genomes. The morphology of the plastid in a given plant tissue, however, can be quite different to that of the plastids in another tissue from the same plant. For example, in green tissues such as leaves, the plastids tend to be in the form of chloroplasts, containing highly organized mem-

brane systems, while in storage organs such as tubers, the plastids contain few, if any, membranes, but are replete with starch storage bodies, termed starch grains. This indicates that the morphology of plastids is controlled from outside and is not dictated directly by the plastid genome.

Plastids proliferate by division of pre-existing plastids and are passed onto daughter cells at cell division and to daughter organisms by maternal inheritance. The information contained in the plastid genome is relatively limited, consisting as

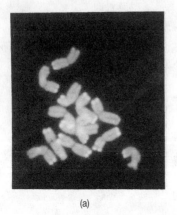

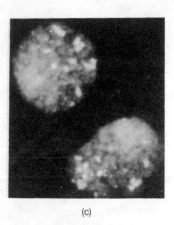

(a)   (b)   (c)

Figure 9.2 DAPI (4′,6-diamidino-2-phenylindole)-stained chromatin from (a) early metaphase, (b) late metaphase and (c) interphase chromosomes from rye nuclei. Courtesy of Dr J.P. Heslop-Harrison.

it does of several copies of one relatively small chromosome of 120–270 kbp. Due to its smaller size and density, DNA from plastid chromosomes can be separated from nuclear DNA on caesium chloride density gradients. The plastid genomes of several plant species have now been sequenced completely. They range in size from about 121 kbp for the genome of the liverwort *Marchantia polymorpha*, to 156 kbp for the plastid genome of tobacco. Plastid genomes are arranged as double-stranded circular molecules, as shown in Figure 9.3[3].

These chromosomes are characterized by the presence of large inverted repeats. It is known that plastid genomes contain all the chloroplast ribosomal RNA genes (six genes), the transfer RNA (tRNA) genes (approximately 30 genes) and probably all the genes for protein synthesis within the chloroplast (about 39 genes). In addition to the above 'housekeeping' genes, the plastid genome encodes many major protein components of the photosynthetic apparatus (about 20 genes). The majority of these genes encode the membrane-bound subunits of the electron transport and photophosphorylation complex. It is interesting, however, that at least one subunit of each of these protein complexes is encoded on the nuclear genome and that the expression of the plastid genes is strictly regulated by the nucleus.

Finally, a number of unidentified open reading frames in the plastid genome have been identified. Therefore the vast majority of the plastid genome contains genes involved either in the transcription and translation apparatuses of the plastid or in the photosynthetic machinery of the plastid. A list of plastid genes and their products is given in Table 9.1. One interesting and unexpected finding is that all of the plastid genomes so far sequenced contain DNA sequences which are predicted to encode respiratory chain NADH dehydrogenases, similar to those found in human mitochondria. These genes are highly expressed in tobacco plastids, which may suggest the existence of a respiratory chain in the chloroplasts of higher plants.

The plastid genome shows a number of prokaryotic features. These include the clustering and co-transcription of the functionally related genes. The order of such genes is often similar to that in prokaryotes such as *E. coli*. For example, the cluster of 11 ribosomal protein genes in the plastid genome is identical to the gene order in the S-10-spc-α operon of *E. coli*. Another prokaryotic feature is the presence of a number of overlapping genes. Normally the two gene products of the overlapping genes are found to be part of the same protein complex, as in prokaryotes. One non-prokaryotic feature found in the

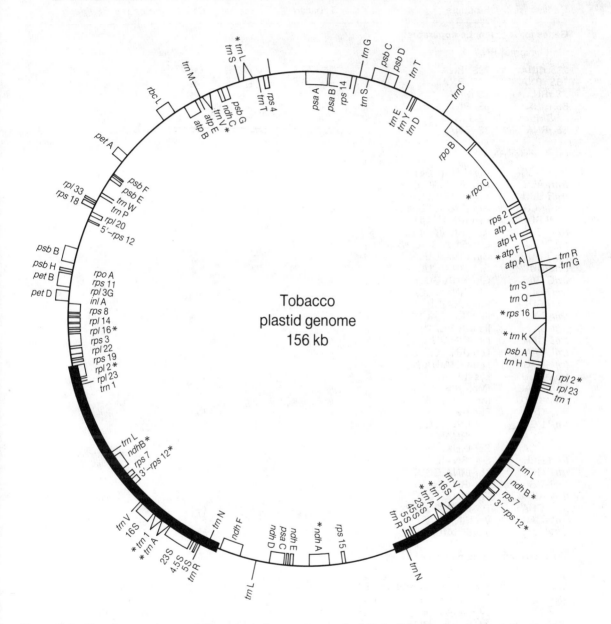

Figure 9.3. Circular gene map of the chloroplast genome of tobacco (*Nicotiana tabacum*). Inverted repeats are shown by bold lines. Genes shown inside the circle are transcribed clockwise and genes shown outside the circle are transcribed counterclockwise. Asterisks indicate split genes. Gene names are explained in Table 9.1. Reproduced with permission from M. Sugiura, 1989, in *The Biochemistry of Plants*, A Marcus (ed.), Academic Press, San Diego, pp. 133–150.

Table 9.1. Plastid genes.

Genes[a]	Products	Tobacco	Pea	*Chlamydomonas*	*Euglena*
Genes for the genetic apparatus					
(i) Ribosomal RNA					
23*S* rRNA	23S rRNA	+ +	+	* *	+ + +
16*S* rRNA	16S rRNA	+ +	+	+ +	+ + + +
7*S* rRNA	7S rRNA	−	−	+ +	−
5*S* rRNA	5S rRNA	+ +	+	+ +	+ + +
4.5*S* rRNA	4.5S rRNA	+ +	+	−	−
3*S* rRNA	3S rRNA	−	−	+ +	−
(ii) Transfer RNA					
trnA-UGC	Ala-tRNA(UGC)	* *	*	+ +	+ + +
trnR-ACG	Arg-tRNA(ACG)	+ +	+		+
trnR-UCU	Arg-tRNA(UCU)	+	+		
trnR-CCGΔ	Arg-tRNA(CCG)	−			
trnN-GUU	Asn-tRNA(GUU)	+ +	+	(+)	+
trnD-GUC	Aps-tRNA(GUC)	+	+		+
trnC-GCA	Cyc-tRNA(GCA)	+			+
trnQ-UUG	Gln-tRNA(UUG)	+			+
trnE-UUC	Glu-tRNA(UUC)	+	+		+
trnG-GCC	Gly-tRNA(GCC)	+	+		+
trnG-UCC	Gly-tRNA(UCC)	*	*		+
trnH-GUG	His-tRNA(GUG)	+	+	(+)	+
trnI-GAU	Ile-tRNA(GAU)	* *		+ +	+ + +
trnI-GAU	Ile-tRNA(CAU)	+ +			
trnL-UAA	Leu-tRNA(UAA)	*		(+)	+
trnL-CAA	Leu-tRNA(CAA)	+ +	+		
trnL-UAG	Leu-tRNA(UAG)	+			+
trnK-UUU	Lys-tRNA(UUU)	*	*		+
trnfM-CAU	fMet-tRNA(CAU)	+	+		+
trnM-CAU	Met-tRNA(CAU)	+			+
trnF-GAA	Phe-tRNA(GAA)	+		(+)	+
trnP-UGG	Pro-tRNA(UGG)	+	+	(+)	
trnS-GGA	Ser-tRNA(GGA)	+			
trnS-UGA	Ser-tRNA(UGA)	+			
trnS-GCU	Ser-tRNA(GCU)	+			+
trnT-GGU	Thr-tRNA(GGU)	+	+		
trnT-UGU	Thr-tRNA(UGU)	+			+
trnW-CCA	Trp-tRNA(CCA)	+	+		+
trnY-GUA	Tyr-tRNA(GUA)	+	+	(+)	+
trnV-GAC	Val-tRNA(GAC)	+ +	+		
trnV-UAC	Val-tRNA(UAC)	*			+
(iii) Ribosomal proteins					
rps2	30S ribosomal protein CS2	+	+		
rps3	30S ribosomal protein CS3	+			
rps4	30S ribosomal protein CS4	+			
rps7	30S ribosomal protein CS7	+		(+)	+
rps8	30S ribosomal protein CS8	+			
rps11	30S ribosomal protein CS11	+			
rps12	30S ribosomal protein CS12	* *			
rps14	30S ribosomal protein CS14	+			
rps15	30S ribosomal protein CS15	+			
rps16	30S ribosomal protein CS16	*			
rps18	30S ribosomal protein CS18	+			
rps19	30S ribosomal protein CS19	+	+		
rpl2	50S ribosomal protein CL2	* *	+	(+)	
rpl14	50S ribosomal protein CL14	+	+		

Table 9.1. *Continued*.

Genes[a]	Products	Tobacco	Pea	*Chlamydomonas*	*Euglena*
rpl16	50S ribosomal protein CL16	*		+	
rpl20	50S ribosomal protein CL20	+			
*rpl21*Δ	50S ribosomal protein CL21	−			
rpl22	50S ribosomal protein CL22	+			
rpl23	50S ribosomal protein CL23	++			
rpl33	50S ribosomal protein CL33	+			
rpl36	50S ribosomal protein CL36	+	+		
rpoA	RNA polymerase, subunit α	+	+	(+)	+
rpoB	RNA polymerase, subunit β	+	+	(+)	
rpoC	RNA polymerase, subunit β'	*	*	(+)	
tufA	Elongation factor TU	−	(+)	*	
infA	Initiation factor 1	+			

Genes for the photosynthetic apparatus

(i) Ribulose bisphosphate carboxylase (Rubisco)

Genes	Products	Tobacco	Pea	*Chlamydomonas*	*Euglena*
rbcL	Rubisco, large subunit	+	+	+	+
rbcS	Rubisco, small subunit	−	−	−	−

(ii) Photosystem I complex (PS I)

Genes	Products	Tobacco	Pea	*Chlamydomonas*	*Euglena*
psaA	PS I, p700 apoprotein A1	+	+	*	*
psaB	PS I, p700 apoprotein A2	+	+	+	(+)
psaC	PS I, 9 kDa polypeptide	+	+		

(iii) Photosystem II complex (PS II)

Genes	Products	Tobacco	Pea	*Chlamydomonas*	*Euglena*
psbA	PS II, D1 protein	+	+	**	*
psbB	PS II, 47 kDa polypeptide	+	+	(+)	(+)
psbC	PS II, 43 kDa polypeptide	+	+	(+)	*
psbD	PS II, D2 protein	+	+	+	
psbE	PS II, cytochrome b_{559} (8 kDa)	+	+		
psbF	PS II, cytochrome b_{559} (4 kDa)	+	+		
psbG	PS II, G protein	+			
psbH	PS II, 10 kDa phosphoprotein	+	+		

(iv) Cytochrome b/f complex

Genes	Products	Tobacco	Pea	*Chlamydomonas*	*Euglena*
petA	b/f complex, cytochrome f	+	+		
petB	b/f complex, cytochrome b_6	*	*		
petD	b/f complex, subunit IV	*	*		

(v) ATP synthetase complex

Genes	Products	Tobacco	Pea	*Chlamydomonas*	*Euglena*
atpA	H^+-ATPase, subunit α	+	+	+	(+)
atpB	H^+-ATPase, subunit β	+	+	+	(+)
atpE	H^+-ATPase, subunit e	+	+	(+)	
atpF	H^+-ATPase, subunit I	*	*	(+)	
atpH	H^+-ATPase, subunit III	+	+	(+)	+
atpI	H^+-ATPase, subunit IV	+	+		

(vi) Putative NADH dehydrogenase components

Genes	Products	Tobacco	Pea	*Chlamydomonas*	*Euglena*
ndhA	NADH dehydrogenase, ND1	*			
ndhB	NADH dehydrogenase, ND2	**			
ndhC	NADH dehydrogenase, ND3	+			
ndhD	NADH dehydrogenase, ND4	+	+		
ndhE	NADH dehydrogenase, ND4L	+			
ndhF	NADH dehydrogenase, ND5	+			

+, continuous genes present; *, split gene present; −, gene absent; (+), genes detected by heterologous hybridization. Numbers of + and * symbols indicate copies per genome.
[a]Translation products for *rps2–19*, *rpl2–36*, *rpoA–C*, *tufA*, *infA* and *ndhA–F* were not analysed.

plastid genome is the presence of introns. For example, more than 16 plastid genes have now been found to contain introns. Therefore, plastids must contain an RNA-splicing mechanism.

As well as the genome organization, the transcription and translation machinery of plastids is very similar to that of prokaryotic organisms. The plastid genome contains all 30 tRNA genes used in plastid protein synthesis. These 30 tRNAs are sufficient to read all codons in the plastid genome. All of the plastid tRNA sequences show closer homology to the corresponding bacterial tRNA than to the corresponding cytoplasmic tRNAs.

Plastids contain 70S ribosomes, which are characteristic of prokaryotic organisms, rather than the 80S ribosomes found in the cytoplasm of plant and animal cells. The 70S ribosomes found in plastids are made up of two main subunits of protein–RNA complexes, measuring 50S and 30S respectively. The 50S ribosomal subunit is made up of ribosomal RNA molecules of 23S, 5S and 4.5S, while the 30S subunit is made up of an RNA molecule of 16S. All of these ribosomal RNAs are encoded on the plastid genome. Approximately half of the ribosome by weight is made up of protein. The 50S ribosomal subunit contains between 33 and 35 proteins, while the 30S subunit contains between 25 and 31 proteins. Therefore, ribosomes contain between 58 and 66 proteins, of which 19 are encoded on the plastid genome. Therefore, in contrast to the ribosomal RNA, all of which is encoded on the plastid genome, less than one third of the ribosomal protein is encoded on the plastid genome. The reason for this discrepancy is not known.

The major part of the plastid genome encodes its own genetic apparatus or proteins which are involved in the photosynthetic apparatus in plant cells. Therefore, the vast majority of the plastid genome tends only to be expressed in photosynthetically active cells. Plastid gene expression is light regulated and is associated with the development of non-differentiated plastids into photosynthetically competent chloroplasts. A large proportion of chloroplast ribosomes is bound to thylakoid membranes during protein synthesis. This is especially true if the ribosomes are involved in the translation of hydrophobic proteins. This mechanism facilitates the direct insertion of protein into the thylakoid membranes.

Many plastid genes contain 5′ prime flanking regions in which prokaryotic-like promoter sequences are present. These regions have been suggested to be potential plastid gene promoters. The putative plastid promoters resemble prokaryotic promoters, but also have some unique features. For example, many plastid genes contain 5′ upstream elements which are similar or identical to the −10 and −35 promoter elements found in prokaryotic upstream sequences. Nevertheless, such elements are not found upstream in all plastid genes. These observations indicate that prokaryotic and plastid RNA polymerases do not have completely identical promoter sequence requirements. Other differences between the plastid and *E. coli* RNA polymerases have also been found. The bacterial enzyme is unable to initiate transcription in the presence of the antibiotic rifampicin, whereas the plastid enzyme is unaffected by this antibiotic. Additionally, the plastid RNA polymerase requires a super-coiled template in order to initiate transcription *in vitro*, while the *E. coli* enzyme will transcribe linear and relaxed plastid templates. The similarity of plastid genomes to those of prokaryotes has lent great support to the endosymbiotic hypothesis of the origin of plastids in plant cells. This hypothesis was originally proposed by Schimper in the 1870s and was developed more fully by Mereshkovsky in 1905. These scientists proposed that the green organelles of photosynthetic plant cells arose from a symbiotic association between a photosynthetic bacterium and a non-photosynthetic eukaryotic organism. It is likely that the plastids in present-day plants have at least a dual ancestry. The plastids of green algae and higher plants are probably derived from non-cyanobacterial, oxygenic bacteria exemplified by the present-day organism *Prochloron*. Red algae and cryptomonad plastids are probably derived from cyanobacterial-like organisms. It is suggested that a eukaryotic host cell, which would have contained a normal eukaryotic nucleus plus mitochondria, would have captured the endosymbiotic organisms by endocytosis. This would explain the double envelope membrane which surrounds the plastids of green algae and higher plants. A proposed model for the endosymbiotic origin of plastids is shown in Figure 9.4.

One obvious point that is not explained in the endosymbiotic hypothesis is why the plastid genome is now so much reduced. The original endosymbiotic organism must have had a genome which encoded all enzymes and proteins necessary for the existence of the organism. Almost all of the genes have now been lost from the relic organelle, the plastid. Presumably, there was a selective advantage in the loss of genes from the plastid genome, where these were duplicated by genes already present in the nuclear genome of the host cell. Other genes that must have originally been present in the endosymbiont, for example those encoding the small subunit of Rubisco and the light-harvesting complex proteins, have now been lost to the nuclear genome. This may enable the nuclear genome to exert a greater degree of control over the expression and development of the plastids that would be possible if the plastids were genetically and biochemically autonomous. There is now growing evidence for the existence of a mechanism by which DNA may be transferred between different genomes within a eukaryotic cell. Hence, mitochondria-like DNA sequences have been found in the chloroplast genome and mitochondria-like and plastid-like sequences are present in the nuclear genome. It is likely that DNA transfer occurs bidirectionally between organelles and nuclei. It appears that selection pressure has favoured the net transfer of genes from the plastids to the nuclear genome. A recent estimate is that the net rate of DNA transfer from mitochondria to nuclei is of the order of 100 000 times faster than the transfer rate from nuclei to mitochondria[4].

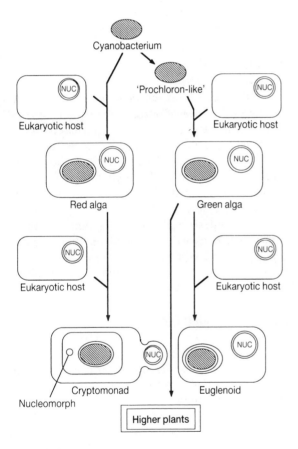

Figure 9.4. Scheme for the endosymbiotic origin of plastids in plants and algae. NUC, nucleus; N, nucleomorph. Plastids and their prokaryotic ancestors are represented by hatched bodies. Note that following each endocytotic event the plastid becomes surrounded by an extra membrane originally derived from the plasmalemma of the host origin. Hence the presence of a triple plastid envelope in euglenoids may be due to two endobiotic events.

MITOCHONDRIAL GENOME ORGANIZATION

In contrast to plant plastid genomes, which are relatively similar to one another, the plant mitochondrial genome is exceptionally diverse in both size and organization. In some plant species, mitochondria contain low-molecular-weight circular and linear DNAs and both single and double-stranded RNAs. There appears to be no relationship between mitochondrial genome size and genetic complexity. The mitochondrial genetic code may also exhibit at least one difference from the universal code.

Studies of the organization of mitochondrial genomes are complicated by their large size and multipartite organization. Mitochondrial genome sizes of up to 2400 kbp have been measured in the case of musk-melon, although a more typical size is between 200 and 500 kbp. Most mitochondrial genomes already studied consist of circular molecules, with a variable number of repeating elements. In addition to their main chromosomes, a number of plant mitochondria contain small circular DNA species referred to as mini-circular DNAs, or extra-chromosomal plasmids. These plasmids can take up three distinct topological states, i.e. super-coiled, relaxed circular and linear. The mitochondrial plasmids are invariably unrelated to the main mitochondrial chromosome with regard to their DNA sequences. This makes their origins and genetic functions uncertain and it is now believed that such plasmids make no genetic contribution either to the mitochondrion or to the plant as a whole. The plant mitochondrial genome, like all mitochondrial genomes, encodes all of the ribosomal RNAs found in mitochondria. It is also probable that the mitochondrial genome encodes a complete set of tRNAs. Also, in analogy with the plastid genome, the mitochondrial genome encodes many membrane proteins involved in electron transport and ATP synthesis. Finally, the mitochondrial genome contains a series of unidentified open reading frames, non-functional genes and non-functional transcribed sequences.

Mitochondrial genome organization and transcription and translation machinery is similar to that of prokaryotes such as *E. coli*. Like the plastid genome, however, the mitochondrial genome has a number of anomalous attributes, which are not found in present-day prokaryotes. For example, at least three mitochondrial genes contain introns.

It is likely that mitochondria, like plastids, arose from an ancestral endosymbiont. Recent evidence from the comparison of DNA sequences for conserved proteins such as the ATP synthetase subunits of mitochondria leads to the conclusion that mitochondrial genomes have a polyphyletic origin. This implies that the eukaryotic host organisms which eventually gave rise to plants may have acquired their mitochondria at different times to the ancestors of the present-day animal and yeast lineages. Mitochondria are closely related to the present-day purple non-sulphur photosynthetic bacteria. The ancestors of such bacteria may have been the original mitochondrial endosymbionts and may have lost their photosynthetic capacity only after capture by the eukaryotic host cell.

PROTEIN SYNTHESIS

As outlined above, the existence of three separate genomes in plants dictates that there are three separate sites of protein synthesis, namely the cytoplasm, mitochondria and chloroplasts. Nuclear-encoded proteins are translated in the cytoplasm by a mechanism which is broadly similar to that found in animals. Reflecting their probable prokaryotic origins, mitochondrial and chloroplast protein synthesis is very similar to that found in eubacteria. A detailed description of protein synthesis will not be carried out here, rather emphasis will be placed on those aspects unique to plants and on the differences between organellar and cytoplasmic modes of synthesis.

Nuclear-encoded Proteins

Messenger RNA for nuclear-encoded proteins is translated by ribosomes found in the cytoplasm. The details of ribosome structure are essentially identical in all eukaryotes. The overall size of the ribosomes is in the range of 80–87S. The large subunit of 60S contains rRNA species of 25S, 5.8S and 5S and 49 different proteins. The small subunit is 40S in size with an 18S rRNA and contains 33 proteins.

Initiation

Translation in all organisms can be divided into three basic steps: initiation, elongation and release. One of the major differences between animal and plant protein synthesis is found in the stages of initiation. The complex series of steps involving multiple accessory initiation factors (eIFs or eukaryotic initiation factors) has best been described in animal systems. This will be described first, and then a brief outline of where plants differ will follow.

In the cytoplasm is a large pool of ribosomes not always involved in active protein synthesis. The inactive 40S and 60S subunits will bind to each other with high affinity to form inactive

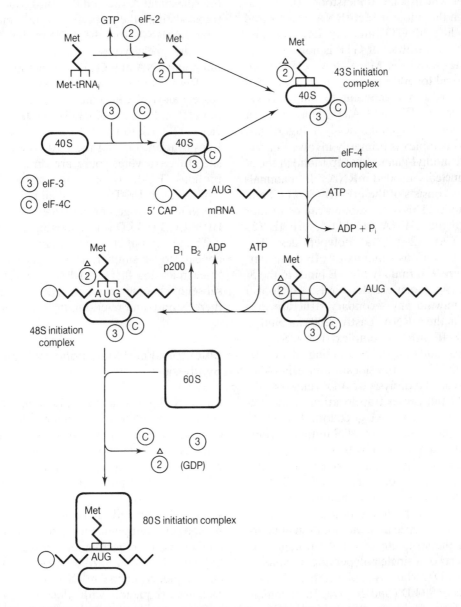

Figure 9.5. The initiation of protein synthesis. The details of the eIF-4 complex are omitted for clarity. 40S and 60S refer to the small and large ribosomal subunits, respectively. Refer to the text for details.

complexes unless kept apart. This is achieved by eIF-3, which binds to the 40S subunit. This initiation factor acts as an anti-binding factor. It may also be involved in termination of translation, again to keep the 40S and 60S subunits apart when translation finishes. The 40S–eIF-3 complex is stabilized by another initiation factor, eIF-4C, in a manner which is not understood. In a separate reaction the initiator Met-tRNA$_f$ is activated by the binding of GTP and another initiation factor, eIF-2. Hydrolysis of GTP is not required. The 40S–eIF-3 and the Met-tRNA$_f$–eIF-2 complexes can bind together to form an activated 43S particle that is now competent to bind mRNA with the help of eIF-4 and ATP. This series of reactions is shown diagrammatically in Figure 9.5.

The eIF-4 complex is primarily involved in recognizing the methyl-guanosine cap found at the 5' end of all nuclear-encoded mRNAs. In mammals this complex consists of the eIF-4B (80 kDa) and eIF-4F subunits. This latter subunit can be further divided into the eIF-4A (46 kDa), eIF-4E (25 kDa) and p200 (200 kDa) polypeptides. The actions of the various subunits of eIF-4 are outlined in Figure 9.6. Initially eIF-4E binds to the 5' cap of mRNA[5]. Then eIF-4A binds, which also appears to unwind any secondary structure present locally in the mRNA. Lastly p200 can bind to form the eIF-4F–mRNA complex. The 43S initiation complex can now bind. Binding of eIF-4B causes the mRNA to translocate and eIF-4A to bind to the cap. Hydrolysis of ATP releases eIF-4B and p200 and causes translocation of the 43S complex to the initiator AUG codon. The whole complex is now known as the 48S initiation complex. The 60S subunit can now bind. This requires eIF-5. Successful binding causes release of all the eIFs and hydrolysis of GTP. The 80S–Met-tRNA–mRNA complex is now ready to enter the elongation phase of protein synthesis.

The major difference in plant initiation appears to reside in the structure of eIF-4. In mammals eIF-4B consists of a single polypeptide of molecular weight 80 kDa, whereas in plants there are two polypeptides of 80 kDa and 28 kDa. In mammals eIF-4F consists of three polypeptides: eIF-4A, eIF-4E and p200. In plants there appears to be no equivalent of eIF-4A.

Elongation and Release

The process of elongation of the polypeptide chain appears to be little different between plants and animals, and mechanistically similar in prokaryotes. There are three elongation factors: two concerned with loading the amino-acyl tRNA into the ribosomal A site and the third concerned with translocating the ribosome and moving the peptidyl-tRNA from the A site to the ribosomal P site. In prokaryotes elongation factor EF-Tu binds aa-tRNA and GTP. This complex loads the aa-tRNA into the A site. Successful loading of the correct aa-tRNA into the A site causes hydrolysis of GTP and release of EF-Tu–GDP. EF-Ts recycles GTP back to EF-Tu. EF-G causes ribosomal translocation, moving the next codon into the A site. In eukaryotes there are directly analogous proteins. EF-1a (molecular weight 53 kDa) is equivalent to EF-Tu. EF-1b (30 kDa) plus EF-1g (53 kDa) are together equivalent to EF-Ts, and lastly EF-2 (80 kDa) is analogous to EF-G.

The process of chain termination in eukaryotes appears to be simpler. In prokaryotes there are three releasing factors (RF), each recognizing a subset of chain termination codons. In eukaryotes there is only one releasing factor, of 56.5 kDa.

Chloroplast and Mitochondrial Protein Synthesis

Little is known about mitochondrial protein synthesis in plants. This is mainly due to the lack of an *in vitro* system by which the process can be fractionated and analysed. Isolated mitochondria can be shown to synthesize a defined set of proteins. The rRNAs are very similar to those found in eubacteria, suggesting that the whole process is similar. Mitochondrial protein synthesis is inhibited by chloramphenicol but resistant to cycloheximide, again reflecting the similarities to prokaryotes. In bacteria, ribosomes recognize a specific sequence in the mRNA, called the Shine–Dalgarno sequence, with which to bind and thus initiate peptide synthesis. Shine–Dalgarno-like sequences have been observed in plant mitochondrial mRNA.

Protein synthesis has been more extensively studied in chloroplasts. One difference between mitochondria and chloroplasts is the site of synthesis of elongation factors: chloroplasts synthesize their own whereas mitochondrial EFs are nuclear encoded. However, research again shows the overall similarities with eubacteria. The ribosomes are 67–70S and there are similar rRNAs, 23S, 5S and 4.5S in the large 50S subunit and 16S in the small 30S subunit. The similarities extend to the ribosomal proteins: antibodies to chloroplast ribosomes cross-react with low-molecular-weight *E. coli* ribosomal proteins. This shows a high degree of functional and structural relatedness.

Similar to mitochondria, chloroplast protein synthesis is sensitive to those antibiotics that inhibit bacteria, such as chloramphenicol, strep-

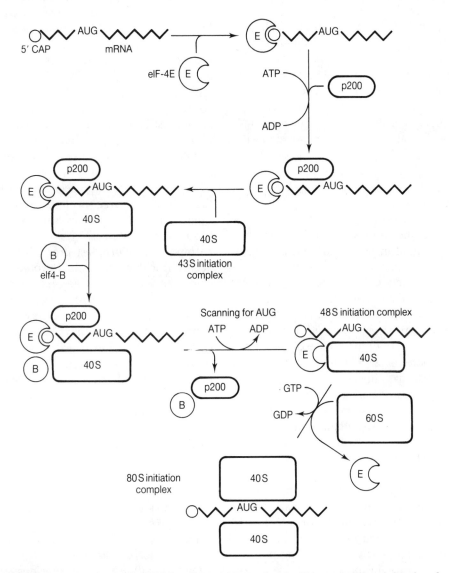

Figure 9.6. The role of eIF-4 in the initiation of protein synthesis. The remaining initiation factors and the methionyl-tRNA are omitted for clarity. The 43S initiation complex is the same as in Figure 9.5. Refer to the text for details. After R.E. Rhoads, 1988, *Trends Biochem. Sci.*, **13**, 52–56, by permission of Elsevier.

tomycin and linomycin. A notable exception to this general rule is the complete resistance of chloroplast RNA polymerases to rifampicin at concentrations far above those needed to inhibit the *E. coli* enzyme. This reflects the much simpler structure of chloroplast RNA polymerase in comparison with the bacterial enzyme.

PROTEIN SORTING

The vast majority of proteins within a cell are synthesized within the cytoplasm from nuclear-encoded transcripts. Yet their final subcellular location can be in one of a whole array of membrane-bound compartments. The means by which these proteins arrive at their final destination is the subject of protein sorting. The mechanisms differ for the various organelles, so each will be taken in turn.

Targeting to Plastids

Chloroplasts are the most structurally complex type of plastid found in plant cells and amongst the most complex organelles in a eukaryotic cell. There are six functionally distinct locations, namely the outer and inner envelope membranes, the intermembrane space, the stroma, the thylakoid membrane and the thylakoid lumen, as shown in Figure 9.7. All of these have distinct sets of proteins. There is also a diverse array of biochemical pathways in chloroplasts besides photosynthesis. Chloroplasts are responsible for fatty acid synthesis, starch metabolism, nitrite reduction and ammonia assimilation and also for some amino acid biosynthesis. To complicate the matter further, chloroplasts are developmentally plastic, hence the name plastid given to chloroplasts and related organelles. In different tissues they can exist as amyloplasts, responsible for starch metabolism, etioplasts in etiolated leaf and steam tissue and chromoplasts in flower petals.

The plastid genome has limited coding potential so most of the proteins present in these organelles are encoded for by nuclear genes and synthesized on cytoplasmic ribosomes. Thus all such proteins must cross at least one and possibly as many as three membranes in order to reach their final subcellular location.

Investigating Chloroplast Protein Import

Both *in vivo* and *in vitro* methods have been used to investigate the correct sorting of chloroplast proteins. *In vivo* studies have been greatly aided by recombinant DNA techniques and the ability to transform plants routinely. The main approach has been to fuse various protein domains from chloroplast proteins with a simpler reporter protein, such as the bacterial protein chloramphenicol acetyl transferase (CAT). This can readily be accomplished via gene fusions. After transformation of these chimeric genes into plants, the subcellular location of the reporter protein is determined by methods such as immunogold labelling of electron microscope sections. In this manner the effects of large protein domains down to single amino acid changes upon correct targeting can be investigated.

Import of chloroplast proteins *in vitro* can be achieved with radiolabelled proteins and purified chloroplasts. The final destination of the protein can again be determined using fractionation techniques coupled with protease protection experiments. Proteins that have successfully entered chloroplasts become resistant to digestion with proteases such as proteinase K. These experiments have also been aided by recombinant DNA methods. Cloned genes for chloroplast proteins can be transcribed and translated *in vitro* to give single radiolabelled proteins of high specific activity, thus greatly increasing the sensitivity of this technique.

Precursors of Chloroplast Proteins

All nuclear-encoded chloroplast proteins are initially synthesized as large precursor molecules. They have N-terminal extensions of 20 amino acids and longer which are not found in the

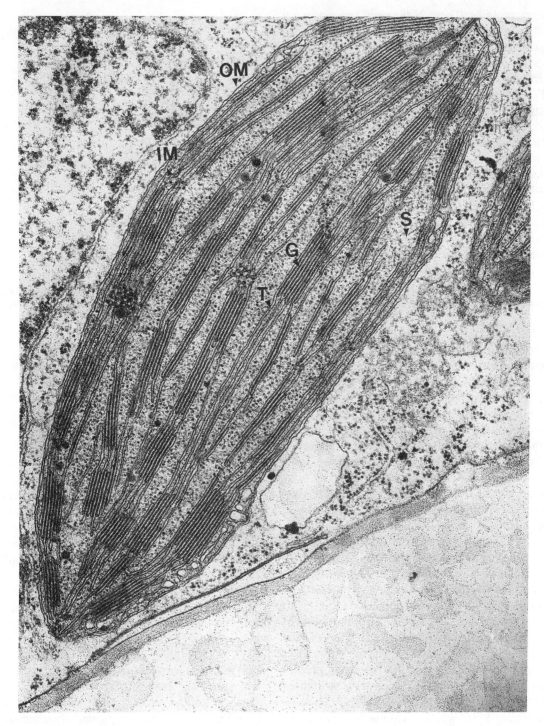

Figure 9.7. An electron micrograph of a chloroplast. OM, outer membrane; IM, inner membrane; S, stroma; T, thylakoid; G, granal stack.

mature proteins. These N-terminal sequences act as chloroplast-addressing domains. To differentiate them from the functionally similar signal peptides of secreted proteins, these chloroplast-addressing domains are called transit peptides (TPs). Upon entry to the chloroplast the transit peptide is removed by proteolysis to produce the mature protein. That these transit peptides are both necessary and sufficient to cause targeting of chloroplast proteins to the stroma was demonstrated by the observation that precursors of the small subunit of Rubisco (pre-ssRubisco) could not be targeted to the chloroplast without the transit peptide. Furthermore the transit peptide alone was sufficient to target foreign reporter proteins to the chloroplast.

Structure and Specificity of Transit Peptides

Analysis of many chloroplast transit peptides shows that they have little overall amino acid homology, although the types and distribution of amino acids show a distinctive distribution. Transit peptides are rich in basic, hydroxylated and hydrophobic amino acids, with an absence of acidic residues. This overall pattern is very similar to that found for mitochondrial transit peptides, so how do plant cells which always have both types of organelle differentiate between the two? The answer to this question was somewhat confused by experiments with the TP of ssRubisco from *Chlamydomonas rheinardii*, a motile unicellular alga. This TP mistargeted ssRubisco in spinach and pea and caused the targeting to mitochondria in yeast. However, recent experiments have clarified the situation. The bacterial reporter protein CAT was correctly targeted to the appropriate organelle by a tobacco mitochondrial signal peptide from the ATPase β-subunit or a chloroplast TP from ssRubisco in transgenic plants. Von Heijne[6] has made predictions of transit peptide secondary structure. Mitochondrial TPs form amphipathic α-helices (see Figure 9.8), whereas chloroplast TPs form amphipathic β-strands. The *Chlamydomonas* TP for ssRubisco is intermediate between these struc-

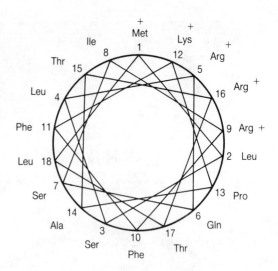

Figure 9.8. A helical wheel of the transit peptide of the mitochondrial enzyme cytochrome oxidase IV. The amphipathic nature of the alpha helix is clearly seen, with all of the basic residues (+) aligned on one surface of the helix.

tures and thus accounts for its mistargeting in higher plants.

Targeting of Proteins to the Stroma

The TP alone will target proteins to the stroma. The proteins enter the chloroplast at defined points where the inner and outer envelope membranes meet, so-called contact sites. Energy in the form of ATP is required for translocation across the contact sites. The initial step in translocation is binding of the precursor proteins to the chloroplast. Protease treatment of intact chloroplasts inhibits binding and prevents uptake of proteins. This suggests that a protein receptor is required for binding. Recently a potential receptor has been identified as an integral membrane protein of 36 kDa by binding of an anti-idiotypic antibody that mimics a transit peptide. The potential receptor protein has eight membrane-spanning domains and an antibody to it will inhibit import of ssRubisco by interfering with binding of the TP. The protein has been localized to the contact sites. A second component of the receptor has been identified by using photoactivatable reagents to bind radioactive pre-ssRubisco to chloroplasts. This procedure identified a protein of 66 kDa. The relationship, if any, between the 66 kDa protein and the 36 kDa protein remains obscure.

The exact mechanism for translocation is unknown. The proteins have to unfold in order to cross the envelope membranes and refold within the stroma. There are cytoplasmic proteins involved in unfolding and these are members of the 70 kDa heat-shock protein family (hsp70). Within the stroma there is another hsp70 and also a 60 kDa hsp60 involved in refolding. The hsp60 is related to the *E. coli* enzyme GroEL and has been given the general name of chaperonin. This protein is also involved in the correct assembly of the large and small subunits of Rubisco into the holoenzyme. Experiments have shown association of chaperonin with seven different proteins, both natural and foreign reporter proteins. Thus chaperonins appear to be involved in general protein assembly in most cells. Once the TP has entered the stroma and during the process of translocation it is cleaved by a stromal protease of 180 kDa.

Targeting to the Envelope Membranes and the Intermembrane Space

To date, no proteins from these compartments have been studied in detail for chloroplasts. The only information available is that taken by reference to mitochondria, which are presumed to function using similar mechanisms. Proteins for the inner envelope membrane which protrude into the intermembrane space (IMS), or those that are soluble within the IMS, are first targeted to the stroma with a normal chloroplast-addressing TP. This is then removed by the stromal protease. A second targeting sequence then sorts the proteins back through the inner envelope membrane. This second targeting sequence is hydrophobic and, if not removed, anchors the protein to the inner envelope membrane, with the bulk of the protein protruding into the IMS

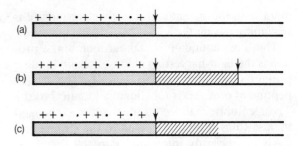

Figure 9.9. Schematic transit peptides for nuclear-encoded chloroplast proteins. Targeting to (a) the lumen. Plus signs represent basic residues, and dots hydroxy residues. The arrow represents the site of proteolytic cleavage: (b) the inter-membrane space (IMS). The striped region represents a helical hydrophobic domain which is cleaved in the IMS by a protease: (c) the inner membrane. The hydrophobic domain is not cleaved off in the IMS, thereby anchoring the protein to the inner membrane.

(see Figure 9.9). If the targeting sequence is removed by a processing protease within the IMS then it becomes a soluble protein. Proteins for the outer envelope membrane (OM) appear to be targeted directly from the cytoplasm. Indeed one OM chloroplast protein has recently been described. It is a 6.7 kDa protein that can be proven to be located to the OM by cell fractionation and protease K sensitivity. The N-terminal domain is in the IMS. There is no cleavable TP and the protein does not require ATP for insertion, unlike stromal proteins. It is thus inserted into the chloroplast by a mechanism different from other proteins.

Targeting to the Thylakoid

Unique to chloroplasts there is a third membrane-bound compartment within the stroma, namely the thylakoids. These are the sites of photosynthetic electron transport and photophosphorylation and have specific proteins for this purpose. Many of these proteins are nuclear encoded and must cross both chloroplast envelope membranes and the thylakoid membrane for correct targeting. Less is known about targeting to this compartment than the stroma but a fairly clear picture is beginning to emerge.

The most abundant thylakoid membrane protein is the light-harvesting chlorophyll a/b binding protein (CAB) of photosystem II. The transit peptide of preCAB is functionally identical to the TP of ssRubisco. Indeed if the two are exchanged then ssRubisco is still targeted to the stroma and CAB is efficiently integrated into the thylakoid membrane. The import into thylakoids can be reconstituted *in vitro*; this requires ATP and a protein of approximately 65 kDa. The relationship of this protein to chaperonin has not been tested. The conclusion from these studies is that a domain within mature CAB must be involved in import to thylakoids. A series of experiments by Kohorn and Tobin[7] allowed the dissection of CAB. They propose that there are three membrane-spanning α-helical domains within CAB. Helix 3 is near the C-terminus which protrudes into the thylakoid lumen. Fusion of helix 3, but not helices 1 and 2, to ssRubisco causes its association with the thylakoid membranes. A deletion mutant of CAB lacking helix 3 was unable to associate with thylakoids. Other deletions of amino acids proximal to the amino terminal end allow thylakoid association, but CAB is inserted incorrectly into the membrane. Thus helix 3 is required for targeting, but other sequences are required for correct assembly within the thylakoid membrane. This proposed structure is debated—others propose that there are more than three membrane-spanning domains, but all agree that one of these is responsible for targeting. Perhaps the best-studied thylakoid luminal protein is plastocyanin (PCy). This protein is a small electron carrier used during photosynthesis. The presequence of PCy is rather long, 66 amino acids, in comparison with TPs of stromal proteins. Analysis of this presequence showed that it could be divided into two domains. The first part is rich in basic and hydroxy amino acids and closely resembles the organization of stromal TPs. The C-terminal part is more hydrophobic and resembles the signal peptides of secretory proteins of both eukaryotes and prokaryotes. This division of the presequence into two domains is supported by

experimental evidence that shows that the presequence is cleaved in two successive reactions. First, the stromal peptidase removes the N-terminal domain that resembles a TP. Indeed this domain functions solely as a normal stromal targeting TP. Second, the C-terminal hydrophobic domain is cleaved by a second peptidase found in the thylakoid membrane. This peptidase has been partially purified and is clearly distinct from the stromal enzyme. This second domain can also act independently of the first and is solely responsible for targeting to the thylakoid lumen. Thus it is now clear that the presequence of plastocyanin acts as a bipartite TP, with two functionally independent domains, the first responsible for stromal and the second for luminal targeting (see Figure 9.10). The mechanism of translocation across the thylakoid membrane is unknown, but may resemble secretion in bacteria.

Unlike plastocyanin, another thylakoid luminal protein, the 33 kDa oxygen-evolving protein (OEP1), is complexed with photosystem II. To investigate if this protein is sorted to the lumen by a similar mechanism to plastocyanin, Ko and Cashmore[8] analysed the 85-amino acid presequence of OEP1 by fusing it to reporter proteins and mutational analysis. Substituting the entire presequence for that of ssRubisco predictably caused the targeting of OEP1 to the stroma. Deletion of

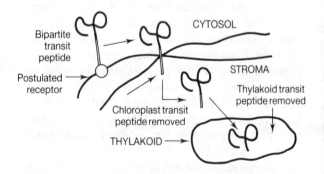

Figure 9.10. The two-stage import process of thylakoid luminal proteins. Reproduced from B. Alberts e al., 1989, *Molecular Biology of the Cell* (2nd edn), Garland Publishing Inc., New York, by kind permission of K. Roberts.

the first 58 amino acids prevented accumulation in chloroplasts, whereas this domain alone formed a chloroplast-addressing, stromal-targeting sequence. The entire presequence alone was capable of targeting the reporter protein, mouse dihydrofolate reductase (DHFR), to the thylakoid lumen. Thus it appears that just as in plastocyanin, the OEP1 contains a bipartite transit peptide.

Several luminal proteins are encoded on the chloroplast genome. These are presented with the same problem as nuclear-encoded luminal proteins which have reached the stroma. One such chloroplast-encoded protein is cytochrome f. This protein is anchored in the thylakoid membrane with a C-terminal α-helical membrane-spanning domain. The bulk of the protein is in the lumen, where it binds haem. There is an N-terminal cleavable signal sequence which presumably acts as the lumen-targeting domain. As for plastocyanin, the targeting to the lumen resembles secretion in bacteria. The lumen-sorting domains resemble bacterial signal sequences. The similarities are further reinforced by the observation that when cytochrome f is expressed in *E. coli* it behaves as a membrane-bound protein which requires the bacterial SecA-mediated secretory pathway for insertion.

The Secretory and Vacuolar Pathways

In animals cells, the secretory pathway forms an elaborate series of membrane-bound structures that account for a large proportion of the total protein and membrane complement of a cell. The system starts with the endoplasmic reticulum and includes the Golgi apparatus, the lysosome and the plasma membrane. The same is true in plants but here the functionally similar vacuole replaces the lysosome, as shown in Figure 9.11. In animals extensive investigation has started to show how this system is organized and also how complex it is. This has been aided by genetic studies in simple eukaryotes such as the yeast *Saccharomyces cerevisiae*. In plants, research is as yet at a much earlier stage.

Secretion

Secretion is the default targeting pathway through the endoplasmic reticulum (ER) and Golgi network in animal cells. Proteins enter the ER cotranslationally on membrane-bound ribosomes. The vast majority of secreted proteins have N-terminal signal sequences of about 20–25 amino acids. These are recognized by a complex ribonucleoprotein called the signal recognition particle (SRP) on nascent polypeptides during translation (Figure 9.12). The SRP then complexes with an ER integral membrane protein called the SRP receptor or 'docking protein'. The signal peptide is translocated through the membrane into the lumen of the ER, whereupon the SRP and docking protein dissociate. Translation continues until all of the protein has entered the lumen of the ER or become inserted into the ER membrane. The signal peptide is removed by a specific signal peptidase. Other modifications include core glycosylation at specific Arg–X–Ser/Thr motifs. In the absence of any other sorting information within the protein, it travels through the ER, then the Golgi apparatus, and finally is secreted. Integral membrane proteins become cell surface proteins bound to the plasmalemma.

A series of recent experiments show that secretion is the default targeting pathway also in plants. The bacterial neomycin phosphotransferase gene (NPTII) was fused to signal peptide-coding regions from several sources. After expression in tissue culture and whole plants, the NPTII activity was found to be secreted. Other reporter genes such as bacterial β-glucuronidase have been found to be both secreted and glycosylated. Thus in most respects the default secretory pathway appears to be the same in both plants and animals, and many details of this pathway found in animals will also hold true for plants.

Sorting to the Vacuole

Trafficking of proteins to the vacuole requires positive sorting information to divert proteins at the Golgi apparatus from the default secretory pathway. The vacuole is the lysosome-like organelle of the plant cell. It also has other functions such as maintaining the turgor of the cell and being a site for the accumulation of secondary metabolites. There are, however, two major differences between vacuoles and mammalian lysosomes. First, vacuolar development is very plastic and varies from tissue to tissue. This plasticity is particularly seen in seed cotyledons. Here the vacuoles are not lytic but serve as a repository for seed storage proteins. These proteins are laid down in the vacuole during seed development.

The vacuole then fragments as it becomes filled with protein, giving rise to protein bodies. The protein serves as an amino acid, nitrogen and energy store for future seed germination. Second, sorting of proteins to mammalian lysosomes is well characterized and depends on the formation of mannose 6-phosphate residues on the glycan side chains of lysosomal glycoproteins. Plants do not use this system, indeed glycosylation *per se* is not necessary for correct sorting to the vacuole. Thus sorting information must reside within the amino acid sequence of the protein itself.

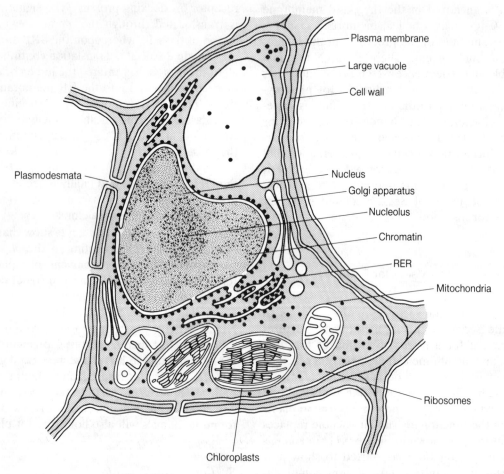

Figure 9.11. A generalized plant cell. The relationship between the outer membrane of the nuclear envelope, the rough endoplasmic reticulum (RER), the Golgi apparatus, the vacuole and the plasma membrane can clearly be seen. These constitute the organelles and membranes of the secretory pathway. Reproduced with permission from P. Sheeler and D.E. Bianchi, 1983, *Cell Biology, Structure, Biochemistry and Function*, John Wiley & Sons Inc., New York.

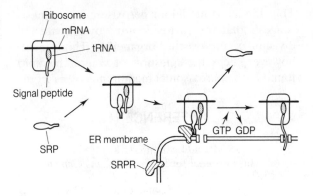

Figure 9.12. The interaction of the signal peptide recognition particle (SRP), ribosomes and the SRP receptor (SRPR).

Yeast also uses protein-sorting domains to target proteins to the vacuole. This prompted Tague et al.[9] to analyse sorting information on the plant storage protein phytohaemagglutinin-L (PHAL) within yeast. The best-characterized yeast vacuolar protein is carboxypeptidase Y (CPY), which has a 91-residue presequence. At the N-terminus end of this sequence is the four-amino acid motif QRPL (glutamine, arginine, proline, leucine), which is both necessary and sufficient for vacuolar targeting. PHAL when expressed in yeast is sorted to the vacuole. Analyses of the amino acid sequence revealed the motif LQRD near the N-terminus end, which is involved in targeting PHAL to the yeast vacuole. Subsequent experiments to test this motif in plant cells have met with mixed results. It thus appears that vacuole sorting is more complex in plants than in yeast. Recent evidence has shown that the 16-amino acid N-terminal propeptide of sporamin from sweet potato is responsible for vacuolar targeting, whereas a C-terminal 15-amino acid sequence is important in the targeting of barley lectin found in protein bodies.

Mitochondria

Little work has been done on protein sorting to plant mitochondria. The basic assumption has been made that the mechanism is similar if not identical to that found in the well-characterized mitochondria of *Neurospora crassa* and yeast (for reviews see Schatz, 1987; Hartl *et al.*, 1989). As in these organisms, mitochondrial proteins are synthesized as higher-molecular-weight precursors. The gene for manganese superoxide dismutase was cloned and sequenced from tobacco. From the DNA sequence it was predicted that an N-terminal transit peptide was present in the immature protein which, based on the work of Schatz in yeast, predicted a mitochondrial-located enzyme. Subsequent cell fractionation confirmed this prediction. Similarly the gene for *Arabidopsis* mitochondrial citrate synthase has been cloned. The predicted 53 kDa preprotein when transcribed and translated *in vitro* is incorporated into isolated pea mitochondria, where it forms a 51 kDa mature protein. Thus all evidence to date confirms the assumption that there is little difference in the mitochondrial targeting mechanisms of yeast, *Neurospora* and plants[10].

Microbodies

Microbodies are small organelles about 0.5 µm in diameter, bounded by a single membrane, that are ubiquitous in eukaryotes. Although differences occur between microbodies from different species and tissues they all possess a primitive respiratory chain which produces hydrogen peroxide and also contain catalase for the decomposition of this hydrogen peroxide. Because of this they are frequently called peroxisomes. Besides the respiratory chain, an inducible fatty acid β-oxidation system is present but its activity varies considerably. In plants β-oxidation is exclusive to peroxisomes and is present at high activity in germinating cotyledons of oil seeds. Some specialized plant microbodies contain all five enzymes of the glyoxylate cycle and are thus called glyoxysomes.

Protein import into peroxisomes was originally thought to be via the ER. It soon became clear, however, that matrix proteins are synthesized on free ribosomes and there is no ER involvement.

A refinement to this hypothesis stated that pero-xisome membrane proteins come via the ER, but matrix proteins are imported directly. However, two microbody membrane proteins have been shown to be incorporated post-translationally: a 22 kDa protein of rat liver peroxisomes and glyoxysomal alkaline lipase of castor beans. Both of these are major membrane proteins and so could not be contaminants. Thus there is no reason to suppose that the ER has any role in the incorporation of protein into peroxisomes. This was confirmed by translation experiments *in vitro*, where a number of glyoxysomal integral mem-brane proteins failed to be incorporated into canine pancreatic microsomes, whereas other integral membrane proteins from protein bodies which are known to enter the ER were incor-porated.

What is the nature of the sorting signals within the amino acid sequence of peroxisomal proteins? Most are not synthesized as precursors and are incorporated as mature proteins. However, a few examples of peroxisomal proteins that do exist as precursors are known. The function of the pre-sequences is unclear, although one example is known where the presequence is involved in targeting. Water melon has two isozymes for malate dehydrogenase. One is mitochondrial and the other is glyoxysomal. While sequences of the mature enzymes are nearly identical, the presequ-ences differ markedly. The mitochondrial enzyme, mMDH, has a 27-amino acid transit peptide. The glyoxysomal enzyme, gMDH, has a 37-amino acid presequence, within which is found the sequence Ala–His–Leu, a potential peroxisome-targeting sequence. A common peroxisomal enzyme in diverse species, including plants, is luciferase. There is a highly conserved C-terminal sequence of Ser–Lys–Leu that appears to be both necessary and sufficient for peroxisomal targeting. An antibody to this sequ-ence recognizes peroxisomes in plant cells as viewed by EM immunogold labelling. The anti-body also binds to numerous peroxisomal pro-teins in Western blots. Analysis of other peroxiso-mal enzymes shows that this sequence can be Ser, Ala or Cys at the first position, Lys, Arg or His in the middle and only Leu in the third position.

Thus the latest model for peroxisomal biogenesis suggests that they arise *de novo* by growth and division of pre-existing organelles. This model, however, begs the question of where the lipid found in the peroxisomal membrane comes from.

REFERENCES

1. Gasser, S.M. and Laemmli, U.K. (1987) A glimpse at chromosomal order. *Trends in Genetics*, **3**, 16–22.
2. Heslop-Harrison, J.S. and Bennett, M.D. (1990) Nuclear architecture in plants. *Trends in Genetics*, **6**, 401–405.
3. Sugiura, M. (1989) The chloroplast genome. In *The Biochemistry of Plants*, Vol. 15 (ed. A. Mar-cus), pp 133–150. Academic Press, San Diego.
4. Thorsness, P.E. and Fox, T.D. (1990) Escape of DNA from mitochondria to the nucleus in *Saccharomyces cerevisiae*. *Nature*, **346**, 376–379.
5. Rhoads, R.E. (1988) Cap recognition and the entry of mRNA into the protein synthesis initiation cycle. *Trends in Biochemical Sciences*, **13**, 52–56.
6. Von Heijne, G., Stepphun, J. and Herrmann, R. G. (1989) Domain structure of mitochondrial and chloroplast targeting peptides. *European Journal of Biochemistry*, **180**, 535–545.
7. Kohorn, B.D. and Tobin, E.M. (1989) A hyd-rophobic, carboxy-proximal region of a light-harvesting chlorophyll a/b protein is necessary for stable integration into thylakoid membranes. *Plant Cell*, **1**, 159–166.
8. Ko, K. and Cashmore, A.R. (1989) Targeting of proteins to the thylakoid lumen by the bipartite transit peptide of the 33 kd oxygen-evolving pro-tein. *European Molecular Biology Organisation Journal*, **8**, 3187–3194.
9. Tague. B.W., Dickinson, C.D. and Chrispeels, M.J. (1990) A short domainof the vacuolar protein phytohemagglutinin targets invertase to the yeast vacuole. *Plant Cell*, **2**, 533–546.
10. Schmitz, U.K., Hodge, T., Walker, E. and Lons-dale, D. (1990) Targeting of proteins to plant mitochondria. *Plant Gene Transfer*, 237–248.

FURTHER READING

Genome Organization

Good general accounts of prokaryotic systems:

Alberts, B. *et al.* (1989) *Molecular Biology of the Cell* (2nd edn). Garland, New York.

Lonsdale, D. The plant mitochondrial genome. In *The Biochemistry of Plants, Vol. 15* (ed. A. Marcus), pp. 230–295. Academic Press, San Diego.

Palmer, J.D. (1992) Comparison of chloroplast and mitochondrial genome evolution in plants. In *Plant Gene Research Vol. 6: Organelles* (ed. R.G. Herrmann). Springer-Verlag, Berlin.

Watson, J.D. *et al.* (1987) *Molecular Biology of the Gene* (4th edn). Benjamin/Cummings, Menlo Park, California.

Protein Synthesis

Useful general descriptions of prokaryotic systems:

Alberts, B. *et al.* (1989) *Molecular Biology of the Cell* (2nd edn). Garland, New York.

Lewin, B. (1990) *Genes IV*. Oxford University Press, Oxford.

Chloroplast Targeting

A short readable account of chloroplast targeting:

Smeekens, S. *et al.* (1990) Protein transport into and within chloroplasts. *Trends in Biochemical Sciences,* **15**, 73–76.

Mitochondrial Targeting

A well-written short account of mitochondrial targeting:

Schatz, G. (1987) Signals guiding proteins to their correct locations in mitochondria. *European Journal of Biochemistry,* **165**, 1–6.

For an extensive detailed review:

Hartl, F.-U. *et al.* (1989) Mitochondrial protein import. *Biochimica et Biophysica Acta,* **988**, 1–45.

Protein Targeting

A detailed and well-written book:

Pugsley, A.P. (1989) *Protein Targeting*. Academic Press, London.

10

REGULATION OF GENE EXPRESSION

Nigel J. Robinson, Anil H. Shirsat and John A. Gatehouse

Department of Biological Sciences, University of Durham, UK

Plant Biochemistry and Molecular Biology. Edited by P.J. Lea and R.C. Leegood
© 1993 John Wiley & Sons Ltd

INTRODUCTION

The generation of a phenotype from a genotype is the manifestation of gene expression. This chapter considers some of the processes responsible for the accumulation of either protein or RNA following gene transcription. Phenomena such as protein turnover and post-translational protein modification are dealt with in other chapters.

The expression of specific plant genes increases or decreases in response to environmental factors and an inherent developmental programme. Environmental factors affecting plant gene expression include interactions with pathogens and abiotic factors such as light, temperature, anoxia, drought, nutrient excess, nutrient deficiency and non-essential chemicals. Developmental cues may be temporal and/or spatial and coincide with processes such as seed germination, organ development (root, leaf), the transition to floral morphology, organ senescence, embryogenesis, seed development and fruit ripening. How are these changes in gene expression regulated?

This chapter reminds the reader of the structure of plant genes and outlines the different mechanisms by which plant gene expression is modulated. Selected examples of the responses of plant genes to specific developmental and environmental factors are described, and some of the technology which has been used to examine the regulation of plant gene expression is explained.

A REVIEW OF PLANT GENE STRUCTURE

All genes include a transcribed sequence bounded by sequences at which transcription starts and stops (Figure 10.1). The transcribed region is flanked by elements necessary for transcription initiation and regulation. The transcribed sequence is 'read' by RNA polymerase. Nuclear genes are transcribed by any one of three RNA polymerases. RNA polymerases I and III are responsible for the synthesis of rRNA and tRNA respectively. RNA polymerase II is responsible for the transcription of structural genes encoding proteins. Within the transcribed region of structural genes are one or more exons which contain the protein coding sequence (called an open reading frame, ORF) and non-translated 5' and 3' flanking sequences which are involved in different aspects of gene expression. The transcribed sequence may contain introns which can be present in the transcript flanking sequences as well as the protein coding region. Introns are subsequently removed from transcripts and the exons and flanking sequences are then covalently joined in order to form mature mRNA.

In addition to the genes present in the nucleus, plants also contain mitochondrial and plastid genomes which encode a small proportion of the proteins found in these organelles (see chapter 9). These genes have features in common with prokaryotic genes (often polycistronic; Shine–Dalgarno ribosome binding sites) and certain nuclear and organellar genes are coordinately regulated.

Genes are only a Fraction of Plant DNA

Growth and development processes in all higher plants are very similar. Typically, the plant life cycles goes through stages of seed germination, enlargement and differentiation of the mature plant body, development of reproductive floral structures and subsequent fertilization, leading to the formation of a mature seed. The similarity of these processes between plants might lead one to expect that the total DNA content of most higher plants would be roughly equal. However, as Table 10.1 shows, this is far from being the case. *Arabidopsis thaliana* has one of the smallest plant genomes, with 145 Mbp of DNA per haploid genome, whereas *Tulipa* sp. has one of the highest, with over 24 000 Mbp per haploid genome. This wide variation in the DNA content between plant species is due to the presence in plant genomes of varying amounts of non-coding repetitive DNA. Coding genes are mostly found in the non-repetitive fraction, with single or small numbers of copies, exceptions to this being the genes cod-

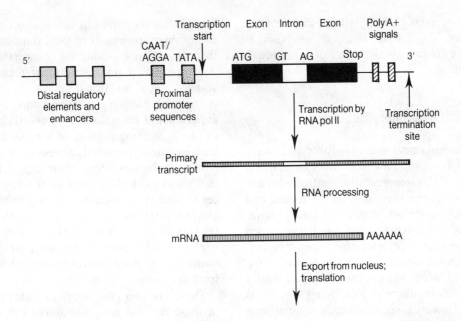

Figure 10.1. Structure of an idealized plant gene and representation of its expression.

ing for the cytoplasmic ribosomal RNAs, which are highly reiterated in plant genomes. A more complete description of the plant genome is provided in the previous chapter.

Role of Proximal Promoter Regulatory Sequences

The regulatory sequences in the 5′ regions of most genes can be divided into two categories: the proximal sequences responsible for the correct initiation of transcription, and the distal sequences which regulate and modulate gene expression. Although both regions are commonly referred to as 'promoters', these regions in some genes can be separated by hundreds of base pairs. The proximal promoter consists of up to three known regions: (a) the start of transcription (TCS); (b) a conserved sequence referred to as the 'TATA box', approximately 30 bp 5′ to the TCS, which is involved in transcription initiation by direction of RNA polymerase II to the correct start site; and (c) often a further conserved sequence referred to

Table 10.1. DNA content (in Mbp) of unreplicated haploid genome of some selected plants.

Scientific name	Common name	DNA content
Arabidopsis thaliana	Arabidopsis	145
Prunus persica	Peach	262
Ricinus communis	Castor bean	323
Citrus sinensis	Orange	367
Oryza sativa ssp. *javanica*	Rice	424
Petunia parodii	Petunia	1221
Pisum sativum	Garden pea	3947
Avena sativa	Oats	11 315
Tulipa sp.	Garden tulip	24 704

as 'CAAT or AGGA box' within 75 bp of the TCS (Figure 10.1), which has been proposed to regulate the frequency of transcript initiation in some genes.

Role of Distal Regulatory Sequences

Distal gene regulatory elements (Table 10.2) have been found to regulate a wide variety of environmentally and developmentally controlled gene systems. There are several ways to locate and identify such upstream distal regulatory sequences. Once a particular gene has been cloned and its DNA sequence determined, putative regulatory elements can be proposed by comparing sequences of genes which are activated under similar conditions, thereby identifying consensus sequences. However, such sequence comparisons must be coupled with functional analyses since

homologies identified in this manner may reflect evolutionary processes of gene duplication rather than functional sequence conservation. The development of transgenic plant technology has allowed such functional analysis of sequences controlling plant gene expression. A standard experimental procedure is to construct deletions of the 5' flanking regions of a gene, and then to transfer such modified genes to a host plant (usually tobacco) where their expression can be easily analysed. The putative regulatory sequences can be systematically deleted either by using existing restriction enzyme sites in the sequence (Figure 10.2), or by using enzymes such as Bal31 nuclease which hydrolyse double-stranded DNA molecules by progressively removing nucleotides from an exposed terminus.

There are two main ways in which the activity of these deleted gene sequences can be investigated. The accumulation of mRNA species can be

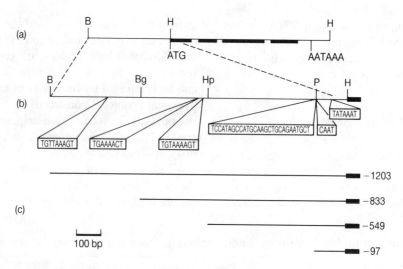

Figure 10.2. Construction of deletions of the 5' flanking region of the pea legumin (*legA*) gene using existing restriction sites. A succession of deletions are produced which remove various regulatory sequences. The effect of loss of these sequences on gene expression can be determined following transient or stable transformation of a host plant, often tobacco. (a) Map of the *legA* gene representing the coding region (exons in solid bars) and flanking sequences. The translation start (ATG) and polyadenylation signal (AATAAA) are indicated. (b) Expanded region of the *legA* 5' flanking sequence showing the location of the TATA and CAAT promoter consensus elements, the 28 bp 'legumin box', and the three sequences homologous to the cereal glutenin gene control elements. (c) The extent of the four deletions. Numbers to the right of each deletion refer to the length in base pairs upstream from the transcription start site of the *legA* gene. Restriction endonuclease recognition sites are represented as follows: B, *Bam*H1; Bg, *Bgl*II; H, *Hind*III; Hp, *Hpa*II; P. *Pst*I. From Shirsat *et al.*, 1989, with permission.

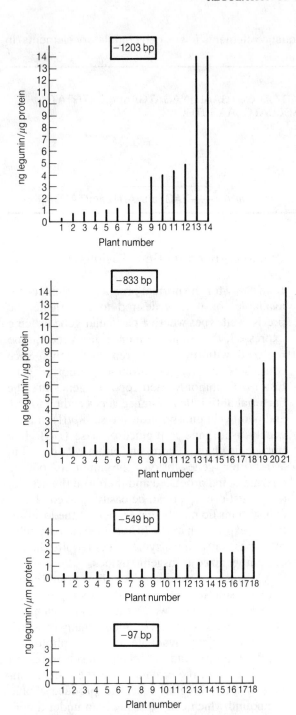

assayed by the technique of Northern blotting. This involves using the radiolabelled coding region of the transferred gene to seek out its transcriptional complement in total RNA extracted from transgenic plants. Alternatively, the protein product of the genes can be assayed if antibodies to the specific protein are available. Figure 10.3 shows the use of the latter approach to identify controlling sequences in the 5′ region of the *legA* legumin gene from *Pisum sativum*. Three deletions and the undeleted wild-type gene were introduced into a series of tobacco plants and expression from the truncated 5′ region was analysed in the transgenic plants. The protein product (legumin) was assayed by enzyme-linked immunosorbent assay (ELISA) in seeds of individual transgenic plants. This indicated that one of the deletions which terminated at −97 bp relative to the TCS abolished the activity of the gene. Synthesis of legumin was first seen in a deletion which had 549 bp relative to the TCS. This indicated that sequences upstream of −97 bp were required for the activation of the *legA* gene. Seeds from plants transformed with the −833 bp and the normal (not truncated) gene showed high levels of legumin expression; in some transgenic plants legumin protein constituted up to 0.5% of the total seed protein. When the transcript abundance was determined using Northern blots, results in agreement with ELISA were obtained.

This *in vivo* transgenic plant experiment illustrates two important facts about plant regulatory elements: (a) sequences necessary for low-level tissue-specific expression are usually located within about 500 bp of the TCS; (b) further upstream sequences can increase the level of gene expression without contributing to tissue specificity. Some of these sequences do not function

Figure 10.3. Analyses of pea legumin gene (*legA*) expression in tobacco plants containing the constructs shown in Figure 10.2. Pea legumin protein was quantified in the mature seeds of four sets of transgenic tobacco plants using a specific antibody. The boxed numbers over each histogram refer to the deletions shown in Figure 10.2. This reveals that the presence of increasing lengths of 5′ flanking region results in an increase in the amount of protein synthesized. Such experiments can be used to identify regulatory sequences.

Table 10.2. Some selected examples of conserved sequence elements (distal gene regulatory elements) in specific plant genes.

Developmental regulatory elements

Prolamin box	ACAtgTGTAAAGGTGAAt/gNAGATGAgt/tgCATGTAT
Legumin box	TCCATAGCCATGCAAGCTGCAGAATGTC
Vicillin box	GCCACCTCaattt

Environmental response elements

Heat-shock	CTNGAANNTTCNAG
Anaerobic response	CGGTTT---TGGTTT
Wound response	Tg/tGTTGAAATAa/tA-----TAGTa/tAAATaa/gtTATGA

when disassociated from their original gene, while others appear to function as 'enhancers' when introduced into unrelated genes (see next section). Thus, in the *legA* experiment, sequences up to −549 bp conferred a low level of tissue-specific expression, while sequences between −549 and 1203 bp increased total expression levels.

Enhancers

Enhancers are sequences which can be located at considerable distances from the TCS, act in a position and orientation-independent manner to elevate gene transcription, and can stimulate the activity of any gene. Such enhancer sequences have been found in the soya bean conglycinin (a seed storage protein) gene, in a region of DNA between −159 to 257 bp upstream of the TCS. This region contains four repeats of the 6 bp A [A/G/C] CCCA sequence. These repeats were inserted in different positions and orientations into a reporter gene system (see next section) consisting of the cauliflower mosaic virus promoter linked to chloramphenicol acetyl transferase (CAT), and transgenic plants regenerated. A 25-fold increase in the levels of CAT expression over controls was found in the seeds of some transgenic plants, indicating that the upstream conglycinin flanking sequences were behaving as enhancers.

Promoter–Reporter Gene Fusions

If an assay for the normal product of a gene is not available, or if it is desired to investigate the precise cell types where a particular gene is being expressed, the coding sequence of a gene can be replaced with that of a 'reporter' gene. Such constructs are termed promoter–reporter gene fusions. Commonly used reporter genes are the bacterial antibiotic resistance genes coding for *cat* and neomycin phosphotransferase (*npt*II); a third widely used gene is β-glucuronidase (*gus*). If a transgenic plant is expressing either *cat* or *npt*II in a reporter gene fusion, the amounts of the enzyme being produced and therefore the activity of the test promoter can be easily assayed. These two systems do not, however, permit the localization of expression of the fusion genes in particular cell types, since the enzyme assays require the use of homogenized transgenic tissues.

A reporter gene which does allow visualization of expression in particular cell types is *gus*. GUS is an *E. coli* enzyme which catalyses the cleavage of a wide variety of glucuronides, many of which are commercially available as spectrophotometric, fluorometric and histochemical substrates. In the presence of the enzyme GUS the histochemical substrate 'X–gluc' reacts to produce a blue compound which can be easily seen under a microscope, and specific cell types expressing a particular promoter–*gus* gene fusion can thus be identified. This can reveal the tissues where promoters of genes of unknown function are expressed.

Processing of RNA

The transcription of a gene initially results in the production of nuclear RNA in the nucleoplasm (Figure 10.4), the population of which is referred to as heteronuclear RNA (hnRNA). The initial transcript has many more nucleotides in it than are required to code for the final protein. These extra sequences consist of the 5′ untranslated region, the 3′ untranslated region and the intervening sequences or introns. Processing of this

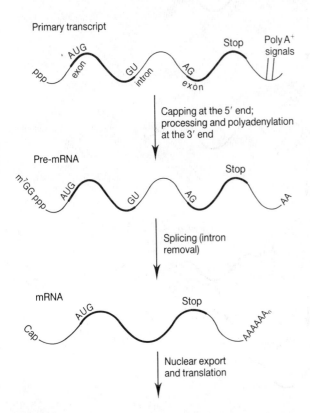

Figure 10.4. Stages involved in the processing of primary transcripts to produce mRNA. These are the addition of a 7-methyl guanosine cap; polyadenylation; and the removal of introns.

molecule involves 'capping' the 5′ terminus of the RNA, the addition of a poly(A) 'tail' to the 3′ terminus, and, finally, the splicing out of the introns to leave a mature, processed mRNA molecule. Unlike the nuclear genes, organellar genes very rarely contain introns. Chloroplast gene transcripts are not capped or polyadenylated, whereas mitochondrial transcripts, though polyadenylated, are not capped.

Capping

At the 5′ end of 'capped' transcripts, a 7-methyl guanosine residue is added through a 5′–5′ linkage (Figure 10.5). This cap structure protects the pre-mRNA from exonucleotic degradation, and is also involved in the binding of the 40S ribosomal subunit during protein synthesis. The 40S subunit recognizes the 5′ cap and travels along the mRNA until the initiator methionine codon is reached.

Polyadenylation

Following capping of the primary transcript, a stretch of adenosine residues—the poly(A) tail—is added to the 3′ end (Figure 10.4). The signal for polyadenylation is usually AAUAAA in plants, and precedes the site of poly(A) addition by 13–23 bp. Plant genes sometimes contain more than one poly(A) addition signal, and the use of alternative polyadenylation signals has been observed in one mRNA species. The poly(A) tail of animal messages protects the mRNA from ribonuclease digestion (a role akin to that played by the cap at the 5′ end), and is involved in the transport of the mRNA from the nucleus to the cytoplasm. Similar roles are expected for the poly(A) tail of plant mRNA species.

RNA Splicing

The final event which takes place before the production of the mature message is the splicing out

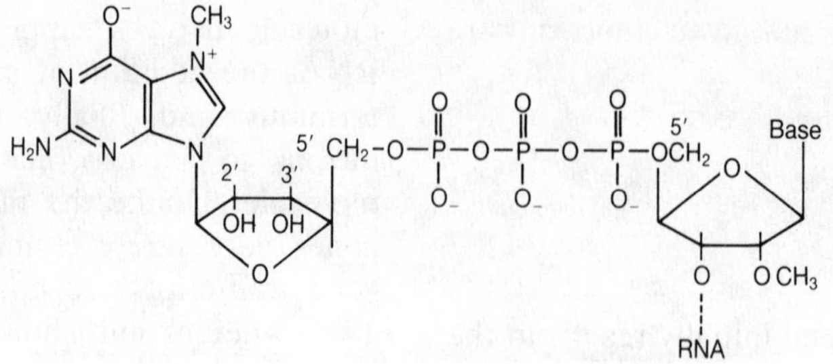

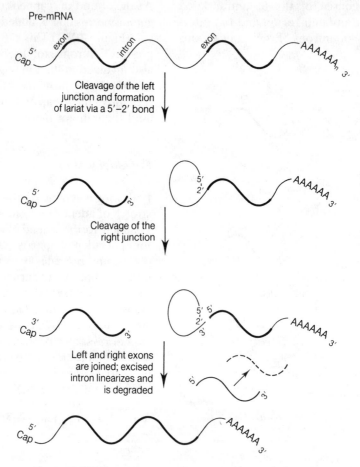

Figure 10.5. 'Capping' the 5' end of the transcript involves the addition of 7-methyl guanosine.

Figure 10.6. Intron splicing leading to the formation of mature mRNA. Introns may occur within the protein coding regions of a transcript or within its untranslated flanking sequences. The left exon–intron junction is cleaved and the intron forms a lariat structure. Following subsequent cleavage of the right junction the two exons are covalently joined. The released intron is assumed to be rapidly degraded. The mature mRNA, containing the translation initiation and termination codons, is exported to the cytoplasm for translation.

of the introns. Splicing involves interactions between intron sequences, U-type small nuclear ribonucleoprotein particles (UsnRNP), as well as other less well-defined proteins, and is an energy-dependent process, requiring ATP. Intron–exon junctions have boundaries which are precisely defined by dinucleotide pairs—the 5′ terminus of an intron begins with GT, and the 3′ terminus ends with AG. The mechanism of splicing is shown in Figure 10.6. The initial event in splicing is the assembly of a ribonucleoprotein complex called the spliceosome in the pre-mRNA. Splicing then proceeds in a two-step pathway. The mRNA is first cleaved at the 5′ intron junction, and a 5′–2′ phosphodiester bond is formed between the free 5′ intron end and a 2′OH of an intron residue near the 3′ intron splice site. This is termed a lariat structure. The 3′ intron junction is then cleaved, releasing the lariat, and the two exon junctions are then covalently ligated together to form the mature mRNA, ready for export to the cytoplasm. The consensus sequence for plant intron lariat sites differs from animals and lower eukaryotes like yeast. For this reason, intron processing of foreign genes from other kingdoms may not occur efficiently in transgenic plants.

REGULATORY MECHANISMS

Signal Transduction Pathways

The responses of plant cells to environmental stimuli, and to the signals passing between tissues or cell types (hormones, metabolites, ions), are often rapid. What are the mechanisms for rapidly transmitting these stimuli from outside to inside the plasmalemma and for transducing the initial signal into a form that can be perceived and amplified within the plant cell? In many cases these pathways lead to direct modification of specific enzymes, for example a change in their phosphorylation status, facilitating stimulation or inhibition of their activity. Such signal transduction pathways which bypass DNA are not considered here. In many cases, however, the response involves changes in gene expression. It is convenient to consider the transmission of information

to the genetic material in stages. First, the detection of the external stimulus by a primary sensor or receptor, and second, the transmission of information to regulatory proteins which interact with specific DNA sequences. These regulatory proteins are often referred to as transcription factors, or *trans*-acting factors. In some cases, secondary messengers convey information between a primary sensor and DNA-binding transcription factors.

Before considering more complex signal transduction pathways it should be noted that in prokaryotes there are also precedents for direct effects of stimuli on gene expression and it is possible that similar mechanisms could operate in plants. For example, *proU* expression in *Salmonella typhimurium* is induced by elevated ionic strength and is positively correlated with accumulation of cytoplasmic potassium ions. It has been proposed that these ions could act directly as a regulatory signal via changes in intracellular ionic strength. Ionic strength may influence the conformation of a regulatory protein, modifying its ability to bind to the *proU* promoter or, at high ionic strength, certain RNA polymerase promoter interactions may be directly altered[1].

Primary Sensors

Primary sensors for many stimuli remain to be identified. It is thought that growth regulators and sugars may have specific receptor proteins. Changes in the oxidative/reductive environment of plant cells may facilitate perception of a number of environmental stimuli including heat-shock. The presence of heat-denatured proteins which become tagged with ubiquitin could also facilitate perception of heat-shock. Plant cell wall-localized β-1,3-glucanase may be one of the sensors involved in pathogen recognition by releasing β-1,3-glucans from the walls of invading pathogens. These diffusible oligosaccharides then act as a signal to elicit transcription of genes involved in the plant defence response. The 'defence' genes include those encoding enzymes such as chitinase and additional glucanase to further hydrolyse cell wall components of the invading pathogens. Some of the induced proteins are referred to as pathogenesis-related (PR) proteins. However, probably the best described primary sensors in plants are those which detect photomorphogenetically active light.

Many changes in plant morphology are regulated by light, often at light levels which are insignificant for photosynthesis. These changes include phototropism, seed dormancy and germination, induction of flowering, growth rate, anthocyanin synthesis, movement of chloroplasts within cells and movement of leaves (see chapter 11 for further discussion of the molecular control of development). The light sensors responsible for mediating these changes include phytochrome and the blue light receptors. In performing their regulatory functions, these receptors modulate the expression of a number of nuclear genes in either a positive or negative manner. For phytochrome this includes the regulation of transcription of its own *phyA* genes. Some other phytochrome responses involve direct modification of the activity of a specific protein kinase or protein phosphatase and hence the phosphorylation status of critical enzymes. The consequent change in enzyme activity occurs without interaction with DNA, and these signal transduction pathways are discussed elsewhere[2].

Many of the phytochrome-regulated responses are inducible by red (r) light and, with certain exceptions, reversible by far-red (fr) light. The phytochrome polypeptide is encoded by a minimum of three distinct genes in angiosperms, often designated *phyA*, *phyB* and *phyC*. The prosthetic group associated with the protein is composed of an open chain of four tetrapyrole rings. Two forms of this chromophore are shown in Figure 10.7. The P_r form has an absorption maximum at 660 nm and absorption of red light converts it to the P_{fr} form. In turn, the P_{fr} form has an absorption maximum at 730 nm and absorption of far-red light converts the P_{fr} form to the P_r form. In the dark, P_{fr} slowly returns to P_r and there is also some degradation of P_{fr} (Figure 10.8). There are a number of intermediate states (not shown in Figure 10.8), with intermediate absorption max-

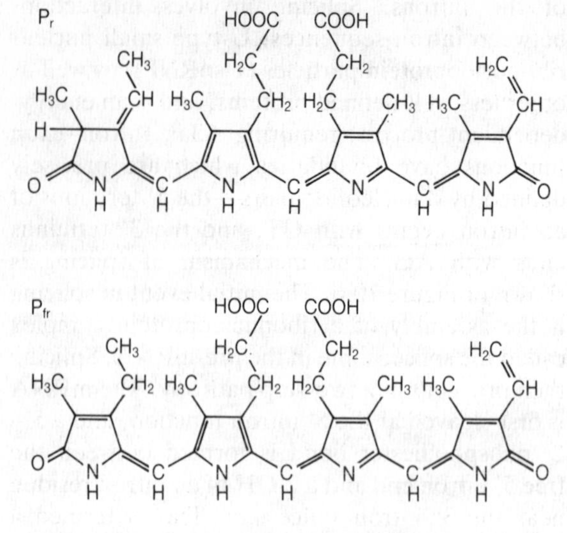

Figure 10.7. The red form (P_r) and the far-red form (P_{fr}) of the prosthetic group of phytochrome.

ima, due to other conformational changes within both the protein and prosthetic group. The interconversion between P_r and P_{fr} allows phytochrome to initiate many light-mediated responses, including modification of gene expression. Some of these responses, such as the positive regulation of chlorophyll a/b-binding protein (*cab*) and ribulose bisphosphate carboxylase small subunit (*rbcS*) genes, have been shown to require *de novo* protein synthesis. This raises the possibility that these genes only respond indirectly to the primary phytochrome signal by way of regulated expression of an intermediary gene, possibly one encoding a *trans*-acting factor. However, other phytochrome-mediated gene responses occur in either the presence or absence of protein synthesis, indicating that all compo-

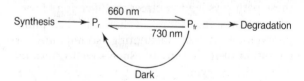

Figure 10.8. A simple scheme representing the light-dependent changes in the phytochrome chromophore.

nents in the signal transduction pathway are present in the cell prior to photoconversion of the sensor molecule. Lissemore and Quail (1988) have shown that transcriptional regulation of *phyA* in oats falls into this latter category and is therefore directly regulated by the primary phytochrome signal[3]. In dark-grown tissue, where only P_r is present, transcription of *phyA* is high but is repressed within 5 min from formation of P_{fr} after a pulse of light, with no dependence on *de novo* protein synthesis. Derepression of *phyA* transcription occurs after some further period in the dark when the P_{fr} concentration drops below some critical level. The duration of repression depends upon the nature of the initial light pulse and hence the amount of P_{fr} formed.

Secondary Messengers

Some primary sensors interact directly with DNA sequences to modulate gene expression. It has been hypothesized that some phytochrome responses could fall into this category, with P_{fr} directly binding to *cis*-regulatory elements with no requirement for a second messenger. However, experimental data suggest that other phytochrome-induced responses require intermediaries such as Ca^{2+} and calmodulin. There is evidence that Ca^{2+}, calmodulin and P_{fr} all play a role in nuclear protein phosphorylation. This has led to the hypothesis that phosphorylation of nuclear proteins may be an intermediate step linking photoactivation of phytochrome with the photoregulation of gene expression. Datta and Cashmore (1989) observed that the binding of a pea nuclear protein, AT-1, to promoters of certain photoregulated genes is modulated by protein phosphorylation[4]. It is also known that phytochrome is itself a phosphoprotein with either its own protein kinase activity (autophosphorylation) or this activity being conferred by a separate protein kinase which is tightly associated with phytochrome.

Protein phosphorylation, Ca^{2+}–calmodulin, phosphoinositide metabolism, pH changes, redox reactions at the cell surface, membrane fatty acids and protein methylation have all been implicated in second messenger pathways in plants. The vacuole is also considered to be a potential store of second messengers in plants. For description of current knowledge of second messengers in plants the reader is referred to Boss and Morre (1989).

Control of Transcription

Activation of transcription is a two-stage process. In the first stage chromatin must be decondensed, since most genes exist at some time in an inert state, tightly packaged with histones into chromatin. The second stage involves the interaction of transcriptional factors with specific DNA sequences. These DNA-binding factors are often a final element in the multi-component signal transduction pathways described in the previous section. For a detailed account of transcriptional regulatory mechanisms general to eukaryotes see Hames and Glover (1989).

Transcription of genes in plants occurs by similar mechanisms to the process in other eukaryotes. Transcription is carried out by a transcription complex, containing RNA polymerase and a series of transcription factors (designated TFIIA, B, D, E, F for RNA polymerase II). The complex is assembled on the immediate 5′ flanking sequence of the gene and its formation is dependent on the presence of the 'TATA' box (or its equivalents for RNA polymerases I and III), which determines the start point for transcription. The 'TATA' box itself is not thought to play a role in control of gene expression, although its presence is necessary for accurate initiation. Current models assume that the bound *trans*-acting factors either aid the assembly of the active transcription complex, or stimulate the activity of the transcription complex, as shown in Figure 10.9.

Measurement of Transcriptional Activity

Until very recently, the transcriptional activity of plant genes could not be measured in an *in vitro* cell-free system; no such systems had been pre-

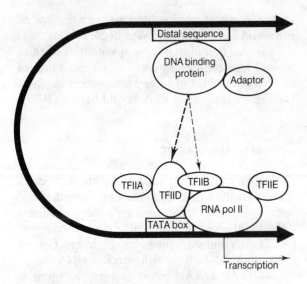

Figure 10.9. Simplified model of the interactions of *trans*-acting DNA binding proteins with the transcription complex containing RNA polymerase II and the transcription factors TFIIA to G (selected representatives of the transcription factors are shown). Current evidence favours TFIID, and possibly TFIIB, as the target for protein–protein interactions which activate the transcription complex. The action of the *trans*-acting DNA binding protein can in turn be affected by interaction with a modifier/adaptor protein, which can act as an activator or repressor. For simplicity, only one *trans*-acting DNA binding protein has been shown; in reality, multiple binding sequences for *trans*-acting factors are present in the 5′ flanking regions of genes.

pared from plant tissues, and animal systems lack the appropriate control factors. However, a transcription system from wheat germ has now been described and is likely to be more widely used in the future. Transcription of specific plant genes has previously been measured in a system containing isolated nuclei, or sometimes (but less satisfactorily) chromatin. The products of such assays are wholly the products of elongation of nascent transcripts, since no reinitiation occurs, so that the 'run-off' products represent the relative transcriptional activities of the genes *in vivo*.

A problem with these assays is the specificity of the probe sequence used to detect the labelled 'run-off' transcripts. Due to the frequent occurr-

ence of multi-gene families in plants, the transcriptional activity measured may be the net result of several genes. For example, assays which aimed to measure the transcriptional activity of the pea seed storage protein gene *legA* in fact measured the net transcription from four genes, *legA/B/C/E*, since the probe used could not distinguish between them, owing to the very high (99%) level of homology. The use of transgenic plants containing the gene of interest as the only foreign gene is one way around this difficulty, but this assumes that regulation of the gene is exactly the same as in the normal environment.

Despite these limitations, it is possible to obtain a good estimate of the transcriptional activity of many plant genes. A general conclusion from these assays has been that modulation of transcription is a primary factor in controlling expression of many plant genes. However, as will be discussed, many other factors are also involved in determining the observed levels of mRNA species and their encoded polypeptides in plant tissues.

Enabling Transcription

For transcription to be possible, the DNA comprising the promoter region of the gene must be available to interact with RNA polymerase and *trans*-acting factors. In turn, this presupposes that the section of DNA containing the gene must form part of a region where the chromatin is in an extended conformation. Direct visualization of regions of extended and condensed chromatin (as observed in insects, such as *Drosophila* chromosomal 'puffs') has not been possible in plants. However, indirect evidence for this phenomenon has been provided by the well-known 'positional effect', where a proportion of transformed plants containing a foreign gene fail to show any expression of that gene whereas others express strongly. This variability in expression has been assigned to the foreign gene being inserted into a region of either 'inactive' or 'active' chromatin. Attempts have been made to measure the accessibility of specific genes by treatment of chromatin with DNase I, a non-specific nuclease. Results have shown that active genes often have so-called 'DNase hypersensitive' sites associated with them, where attack by the nuclease occurs significantly faster than all other areas; in the wheat ribosomal genes, these hypersensitive sites lie in the intergenic spacers, and the degree of hypersensitivity corresponds to the activity of the genes, suggesting changes in chromatin conformation play a controlling role. This is also thought to be the case for other genes.

Plant DNA is also highly methylated, methylation occurring on the cytosine of C-G dinucleotide sequences as in other eukaryotes, but also extensively on the cytosine of C-N-G trinucleotides. The reason for this extra methylation is unknown. A correlation of degree of methylation (as measured by resistance to digestion with selected restriction endonucleases) with gene activity has been observed with many genes; for example, ribosomal genes in wheat and pea, and T-DNA genes in transgenic plants. In animal systems only a small proportion of the total methylation events, usually those occurring in the gene promoter region, lead to gene inactivation, and this has also been shown to be the case in plant genes, e.g. seed storage protein genes. Examples of the progressive inactivation of genes by increasing methylation, as observed in foreign genes in animal cell lines in culture, have not yet been observed in plants.

Activating Transcription

Once a gene becomes available for transcription, it is likely that a basal level of expression will occur. Low levels of transcription have been observed for many genes in most tissues, even those thought to be 'tissue specific', such as the soya bean trypsin inhibitor gene, which is usually regarded as a seed-specific gene. Only in the case of some seed storage protein genes has expression been shown to be truly tissue specific within the limit of sensitivity of the assay; in these cases, the ratio of transcriptional activity in seed and leaf tissue is greater than 300-fold. The increase in transcriptional activity observed when a gene is 'switched on' by endogenous or exogenous factors is assumed to be the result of *trans*-acting proteins binding to *cis*-acting DNA sequences. A previous section has described the *cis*-acting sequences in genes, which are becoming fairly well characterized; in contrast, little is known about *trans*-acting factors in plants, although rapid progress in this area is imminent.

Trans-acting factors are normally assayed by a technique referred to as 'gel retardation assay' or electrophoretic mobility shift analysis (EMSA). The principle of this assay is shown in Figure 10.10. A double-stranded DNA probe, containing a putative *cis*-acting sequence, is end-labelled with ^{32}P. This sequence can either be isolated from a cloned gene, or can be synthesized as complementary oligonucleotides. The labelled probe sequence is then allowed to react with an extract containing nuclear proteins from the plant tissue of choice. If binding takes place between a component of the extract and the probe fragment, a DNA–protein complex is formed which has a different mobility to the free probe (slower, due to the increased size of the complex) on gel electrophoresis. Because the probe is labelled, very small amounts of complex can be detected after autoradiography and the assay is very sensitive. EMSA assays are used to demonstrate the presence of *trans*-acting factors in different tissues, and can show correlations between the presence of binding activity and transcriptional activity of the gene used as probe; for example, binding activity to 5′ flanking regions of seed storage protein genes has been observed in extracts of seed nuclei, but not in extracts of leaf nuclei, in agreement with transcription assays. These assays are also used to map the extent and specificity of *cis*-acting sequences, by producing probe sequences which have been mutated by deletion or sequence alteration, and to investigate the specificity of the *trans*-acting factors by carrying out competition assays between putative *cis*-acting sequences from different genes. Assays of the latter type have been used to show that different genes share common *trans*-acting factors, as would be expected on a hierarchical model of control of gene expression.

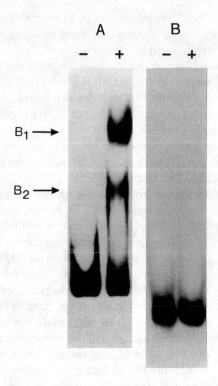

Figure 10.10. Detection of DNA binding proteins (*trans*-acting factors) by electrophoretic mobility shift assay (EMSA). Assays A and B use two different fragments of the 5′ flanking sequence of the pea seed storage protein gene *legA*; fragment A contains a *cis*-element (i.e. a distal DNA sequence interacting with binding proteins), whereas fragment B is a control containing no sequences that interact with specific binding proteins. In each case the probe has been incubated without (−) or with (+) pea seed nuclear proteins before carrying out electrophoresis. Probe A forms two retarded DNA–protein complexes, designated B_1 and B_2.

The precise site of interaction of *trans*-acting factors with DNA sequences can be determined by a 'footprinting' assay. In this technique, a probe labelled at one end only is allowed to form a complex with the *trans*-acting factor, and is then exposed to a reagent which cleaves the DNA at random, such as DNase I, or the *o*-phenanthroline-copper reagent. The partially cleaved probe/factor complex is then dissociated, and the DNA is run on a DNA sequencing gel. The region of probe bound to the *trans*-acting

factor will be protected from the cleaving reagent, and DNA fragments corresponding to the protected region will be absent from the resulting fragments observed on the sequencing gel. Controls of unprotected probe are of course necessary. The optimal situation is where a sequence identified as binding to a *trans*-acting factor corresponds to a *cis*-acting sequence identified in promoter assays, which in turn corresponds to a sequence conserved between different genes sharing a common expression pattern. One example of this is the maize zein storage protein genes[5].

The characterization of genes encoding *trans*-acting factors is just beginning in plants. Several techniques are available to isolate such genes, including the use of heterologous probes, and transposon tagging. Another approach is to make a cDNA expression library, in a vector such as lambda gt11, from the tissue in which the *trans*-acting factor is present (see later section). This library can then be screened, under conditions allowing specific DNA/protein interactions, with a labelled double-stranded probe containing the corresponding *cis*-acting sequence. This method has already been used to isolate tobacco DNA-binding proteins, including a transcription-enhancing factor termed ASF-1. The *trans*-acting factor genes will probably encode DNA-binding proteins similar to the types identified in other eukaryotes (Zn-fingers; Zn-clusters; Zn-twists; leucine zippers), but types unique to plants cannot be ruled out.

In other eukaryotic gene systems, multiple *trans*-acting factors have been shown to bind to the flanking regions of genes. Work on a number of plant genes including, for example, extensin genes, and maize zein genes, has already shown that similar results will be obtained for plants. Some *trans*-acting factors have a non-specific effect on expression, whereas others give specific control of expression in relation to development or environmental. The combination of *trans*-acting factors allow for precise and flexible control of gene expression, and for responses to different stimuli. An overall description of transcriptional control is likely to be very complex, but reductive analysis should make it possible to describe parts of the system (e.g. response to light) in some detail.

Post-transcriptional Control of Gene Expression

Post-transcriptional control comprises the processes that cause the relative steady-state levels of mRNA species to differ from the relative transcription rates of the corresponding genes. The relative contributions of different processes, i.e. processing of the primary transcripts (intron removal, capping, polyadenylation), export of mRNA to the cytoplasm, and stability of mRNA in the cytoplasm, are not fully understood, but mRNA stability seems to play a major role.

Evidence for post-transcriptional control is primarily at the quantitative level, but it is clear that in this respect it can be very significant. A mutant form of the soya bean Kunitz trypsin inhibitor gene containing three base changes in the protein coding region, leading to nonsense codons, was transcribed at the same rate as the normal gene, but steady-state mRNA levels were suppressed, and protein accumulation decreased by at least 100-fold. More spectacularly, a mutant form of the *Phaseolus vulgaris* lectin gene, containing an in-frame stop codon due to a frame shift mutation, was repaired by *in vitro* mutagenesis, and introduced into a transgenic plant. Whereas the mutant gene was not expressed at detectable levels, the repaired gene was fully active, and it was estimated that the 'repair' to the coding sequence had increased steady-state mRNA levels at least 40-fold. These effects are most likely to be at the level of mRNA stability, and to be mediated by the 'translatability' of the message. Similar, if less spectacular, discrepancies have been observed when the relative transcription rates of different seed storage protein genes are compared to the relative steady-state levels of the corresponding mRNA species. For example, for two subfamilies of pea legumin genes (*legA* and *legJ*) the ratio of transcription activities was 1.1 : 1 during the cell expansion phase of seed development,

whereas the ratio of the steady-state levels of the corresponding mRNA species was 5.0 : 1.

Post-transcriptional control also plays an important role in control of gene expression by environmental or genetic factors; the suppression of legumin synthesis by the 'r' locus (which encodes starch branching enzyme) in wrinkled peas has been shown to be due to post-transcriptional control, and post-transcriptional control amplifies the response of pea legumin genes to sulphur deprivation, with mRNA stability apparently changing under different conditions.

Although results from animal systems have provided abundant evidence for the role of the poly(A) tail in determining mRNA half-life, and hence steady-state levels, very little work has been done on comparable plant systems, although seed protein mRNA poly(A) tails have been shown to shorten as seed development proceeds into the desiccation phase, when mRNA levels decline.

A Note about Regulation at the Level of Translation

Protein synthesis has been discussed in chapter 9. The rate of protein synthesis from specific transcripts can vary, thereby further modifying gene expression. The expression of heat-shock (HS) genes during carrot somatic embryogenesis represents one example which has been described in some detail[6]. Somatic carrot embryos show different responses to heat-shock at different stages of development. The abundance of transcripts encoding a small HS protein was significantly less at the mid-globular stage of embryo development when compared to other stages. Furthermore, during the mid-globular stage there is no transcriptional induction of the corresponding HS gene, with its transcripts showing the same low level of abundance both before and after heat-shock. In spite of the low level of HS mRNA available, these globular embryos synthesize the full complement of HS proteins in response to heat treatment. This is accomplished by an elevation in the rate of translation of existing HS transcripts, which compensates for the low level of available mRNA.

Other Regulatory Mechanisms

Besides the universal mechanisms of gene control described above, several other mechanisms exist which apply to specialized cases of gene expression. Of these, perhaps the most significant are gene inactivations and reactivations caused by the insertion and excision of transposable elements. The best-studied cases are genes encoding proteins involved in pigment biosynthesis, such as the chalcone synthase (CHS) gene of petunia and Antirrhinum. The phenotype is the well-known 'streaky' flowers, with zones of pigmentation corresponding to excision of the element from an inactivated CHS gene in some cells during floral development. This gives rise to active CHS gene, and hence pigmentation, in the progeny of these cells produced by cell division. The effect of a transposable element on gene expression will vary according to the site of insertion, and the element; most elements cause a duplication of target site DNA, of a variable number of base pairs, which can permanently disrupt coding sequence or other structural features. For example, in Antirrhinum majus, the phenotype of white flowers with red flecks is given by insertions of transposable elements Tam1 and Tam3 in the promoter region of the CHS gene upstream of the TATA box (where element excision leaves an active promoter), but insertion of Tam2 at the first exon/intron boundary gives a phenotype of stable white flowers, since the splicing site is permanently disrupted even after excision of the element. Thus, the mechanism of inactivation also depends on the position of insertion; insertion into a promoter region prevents transcription, but insertion in the transcribed region leads to inactivation by the post-transcriptional control mechanisms discussed above. The effect of a transposable element on expression of the bronze gene in maize, giving

spots or zones of pigmentation in the seed coat, has also been extensively characterized.

Alterations in copy number of repetitive genes, such as the 5S RNA genes, have been observed in different cell types, and in response to environmental factors such as certain 'stresses'. This may also be a mechanism for controlling their activity, although no direct evidence for such control exists. The possibility also exists that anti-sense RNA may exist naturally in plants as a control mechanism.

Anti-sense RNA is an RNA species complementary to an mRNA species which suppresses expression of the 'sense' mRNA both by interfering with its translation and its transcription. Examples of naturally occurring anti-sense RNA species have been observed in prokaryotes, and anti-sense RNA has been engineered into transgenic plants, and has been shown to suppress expression of the 'sense' gene. One such example is transgenic tomatoes containing an anti-sense construct which suppresses expression of the gene encoding polygalacturonidase, thereby inhibiting softening of the fruit due to reduced degradation of cell wall pectin. Since this control mechanism is practically feasible in plants, it is possible that it may also be exploited in nature.

A final possibility for general control of gene expression is via modification of RNA polymerase, particularly RNA polymerase II, or its associated transcription factors. These possibilities are yet to be investigated owing to the difficulties involved in purifying the proteins from plant tissues.

INTEGRATION OF REGULATORY CIRCUITS

A gene can respond to multiple factors, both environmental and developmental. It can possess several different cis-acting regulatory sequences specific to different DNA-binding proteins. How are the different signals integrated to give appropriate expression in response to conflicting signals?

The answer to this question is likely to be both complex and different for each plant gene.

HOW THE 'REGULATED GENES' ARE ISOLATED

For a more detailed explanation of many of the techniques used in gene isolation the reader is referred to Sambrook *et al.* (1989) and Old and Primrose (1985).

Screening cDNA Libraries for a Gene Encoding a Known Protein

Production of cDNA Libraries

The polyadenylated (poly(A^+)) RNA expressed in plant cells (cells at a particular developmental stage or exposed to an environmental stimulus) can be purified by methods usually involving selective binding and release from oligo(dT) cellulose. The enzyme reverse transcriptase is then used in the production of complementary DNA (cDNA) from the poly(A^+) RNA template (Figure 10.11). Reverse transcriptase can be isolated from cells infected with retroviruses where the viral RNA is used as a template to produce DNA (retroviruses contain only RNA, no DNA). A second strand of DNA is produced following alkali treatment to hydrolyse the RNA in the initial RNA–DNA hybrids. The double-stranded DNA is then introduced into a vector, such as a plasmid or virus, which can be replicated in a bacterial host cell such as *E. coli*. The choice of the appropriate vector will depend upon the method to be used to detect genes of interest. Bacteria containing such recombinant plasmids, or phage, are cultured and all of the cells within each individual colony, or plaque, will carry the same plant gene sequence. The next step is to detect the colonies, or plaques, which are carrying the genes of interest.

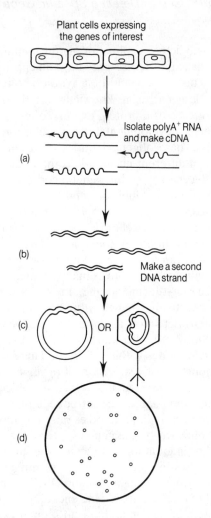

Figure 10.11. Production of a cDNA library. Poly (A^+) RNA is isolated from cells expressing the gene(s) of interest. (a) Reverse transcriptase and oligo(dT) primers are used to produce a first strand of complementary DNA (cDNA). (b) Alkali is used to remove RNA from the DNA–RNA hybrids and a second strand of DNA is synthesized using DNA polymerase. (c) The double-stranded cDNA is ligated to a vector (viral or plasmid) and introduced into *E. coli*. (d) The cells are plated out to produce individual colonies or plaques. Each colony or plaque will contain copies of a DNA sequence which will correspond to part of a single plant gene.

Methods Used to Detect a Specific Gene in a Library

In order to detect a colony or plaque containing a specific gene it is necessary to have an appropriate probe. This probe could be a synthetic oligonucleotide designed using knowledge of the amino acid sequence of a protein expressed in response to a specific developmental cue or environmental stimulus. These oligonucleotides must contain a number of different nucleotide combinations due to the degeneracy of the genetic code. Libraries are plated out in duplicate and the DNA from colonies, or plaques, from one of the replicates is transferred to nitrocellulose membrane (filters). After denaturation of the bound DNA, these membranes can then be incubated with the probe which will hybridize to single-stranded DNA sequences on the membrane. Probes are generally radioactively labelled with ^{32}P to allow their subsequent detection as spots when exposed to X-ray film. Antibodies can also be used to identify specific clones within a library. This requires the production of expression libraries in which the cDNA is inserted into a vector which will facilitate expression of the introduced plant gene sequence in *E. coli*. Antibodies raised against the protein encoded by the gene of interest can then be used to immunologically screen the library for clones expressing the corresponding antigen.

The polymerase chain reaction (PCR) can also be used to characterize a gene encoding a regulated protein of known amino acid sequence. Two redundant oligonucleotides, referred to as primers, are designed to hybridize to opposite strands of the corresponding gene such that they will have proximal 3' ends. A thermostable DNA polymerase is then used to synthesize DNA from the 3' ends of these primers using a template of total genomic DNA, or total cDNA, from the cells of interest. Repeated cycles of primer annealing to the template DNA, DNA synthesis and melting of the complementary strands facilitate amplification of a specific fragment. This fragment can be ligated to a vector, cloned, sequenced and possibly used as a probe to isolate the entire gene from a gene library.

Differential Screening and Subtracted Libraries

Differential Screening

This approach allows sequences that correspond to poly(A$^+$) RNA which is unique to cells at a particular stage of development, or exposed to an environmental stimulus (induced), to be differentiated from other sequences within a library (Figure 10.12). Radiolabelled probes are prepared

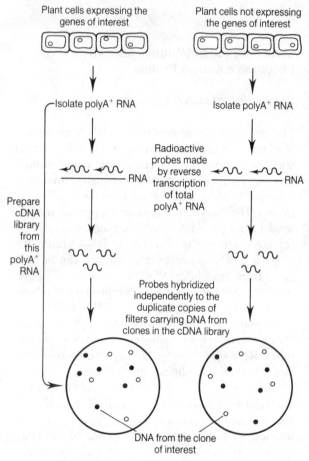

Figure 10.12. Differential screening of cDNA libraries facilitates the identification and characterization of mRNA sequences which are abundant in one cell type but not abundant in another.

from total poly(A⁺) RNA isolated both from induced and non-induced cells. These probes are the independently hybridized to duplicate copies of nitrocellulose membranes on which have been plated a cDNA library prepared from induced cells. DNA from some clones will only hybridize to probe prepared from the induced cells and these contain the unique induced sequences. This technique has been extensively used to isolate both developmentally and environmentally regulated plant genes.

Subtracted Libraries

It is possible to enrich cDNA libraries for sequences which are unique to induced cells. Total cDNA is first prepared from poly(A⁺) RNA purified from these cells. RNA is then removed by alkali hydrolysis and the remaining cDNA annealed to poly(A⁺) RNA isolated from non-induced cells. The cDNA that does not hybridize is purified on hydroxyapatite, which allows separation of single-stranded from double-stranded molecules. The purified cDNA is then used to make a library which can be differentially screened.

Coupling Two-dimensional Gel Electrophoresis of Proteins with Gene Isolation

Two-dimensional Polyacrylamide Gel Electrophoresis

Many of the proteins expressed in a population of cells can be resolved using two-dimensional polyacrylamide gel electrophoresis (Figure 10.13). Proteins are first separated on the basis of their charge, often by isoelectric focusing in thin gel tubes. In this method the tube gels contain a pH gradient and proteins therefore migrate towards either the anode or cathode at opposite ends of the tube, until they reach their isoelectric point (the pH at which they have no net charge). The tube gel is then placed on the surface of a

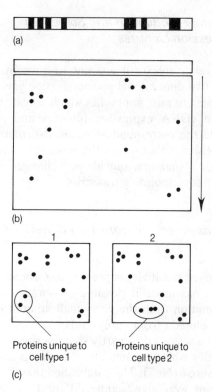

Figure 10.13. Two-dimensional polyacrylamide gel electrophoresis (2-D PAGE) to identify differentially expressed proteins. (a) Proteins are first separated by isoelectric focusing, in tube gels, on the basis of their charge. (b) The tube gel is placed on the top of a slab gel and further proteins are separated on the basis of their size. The proteins can then be visualized, by one of a number of alternative methods, and appear as spots on the gel. (c) Comparison of such gels from extracts of cells exposed to two different sets of environmental conditions (shown as 1 and 2) allows the identification of differentially expressed proteins.

polyacrylamide slab gel which separates proteins on the basis of their size. By repeating this for extracts from both cells expressing proteins of interest, and those not expressing, it is possible to identify the regulated proteins. If the method is being used as an analytical tool the sensitivity can be improved by radiolabelling the proteins *in vivo*. The same technique can also be used to resolve proteins produced by *in vitro* translation of poly(A⁺) RNA, which will of course identify transcripts which change in abundance.

Antibody Production and Screening Expression Libraries

The relatively small amount of protein eluted from two-dimensional polyacrylamide gels can be sufficient to raise antibodies which may be used to screen cDNA expression libraries and thereby isolate the corresponding genes. Northern blots establish whether or not the observed changes in protein abundance coincide with changes in levels of the corresponding transcript.

Microsequencing from Two-dimensional Gels

Improved sensitivity of protein-sequencing instruments has made it possible to obtain sequence information from the very small amount of protein eluted from polyacrylamide gels. This method was used to partly characterize two polypeptides which accumulate in barley plants after exposure to NaCl. This established that the barley proteins were significantly different from osmotin, an NaCl-induced protein which had previously been isolated from cultured tobacco cells. Protein sequence information can be used to design and synthesize oligonucleotides suitable for isolation and characterization of the corresponding genes by methods which have already been described.

Promoter Tagging

Certain plant phenolic compounds activate the virulence genes carried on the tumour-inducing (Ti) plasmid of soil agrobacteria. The virulence genes direct the integration into the plant genome of certain DNA sequences which are present on the Ti plasmid (see chapter 12 for more detailed discussion). The DNA which is transferred (T-DNA) is flanked by specific short DNA sequences. By introducing a promoterless reporter gene between these flanking sequences it is possible to 'tag' active plant promoters since the reporter gene will only be expressed when the T-DNA has inserted itself behind a promoter which is active in the tissue under examination. The 'tagged' gene which has been interrupted by the T-DNA can be isolated by any one of a number of methods, thereby facilitating the identification of genes expressed in a particular tissue.

REFERENCES

1. Higgins, C.F., Cairney, J., Stirling, D.A., Sutherland, L. and Booth, I.R. (1987) Osmotic regulation of gene expression: ionic strength as an intracellular signal. *Trends in Biochemical Sciences*, **12**, 334–339.
2. Budle, R.J.A. and Randall, D.D. (1990) Light as a signal influencing the phosphorylation status of plant proteins. *Plant Physiology*, **94**, 1501–1504.
3. Lissemore, J. and Quail, P.H. (1988) Rapid transcriptional regulation by phytochrome of the genes for phytochrome and chlorophyll a/b-binding protein in *Avena sativum*. *Molecular and Cellular Biology*, **8**, 4840–4850.
4. Datta, N. and Cashmore, A.R. (1989) Binding of pea nuclear protein to promoters of certain photoregulated genes is modulated by phosphorylation. *Plant Cell*, **1**, 1069–1077.
5. Shewry, P.R. and Tatham, A.S. (1990) The prolamin storage proteins of cereal seeds: structure and evolution. *Biochemical Journal*, **267**, 1–12.
6. Zimmerman, J.L., Apuya, N., Darwish, K. and O'Carroll, C. (1989) Novel regulation of heat shock genes during carrot somatic embryo development. *Plant Cell*, **1**, 1137–1146.
7. Shirsat, A.H., Wilford, N.W., Croy, R.R.D. and Boulter, D. (1989) Sequences responsible for the tissue specific promoter activity of pea legumin gene in tobacco. *Molecular and General Genetics*, **215**, 326–331.

FURTHER READING

Boss, W.F. and Morre, J.D. (1989) *Second Messengers in Plant Growth and Development*. Liss, New York.
Hames, B.D. and Glover, D.M. (1989) *Transcription and Splicing*. IRL Press, Oxford.
Leaver, C.J., Boulter, D. and Flavell, R.B. (eds) (1986) Differential gene expression and plant development. *Philosophical Transactions of the Royal Society of London*, **B314**.
Old, R.W. and Primrose, S.B. (1985) *Principles of Gene Manipulation: an Introduction to Genetic Engineering* (3rd edn). Blackwell Scientific Publications, Oxford.
Sambrook, J., Fritsch, E.F. and Maniatis, T. (1989) *Molecular Cloning: A Laboratory Manual*. Cold Spring Harbor Laboratory Press, New York.

11

MOLECULAR CONTROL OF DEVELOPMENT

J.A. Bryant

Department of Biological Sciences, University of Exeter, UK

A.C. Cuming

Department of Genetics, University of Leeds, UK

INTRODUCTION

Development in multicellular eukaryotes commences at fertilization. The zygotic cell proliferates to form a multicellular embryo within which polarity and axes of symmetry are established to provide a framework for groups of cells to differentiate and develop into the tissues and organs of the developing organism. Much of our conceptual framework for understanding the processes which

Plant Biochemistry and Molecular Biology. Edited by P.J. Lea and R.C. Leegood
© 1993 John Wiley & Sons Ltd

control and coordinate developmental processes are derived from studies of animal development, based as they are on the genetic and biochemical analysis of early events in embryogenesis in a few well-characterized model systems. These studies provide us with certain paradigms which may or may not be relevant to the analysis of plant development, and any models which seek to describe how development and pattern formation occur in the developing plant embryo should consider their potential applicability.

The establishment of polarity and the axes of symmetry within developing embryos appears to be a very early event, dictated by the maternal environment. Egg cells are produced in ovaries, which are themselves polar structures, and the establishment of polarity in the unfertilized egg, in the form of morphogenetic cytoplasmic determinants which are unequally distributed, is a well-established concept, typified, for example, by the distribution of *bicoid* gene products within *Drosophila* egg cells. Further development within a maternal organ exposes the embryo to additional maternally derived morphogenetic influences and these, in coordination with internal morphogens, may activate the expression of subsets of genes in a spatially and temporally discontinuous manner. The establishment of segmental identity during *Drosophila* development provides a striking example of the way in which the activation of individual genes by such morphogens coordinates the expression of others to achieve differentiation of structure and function.

Additional features of the development of animal embryos are the movement of cells relative to each other—gastrulation provides a striking example of this process—and the association of groups of cells to form organizing centres which, through the production of diffusible morphogens, provide positional information to which the recipient cells respond accordingly.

Our knowledge of the molecular mechanisms underpinning the establishment of plant embryos is much less extensive. Virtually nothing is known of the way in which cellular organization is coordinated during embryogenesis. In general, plant development is less determinate than animal

development: plant cells retain a high degree of totipotency, exemplified by the ease with which developmental identity can be manipulated in tissue culture. Organized and apparently terminally differentiated tissues can be induced to dedifferentiate to generate disorganized callus tissue, and this can in turn be caused to redifferentiate into recognizable organs or whole plants, via somatic embryogenesis, by relatively simple alterations in the hormonal environment to which the cells are exposed. Such manipulations are now being exploited in order to identify genes and gene products whose expression is associated with developmental decisions in plant tissues, but considerable work remains to be done before the regulation of these genes, during normal development, is understood.

Another important contrast between events occurring in animal and plant embryogenesis is in the contribution made by the movement of cells relative to one another. While this is a key feature of animal embryogenesis, cell movement in plant development is prevented by the rigidity of plant cell structure—a characteristic imposed by the presence of the cell wall. Morphogenesis in plants therefore entirely reflects the shape of their constituent cells, and the orientation of their planes of division in space. It is therefore hypothesized that many of the critical events in determining morphogenesis will involve the regulation of cytoskeletal orientation, as this is responsible both for determining the orientation of nuclear division, through its action in forming the mitotic spindle, and also for controlling the pattern of deposition of cellulose microfibrils within the cell wall. The orientation of these microfibrils constrains the direction in which turgor-driven cell expansion can occur, and changes in the extent and orientation of cellulose deposition may ultimately act to limit the extent of cell growth. It is a longstanding observation that the orientation of newly synthesized cellulose fibrils is underlain by a corresponding orientation of cytoskeletal microtubules beneath the plasmalemma.

Finally, one of the most striking characteristics of plant development is the limitation of those cells able to divide and thus contribute to the growth and pattern of the plant. The ability to undergo continued cell division is restricted to meristematic zones. Those at the root and shoot apices, and in lateral buds and secondary roots (functionally equivalent to primary shoot and root apices) divide to produce additional cells which subsequently expand longitudinally and contribute to root and stem elongation (a variation on this pattern is seen in monocotyledonous species, such as those of the Gramineae, where the meristems responsible for stem and leaf elongation are located basally as intercalary meristems). Cells in the cambium divide perpendicular to the radius of the stem and root, causing thickening of these organs.

Meristematic activity is also associated with cellular differentiation, the internal and external products of cambial cell division forming the xylem and phloem tissues, respectively. The shoot apical meristem is capable of being reprogrammed in the course of plant development. During vegetative growth of the plant, its principal activity is to contribute to stem elongation, and to generate leaf primordia at subapical positions, usually in a highly regular manner giving a typically invariant pattern of leaves (phyllotaxy) around the growing shoot, in a generally indeterminate way. Upon the receipt of an appropriate environmental or developmental signal this pattern of development is disrupted and the vegetative apex undergoes a terminal differentiation in function, producing the reproductive structures, the flowers. The growth pattern of the meristem also changes to a determinate growth habit: the individual flowers are produced when the meristem differentiates to produce the floral organs, and then ceases to proliferate further. The molecular control of floral morphogenesis is one of the few developmental processes occurring in plants for which a molecular explanation is beginning to emerge.

EXPERIMENTAL METHODS IN THE STUDY OF DEVELOPMENT

The objective of developmental biologists is to understand the mechanisms by which the single,

undifferentiated zygote becomes a complex, organized multicellular organism. Within the terms of reference of this chapter, this objective is the elucidation of the pathway by which the developmental programme, encoded within the organism's genome, is interpreted by the cellular biochemical machinery to specify the functions of individual cells and to coordinate these individual functions.

Implicit in this definition is the belief that development is primarily genetically controlled: that there are a relatively small number of genes which act as 'switches', their products regulating the activity of other genes in a spatially or temporally distinctive manner. In very crude terms, we may imagine three basic classes of gene: genes which regulate development, genes which are differentially regulated during development by the action of the first class, and 'housekeeping' genes, whose functions are essential for the viability of all cells (although it should not be thought that 'housekeeping' genes are necessarily unregulated).

The existence of genes which play a central part in coordinating development has long been inferred from studies of mutants showing aberrations in the normal course of development. However, although the inheritance of the mutant genes could be followed, and their positions mapped by classical genetic analysis, the precise nature of the genes and their products has only been amenable to determination since the advent of recombinant DNA technology enabled their isolation and the analysis of their products.

Recombinant DNA techniques permit the molecular cloning of individual DNA sequences. It is a relatively straightforward task to construct a 'library' of DNA fragments, corresponding to the entire genome of a multicellular eukaryotic organism, cloned with a suitable bacterial host–vector system. However, such libraries may comprise up to 10^6 individual clones, and the identification of a clone containing a particular sequence can be an arduous task. Typically, such clones are identified by hybridization with a defined, labelled, probe sequence. However, this approach can only be applied where at least some of the sequence contained with the 'target' gene is already known. In the case of the unknown genes regulating development, this prerequisite is not fulfilled, and alternative strategies have to be devised.

The approaches which have been used to isolate genes controlling development utilize recombinant DNA techniques in concert with classical genetic analysis. Three approaches have proved most fruitful. These are, respectively *chromosome walking, transposon tagging* and *subtractive* or *differential hybridization*.

Chromosome Walking

Chromosome walking is the isolation of overlapping fragments of genomic DNA from a library of cloned fragments by the use of terminal fragments of each clone as successive hybridization probes. The process is illustrated in Figure 11.1. This technique can be applied to the isolation of genes regulating developmental events when they are known to be closely linked to genetic loci for which cloned hybridization probes are available. If mutations defining a developmental process are known to lie between two such markers on a genetic linkage map, it is possible to isolate a collection of clones containing the intervening DNA sequence. In practical terms, this is only the beginning of the procedure. The intervening sequence may be physically large—several hundred kilobase pairs—even when the genetic distance between the markers (defined by recombination frequency) is small. Additionally, because the searcher does not know the nature or extent of the sequence sought, it is necessary to determine, as accurately as possible, the smallest fragment of DNA on which the 'developmental gene' lies. Usually, this must be discovered by reintroducing the candidate DNA sequences into mutant host organisms, in order to determine their ability to correct the mutant phenotype by genetic complementation. Only when this criterion has been fulfilled can the sequence determination of the DNA fragment commence.

Clearly, this is a laborious task, and requires

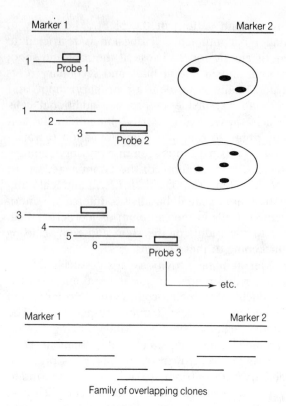

Figure 11.1. Chromosome walking from one genetic marker to another by generation of a series of overlapping clones.

that several prerequisites be met. First, that the organism under study has a comprehensive linkage map, in which the markers correspond to available hybridization probes; this has been achieved for a number of organisms by the use of probes defining restriction fragment length polymorphisms (RFLPs). Second, that the genome of the organism in question is relatively small, thereby reducing the number of clones that must be screened, and that the cloning vector used must accept very long fragments; the development of yeast artificial chromosomes (YACs), in which fragments several hundred kilobases in length may be cloned, has facilitated this task. Finally, it is essential that the organism under study be readily amenable to genetic transformation by foreign DNA to enable the necessary complementation test to be carried out.

Transposon Tagging

Many organisms have been found to harbour mobile genetic elements within their genomes. Such elements, initially discovered in maize, are DNA sequences which exhibit the property of transposition. These sequences encode a 'transposase' function which is responsible for excision and reintegration of the DNA, the extent of the transposable sequence being defined by its possession of terminal repeat sequences which act as the target sequences for the recombinational properties of the transposase functions. During the course of development, transposable elements may become excised from their initial genetic locus, and reappear integrated into another part of the genome. This may occur either in the production of gametes ('germinal excision') or at other times giving rise to a genetically chimeric organism ('somatic excision'). The phenotypes of revertants due to excision can often shed light on the nature of the originally mutated gene.

The integration of such elements is essentially random, and if their insertion occurs within a gene-coding sequence the disruption of the target sequence causes the loss of function of the gene. These transposable elements can thus be used as mutagenic agents—transposon mutagenesis is an established and powerful technique in bacterial genetic analysis. Eukaryotic transposable elements thus provide the developmental biologist with a powerful tool, both for the creation of new mutants and for the isolation of the sequences into which the transposon has integrated. Because a number of transposable elements have been isolated by molecular cloning, it is possible to use their cloned sequences as hybridization probes to identify their sites of integration within a library of cloned DNA derived from a transposon-induced mutant. This approach, known as transposon tagging, is illustrated in Figure 11.2.

In order to utilize this approach, it is necessary that the organism studied be either the host of a previously characterized transposon or accept a foreign transposon introduced by genetic trans-

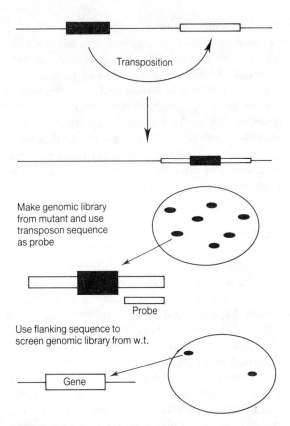

Make genomic library from mutant and use transposon sequence as probe

Probe

Use flanking sequence to screen genomic library from w.t.

Gene

Figure 11.2. Transposon tagging. Transposition of a transposable element into the gene of interest causes a mutant phenotype. The mutated gene is then detectable because of the presence in it of the transposon. The gene sequences can then be used to detect the unmutated gene in wild-type (w.t.) plants.

formation; moreover the behaviour of the transposon should be well understood, so that its activity can be controlled by the experimenter. Ideally, the organism will possess a small genome, reduced in its content of repetitive DNA, thereby increasing the frequency with which transposition results in the interruption of coding sequences. The amenability of the organism to transformation enables the wild-type allele to be introduced to a mutant background, in order to prove its function by complementation testing. However, although desirable, this is not a necessary experiment to prove that the 'tagged' gene encodes the developmental control function in question: because the mutant phenotype depends upon the insertion of a defined sequence, it is possible to demonstrate a precise correlation between the acquisition of the mutation and the insertion of the element by Southern blot analysis of total genomic DNA. As a corollary to this, it is also possible (and essential) to demonstrate that when revertants are isolated, due to germinal or somatic excision of the element, the transposon is lost from the target sequence.

Subtractive/Differential Hybridization

This procedure discriminates between populations of sequences derived from different tissues. It is most readily applicable to the screening of cDNA libraries, particularly in the context of isolating developmentally important sequences. The principle of the technique is simple. It is supposed that tissues undergoing a developmental transition are expressing, as mRNA, both genes which regulate the transition and genes which are regulated during the transition. Tissues in which the transition is prevented, or is abnormal due to a mutation in a regulatory gene, will not express these genes. Messenger RNA is isolated from both classes of tissue, and cDNA is synthesized. This can be used either in a *subtractive* procedure or a *differential* screening.

Subtractive Hybridization

Following cDNA synthesis, the two populations can be hybridized with one another. Conditions are chosen so that one population of sequences is in large excess, i.e. one population acts to drive the hybridization. (This 'driver' population can either consist of cDNA amplified by the polymerase chain reaction, or of the initial mRNA preparation.) Sequences present in both populations will form double-stranded hybrids; sequences present only in one population will remain as single-stranded molecules (Figure 11.3). The double and single-stranded sequences can be physi-

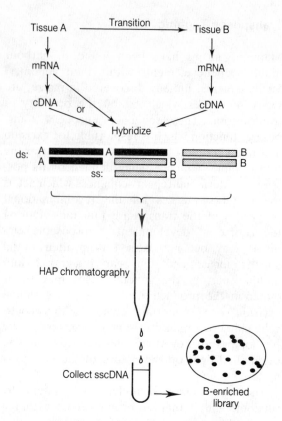

Figure 11.3. Subtractive hybridization.

cally separated, for example by hydroxyapatite chromatography, and the single-stranded component recovered as a population from which the common sequences have been subtracted. (Efficient subtraction usually requires more than one round of hybridization.) The subtracted population can then be cloned and screened— usually by differential hybridization.

Differential Hybridization

As the term implies, this approach relies on identifying sequences in a population by their differential abundance with respect to another population. A cDNA library is constructed, corresponding to the population in which the sought-for sequence is expected to be present (a 'plus' library). The library is then probed by hybridization

DIFFERENTIAL HYBRIDIZATION

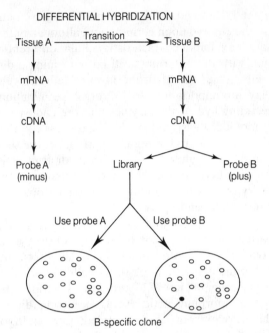

Figure 11.4. Differential hybridization.

with the otal population of sequences comprising the original template mRNA—a 'plus' probe, and also with a 'minus probe'—a population of sequences corresponding to mRNA isolated from a corresponding tissue not expressing the sequence of interest. Individual clones are selected by their ability to hybridize with the 'plus' probe, but not the 'minus' probe (Figure 11.4).

If this approach is to be used to isolate sequences corresponding to genes which regulate development, by subtractive or differential screening of sequences from a wild-type and a mutant, it is only likely to succeed if there is a good expectation that the mutant is a transcriptionally inactive mutant. This is unlikely to be the case if the mutant is a single-base substitution, such as is derived by common chemical mutagenesis procedures, or such as occur spontaneously. However, transposon-induced mutants can fruitfully be screened in this way, as there is a good chance that the insertion of a large genetic element within a coding region will cause the mutant to produce a messenger unrecognizable by a wild-type probe sequence. Once a collection of differentially isolated sequences has been obtained, however, a considerable effort is required to identify which clones correspond to genes regulating developmental change, and which correspond to genes regulated during the change.

GENE REGULATION IN PLANT DEVELOPMENT

Having looked at the range of techniques being employed to investigate the molecular basis of plant development, we now turn to discuss particular facets of plant development which provide interesting examples of gene regulation. In choosing these examples we have concentrated on the events of reproductive development, starting with the fusion of the gametes to form a zygote, moving on to consider the development of the seed and its surrounding fruit, then dealing briefly with germination of the seed before finally discussing the process which starts the cycle over again, floral development. The various phases of development are not treated with equal weight, but are used in a way which gives information on particular facets of the molecular basis of development.

FERTILIZATION AND SEED DEVELOPMENT

Introduction

The seed is the characteristic product of early development in flowering plants. It is a genetically chimeric structure—a consequence of the nature of the fertilization event in angiosperms. Fertilization is the result of the interaction between the female and male gametophytes, and this process is illustrated diagrammatically in Figure 11.5. The female gametophyte—the embryo sac—is a haploid structure derived from a single member of the tetrad of cells produced by meiosis of the 'megaspore mother cell'. This cell undergoes mitotic division and differentiation to

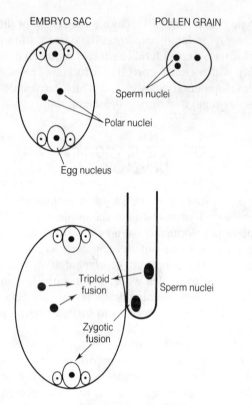

EMBRYO SAC   POLLEN GRAIN

Sperm nuclei

Polar nuclei

Egg nucleus

Triploid
fusion
Sperm nuclei

Zygotic
fusion

Figure 11.5. Diagrammatic representation of the major events of fertilization in angiosperms.

yield a single haploid egg cell and two haploid polar nuclei which participate in fertilization, in addition to a number of associated haploid cells. The embryo sac is fertilized by the pollen grain, which contains a vegetative nucleus and two sperm nuclei. Fusion between one sperm nucleus and the egg nucleus produces the diploid zygote from which the new plant ultimately develops. Fusion between the other sperm nucleus and the two polar nuclei of the embryo sac produces a triploid cell, which proliferates to form the endosperm. This tissue may comprise the bulk of the mature seed, as a storage tissue, as in the case of cereal grains, or it may be reduced in size and importance in seeds where the principal storage reserves are held within the cotyledons which form part of the diploid embryo. In either case, the endosperm may be regarded as a develop-

mental dead end, its sole function being to nourish the embryo upon germination, although in this case it exhibits a characteristic cellular differentiation with the outermost cell layer forming a discrete tissue, the aleurone layer. Aleurone cells play an important role in cereal germination, secreting hydrolytic enzymes into the endosperm to mobilize the stored reserves, a process coordinated by plant growth regulators. Finally, the seed may be surrounded by a pericarp, or fruit wall, which is derived from the maternal tissues of the ovary, and is thus genetically distinct from both embryo and endosperm

Cellular differentiation takes place from the earliest moments, the initial division of the zygote giving rise to two unequally sized daughter cells. Subsequent divisions of these cells define their developmental fates. Figure 11.6 illustrates the typical morphogenetic events occurring during the development of a dicotyledonous embryo. The daughter cell proximal to the maternal tissue divides relatively few times, to form a structure called the suspensor. The distal daughter cell proliferates to form a ball of cells, within which cellular differentiation produces the primordial organs of the new plant. The suspensor is thought to perform a placental function in assisting the conduit of nutrients from the maternal plant into the developing embryo. It is a tissue which is relatively restricted in its development, comprising only a few cells. Following the cessation of the cell divisions necessary for its formation, several further rounds of DNA replication may occur to produce polytene chromosomes in the suspensor cell nuclei.

It is characteristic of angiosperm development that it produces a seed containing an embryo which is essentially a fully differentiated plant in miniature. This comprises anatomically recognizable leaves (cotyledons) or leaf primordia, a shoot meristem and primary and secondary root apical meristems. In monocotyledonous seeds, the shoot and root primordia are enveloped by sheathing structures—the coleoptile and coleorhiza, respectively.

Development of the seed takes place over a lengthy period of time, during which cellular dif-

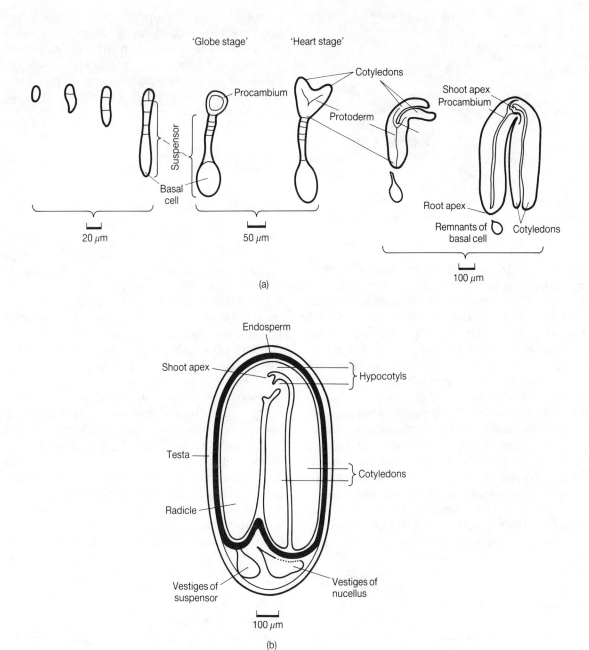

Figure 11.6. (a) Diagram of seed development in a 'typical' dicot plant. (b) Diagram of mature seed structure in *Brassica*, a 'typical' dicot. Reproduced with permission from J.A. Bryant, 1985, *Seed Physiology*, Edward Arnold, London.

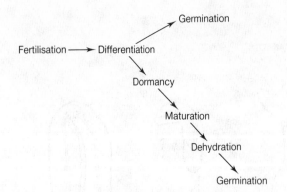

Figure 11.7. Flow diagram illustrating patterns of seed development and growth after fertilization.

ferentiation takes place within the embryo to form the primordial organs of the new plant, and the specialized storage tissues accumulate and package food reserves translocated from the maternal plant. The events occurring during embryonic development can be considered to comprise a series of distinct stages, which are depicted in the flow diagram (Figure 11.7).

The principal events are *fertilization*, which is followed by a phase of cell proliferation and *differentiation*, ending in the production of a functionally mature embryo. This embryo faces a developmental choice. It can germinate immediately, or it can remain in a state of suspended growth (*dormancy*). In almost all plants, dormancy ensues. However, this is not a period of metabolic inactivity. Indeed, it is characterized by the massive accumulation of storage reserves in the appropriate tissues and organs, causing an increase in size and weight, during a phase of development referred to here as *maturation*. The dormant condition is relieved only by the final stage in seed development, *dehydration*. It is perhaps the most remarkable feature of the seeds of flowering plants that not only are they capable of tolerating a period of extreme desiccation (to levels as low as 5% of the original moisture content) but that such dehydration is an essential part of their developmental programme, being both necessary to maintain seeds in a viable form for extended periods (years) and to effect a switch in

their pattern of development. From being a dormant organism, accumulating reserve compounds in an effectively anabolic lifestyle, the embryo is potentiated to germinate immediately upon the uptake of water, its metabolic activity reorientated towards the massive catabolism of these reserves to support growth.

Fertilization

The plant kingdom exhibits clearly the alternation of generations: the sporophyte generation bears 'spores', which germinate to give a gametophyte generation, which bears gametes. In some groups of plants both generations are well developed. Thus in ferns the sporophyte generation is the familiar fern plant; the spores germinate to form the prothallus—very much smaller and very different in morphology from the sporophyte but clearly a plant. The independence of the gametophyte generation has been lost with evolutionary advancement, and this is seen as its most extreme in flowering plants, where the gametophyte generation in the female is reduced to an embryo sac containing eight cells and surrounded by maternal tissues (Figure 11.8), and in the male to two nuclei: the vegetative nucleus and the generative nucleus of the pollen grain. The latter divided either at the time of pollen dispersal or immediately following pollination, to form two sperms or male gametes. In order to achieve pollination, the pollen grain must reach the receptive surface of the stigma, and from that point successful pollination occurs in several steps.

The first step is the adhesion of the pollen grain to the stigmatic surface. This may involve a specific interaction between the surface of the pollen grain and the stigma (see below). The capture of the pollen grain leads to its hydration, followed by germination of the vegetative cell to form the pollen tube. Although this process requires active protein synthesis, there does not appear to be the need for synthesis of any RNA, suggesting that the complete protein synthetic machinery, including mRNA, is present in the ungerminated pollen

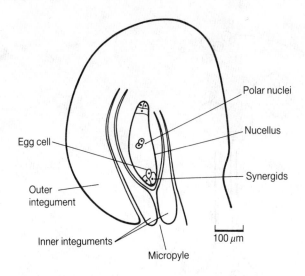

Figure 11.8. Diagram showing structure of embryo sac. Reproduced with permission from J.A. Bryant, 1985, *Seed Physiology*, Edward Arnold, London.

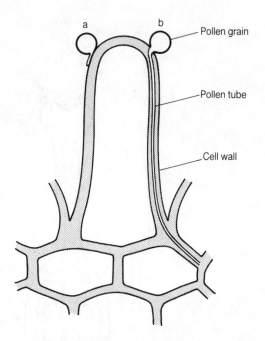

Figure 11.9. Diagrammatic representation of pollen tube growth in *Brassica*. Pollen grain 'a', which is incompatible, has started to grow, but growth is soon aborted. Pollen grain 'b' has grown into the cell wall of a stigmatic papilla, and the pollen tube growth continues towards the embryo sac. The shading represents the presence of the *S*-glycoprotein in the cell wall.

grain, having been laid down during microsporogenesis.

The penetration of the pollen tube and its subsequent growth in the style (Figure 11.9) mean that the gametophyte is essentially acting parasitically within female sporophyte tissues, and again correct intercellular recognition is very important. Finally, on reaching the ovary, the two male gametes (see above) are released, one of which fertilizes the egg cell, and one of which undergoes a triple fusion with the two polar nuclei of the central cell. At this point the cellular recognition system involves direct contact between gametes (or between a male gamete and the polar nuclei).

The series of events just described does not usually come to completion if 'incorrect' or incompatible pollen lands on the stigmatic surface. In other words, there are mechanisms which prevent self-fertilization in obligately outbreeding species and there are mechanisms which prevent fertilization by pollen from a different species. In flowering plants there are two general phases during which compatibility or incompatibility may be manifest. The first is in the interaction between the pollen and the stigma and/or style. The second phase is the interaction between the gametes themselves.

Gametophytic Incompatibility

The best-researched compatibility/incompatibility systems are those which prevent selfing in outbreeding species. The most widespread forms of self-incompatibility are governed first by the genotypes of the pollen grains and second by the genotypes of the diploid stigmatic or stylar tissues. These systems are known as gametophytic, because of the involvement of the pollen grain, and in the commonest type of gametophytic control of compatibility one highly polymorphic genetic locus, the *S*-locus, is involved. The *S*-gene codes for an antigenic protein which, like so many

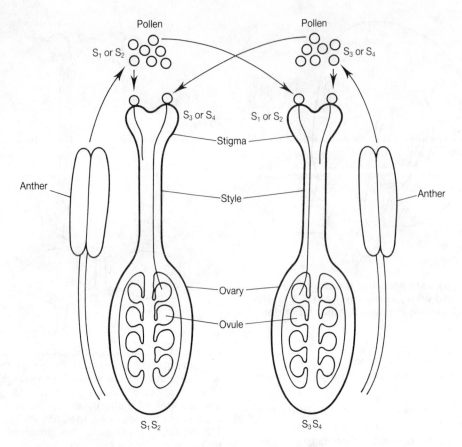

Figure 11.10. The gametophytic system, of self-incompatibility. Redrawn from H.G. Dickinson, 1990, *BioEssays*, **12**, 155–161, by permission of ICSU Press.

proteins involved in cell recognition, is glycosylated. Since only one locus is involved, the *S*-gene product must incorporate both the pollen and stylar components of the recognition system. Growth of a pollen tube carrying a given *S*-allele is prevented if that same *S*-allele is also carried by the stylar tissues (Figure 11.10). If the allele carried by the pollen is not carried by the stylar tissues, then successful pollen tube growth occurs.

As already pointed out, there are a number of possible points at which the incompatibility system may work. In practice, however, it seems that recognition either occurs very early, on the stigmatic surface, or when the pollen tube is growing through the stylar tissues. Examples of the former are seen in the grasses and in evening primrose

(*Oenothera*). In grasses, the stylar papillae are covered with a layer of proteinaceous material. Pollen grains land on this and start to germinate, but in incompatible pairings tube growth stops almost immediately. The protein layer is thus very likely to be part of the recognition system. In *Oenothera*, the stigmatic surface is covered with a fluid secretion; incompatible pollen grains fail to penetrate this, again indicating that the fluid layer is part of the recognition system.

In many other species exhibiting gametophytic self-incompatibility, the recognition of incompatible pollen does not occur until after pollen tube germination. Members of the family Solanaceae, including *Petunia*, tobacco and tomato, provide good examples of this. In these species, the major molecular recognition event occurs while the pollen tube is growing through the stylar tissues. The inhibition of pollen growth shows in a number of different ways even within the same species. The commonest are a swelling and bursting of the tip of the pollen tube, or a cessation of growth accompanied by occlusion by callose, a β1,3-linked glucan which is very characteristic of wound responses in higher plants. The molecular basis of the incompatibility reaction is slowly becoming clear, aided by cloning of the S-gene from tobacco and tomato. On the pollen side, it seems likely that a recognition factor (i.e. the pollen's S-glycoprotein) is exposed after germination of the pollen tube. On the female side, the stylar tissues of *Petunia* and tobacco have been shown to secrete glycoproteins that inhibit the *in vitro* growth of pollen tubes which carry the same S-allele as one of the stylar S-alleles. These glycoproteins have now been shown unequivocally to be the products of the S-genes, and it is now also known that on the female side the messenger RNA molecules for these proteins are only synthesized in stylar tissues. Unfortunately, as yet there is no information on the transcription of the S-gene on the male side.

Analysis of the cloned S-gene from tobacco and tomato indicates that it has three major regions. Two of these, one coding for pollen activity and one coding for stylar activity, are conserved, while the third, representing the region of allelic differences, is highly polymorphic. Exactly how the gene transcripts are processed to give style or pollen-specific antigens is not known; indeed, it is possible that both the pollen-specific and the stylar-specific regions are present in both male and female antigens, and that the differences between the two lie in the patterns of glycosylation. Furthermore, despite quite a detailed knowledge of the gene structure, it is not at all clear how the antigens recognize each other as similar or different. What is now clear, however, at least for tobacco (but likely to be so for tomato too), is the mechanism by which the female S-allele proteins arrest the growth of incompatible pollen. The sequence of the S-allele cDNAs indicates a strong homology with fungal ribonuclease genes,

and it has now been established that the stylar tissues do indeed produce ribonuclease. The ribonuclease enters the growing pollen tube of incompatible pollen, completely breaking down the pollen's RNA. The RNA of compatible pollens, and of the stylar cells themselves, is not degraded at all.

Sporophytic Incompatibility

The second type of incompatibility is based, on the male side, *not* in the genotype of the pollen grain but in the genotype of the paternal parent. For this reason, the system is known as sporophytic incompatibility. Again, the alleles concerned are termed S-alleles, although analysis of cloned S-genes has shown that these genes are not homologous with the S-genes from plants exhibiting the gametophytic system. In the sporophytic system, pollination fails if either of the S-alleles carried by the female parent is matched by either of the alleles carried by the male parent (Figure 11.11). In the best-investigated examples of this, in the families Compositae and Cruciferae, recognition of incompatible pollen occurs very early, usually immediately after hydration of the pollen grain: the stigmatic surface is able, almost instantly, to recognize the genotype of the parent whence the pollen came. Several years ago, Heslop-Harrison, a major contributor to our knowledge of this topic, showed that the mutual recognition depended on substances present in the cell walls of both the stigmatic papillae and in the pollen grain (see below). The involvement of the pollen's parental genotype suggests very strongly that the substances secreted from the pollen come from the exine, since this layer is deposited from parental tissue (the tapetum) and thus contains products from the parental S-alleles. If the interaction is compatible, the pollen tube grows down in the stylar tissues, having dissolved the cuticle of the papillae by secretion of hydrolytic enzymes. In an incompatible reaction, pollen tube tip growth ceases very quickly, and as in the gametophytic system the cell becomes completely occluded by callose. In the sporophytic system,

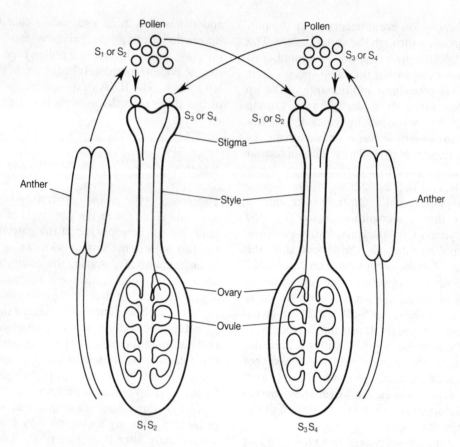

Figure 11.11. The sporophytic system of self-incompatibility. Redrawn from H.G. Dickinson, 1990, *BioEssays*, **12**, 155–161, by permission of ICSU Press.

callose is also laid down in the stylar papilla in contact with the pollen grain.

The antigenic substances mentioned above are now known to be glycoproteins. Glycoprotein products from a number of *S*-alleles from *Brassica* have been sequenced and their genes have been cloned. As expected from the lack of homology at the gene level, these glycoproteins are not homologous with those involved in the gametophytic recognition system. In *Brassica*, the *S*-proteins have seven sites at which they may be glycosylated, all clustered in the N-terminal half of the protein (Figure 11.12), and glycosylation involves the addition of relatively complex branched oligosaccharides (Figure 11.12). Analysis of the proteins coded for by different *S*-alleles indicates that variation is not confined to one polymorphic region of the protein, but occurs as individual amino acid changes throughout the molecule. There are also differences in the number of potential glycosylation sites which actually are glycosylated in the mature protein. As with the gametophytic system, we do not yet understand how a particular *S*-glycoprotein recognizes another as being similar or different.

Barriers to Interspecies Pollination

Some closely related plant species are interfertile, but in the vast majority of instances pollen from one species is incompatible with stigmatic or sty-

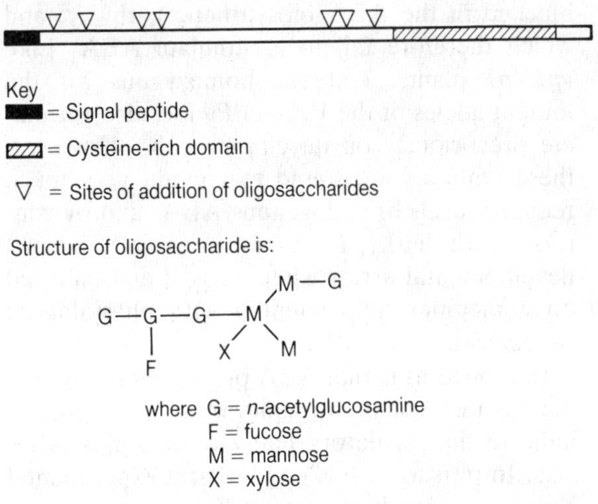

Key

■ = Signal peptide

▨ = Cysteine-rich domain

▽ = Sites of addition of oligosaccharides

Structure of oligosaccharide is:

```
                        M—G
                       /
       G—G—G—M
       |       X \
       F             M
```

where G = *n*-acetylglucosamine
       F = fucose
       M = mannose
       X = xylose

Figure 11.12. Diagram of the structure of the S-locus glycoprotein of *Brassica*. The protein is just over 400 amino acids in length and is glycosylated at seven sites (indicated by the triangles). The 31-amino-acid signal peptide is involved in transport of the glycoprotein into the cell wall and is cleaved from the protein during transport.

lar tissues from another. Very often, this incompatibility is governed by rather non-specific morphological or physiological barriers; for example, the osmotic potential of a stylar secretion may be wrong for the foreign pollen. Obviously these interactions are a result of the normal growth and development of the plant concerned and in that sense are under genetic control. However, there are groups of species in which there is evidence for a more specific type of fertility barrier, somewhat analogous to self-incompatibility systems. In the Compositae and the Cruciferae (two families with a sporophytic mode of self-incompatibility), pollen from other families simply fails to grow. However, pollen from other members of the same family elicits a response similar to that elicited by *self*-incompatible pollen, including the deposition of callose. The molecular basis of this particular interaction is not understood, particularly since there is evidence that the system can be 'fooled': provision of compatible pollen alongside the incompatible will often lead to the acceptance of the incompatible pollen, resulting in the develop-

ment of some hybrid seeds. Traditional plant breeders have known this for many years and often use such pollen mixtures to achieve hybridization between two otherwise incompatible species.

Seed Development

Having achieved a successful landing, germination and tube growth, the compatible pollen eventually delivers its two gametes to the embryo site and fertilization occurs as described in the Introduction to this section. Although the time taken from fertilization for the formation of a mature, dry seed may be between 40 and 60 days, the time taken for an embryo to become a fully differentiated, functionally mature structure is very much shorter. The acquisition of functional maturity can be recognized by the ability of an immature embryo, removed from the confines of the developing seed, to undergo premature or *precocious germination* when supplied with the most basic nutrients. For example, immature wheat embryos will germinate, in culture, between 10 and 14 days following fertilization, if removed from the grain and provided with a simple medium containing sucrose, amino acids, salts and vitamins. Such precocious germination does not occur while the grain is attached to the parent plant. Nor does the embryo germinate precociously within the grain if the entire immature grain is removed from the developing ear and placed on such a growth medium. The ability of the embryo to germinate within the grain is acquired only following dehydration of the grain.

Such a contrast between the short time of acquisition of functional maturity, assessed by potential for precocious germination, and the long period of time that elapses before evident maturity, signalled by the germination of the embryo from within a whole seed, is observed in the majority of angiosperm species. That the differentiation of the embryo and the acquisition of germinability is attained far in advance of the achievement of the maximum size and dry weight of the whole seed is due in large measure (particularly in the cereals, where the embryo constitutes only a small proportion of the overall mass of the mature grain) to the necessity to accumulate storage reserves: triglycerides, carbohydrates and storage proteins. Since the function of these reserves is to supply the germinating embryo with a sufficient stock of nutrients to fuel vigorous growth prior to seedling establishment and the adoption of an autotrophic lifestyle, it would clearly be highly disadvantageous if the potential of the embryo to germinate was expressed immediately upon its acquisition. In that event, the embryo would rapidly exhaust its resources.

Dormancy in immature seeds is believed to be imposed through the action of a plant growth regulator, abscisic acid (ABA). Three lines of evidence support this contention. First, measurements of ABA within developing seeds indicate a general increase in the endogenous levels of this substance which is correlated with the onset of dormancy. More critically, immature embryos dissected from developing seeds can be prevented from undergoing precocious germination, in culture, if the culture medium is supplemented with physiological concentrations of ABA. Moreover, under such conditions the embryos can be observed to undergo a series of metabolic changes characteristic of those which occur normally within the developing seed during the maturation period. Finally, there is genetic evidence to link the maintenance of embryonic dormancy with the action of ABA. A series of maize mutants have been characterized which display a viviparous phenotype. Their embryos commence germination from immature kernels attached to the developing cob. These mutants, designated *vp* (for *viviparous*) fall into two classes. The first class consists of an ABA non-responsive mutant: embryos which are homozygous for the recessive mutant allele at the *VP1* locus (*vp1/vp1*) contain normal levels of ABA, but fail to respond to it. Such embryos germinate vigorously when dissected from the developing grains even when cultured in the presence of high concentrations of exogenous ABA, whereas wild-type or *vp1/+* heterozygous embryos are arrested. The second class consist of a number of mutants which are

blocked in the ABA biosynthetic pathway, and which therefore fail to accumulate ABA. Like *vp1/vp1* plants, embryos homozygous for the mutant alleles of the *VP2—VP9* loci also germinate precociously on developing cobs. However, these embryos are found to contain very much reduced levels of endogenous ABA, and by contrast with *vp1/vp1* embryos exhibit normal developmental arrest when excised and cultured on a medium supplemented with physiological concentrations of ABA.

Demonstrating that ABA plays a central role in the maintenance of dormancy does not, however, indicate the regulatory pathway by which ABA acts. In pursuit of this end, several experimental approaches are being taken. The most common approach has been to identify gene products which are synthesized in isolated immature embryos, in response to exogenously applied ABA. Such products are the mRNA and polypeptide products of genes which are expressed naturally towards the end of the period of embryonic dormancy—the phase during which the dehydration of the seed occurs. These gene products accumulate to high levels in dry embryos, and in consequence of this, and their temporal appearance during the later phases of embryogenesis, have been designated 'late-embryogenesis-abundant' or *Lea* gene products.

The *Lea* proteins characterized thus far constitute an interesting subset of proteins which exhibit a high degree of homology across species boundaries. Theoretical consideration of their physical nature—they may form either amphipathic α-helices capable of associating with both the hydrophobic components of cellular membranes and the polar regions of cytosolic macromolecules, or be essentially unstructured random coil polypeptides with a high affinity for water molecules—suggests that their principal role may be to protect macromolecular structures within cells from irreversible denaturation upon dehydration. This has been projected to occur by close association between the *Lea* proteins, retaining a shell of hydration in the form of tightly bound water, and other more susceptible macromolecules, thereby providing these latter with a sufficient level of hydration within their immediate molecular environment for them to retain their three-dimensional structure.

In addition to being temporally expressed during late embryogenesis, *Lea* genes are activated by exogenous abscisic acid, both in immature embryos and in the vegetative tissues of seedlings. Additionally, the accumulation of the principal *Lea* proteins is induced in response to osmotic stress. Since in vegetative tissues the imposition of stress appears to cause a consequent increase in the levels of endogenous ABA, it is easy to see how both stimuli may be related in the regulation of *Lea* gene expression. In immature embryos, the causative link between osmotic stress and ABA accumulation is less clear-cut, and there is also some evidence to indicate that the effect of ABA, in preventing precocious germination, is to modify the permeability of the cell membrane to water, thereby preventing water uptake-mediated turgor pressure from driving the cell expansion, which is the characteristic feature of germination. Notwithstanding the apparent circularity of the arguments which link stress and ABA responses in plants, the regulation of *Lea* genes provides a model wherein the consequences of two factors, one developmental and one environmental, but each instrumental in the developmental transitions occurring during embryo development, may be dissected by experiments at the molecular level.

Progress is now being made towards the isolation and analysis of genes whose role is to coordinate embryonic dormancy, and the molecular responses to ABA within embryos. The isolation of a number of *Lea* genes has enabled the identification of proteins which interact with their promoter sequences, and which are thus potential transcriptional regulators. This in turn has permitted the isolation of the DNA sequences encoding these proteins, amongst which one might expect to be numbered genes intimately connected with the regulation of embryonic dormancy.

The most significant results have attended the recent isolation of the maize *VP1* gene, by transposon tagging. Simple inspection of the DNA sequence of this gene, and hence deduction of the amino acid sequence of the predicted gene product, provided few clues to the functional nature of the *VP1* protein in bringing about the ABA response of maize embryos. However, functional tests have provided valuable insights. First, the *VP1* protein contains a sequence encoding a protein domain with a distinctly acidic character. Such domains have been identified in a number of transcriptional regulators, and when the *VP1* acidic domain was fused with the DNA-binding domain of a yeast transcription factor (GAL4), using recombinant DNA techniques, the hybrid protein proved capable of activating the transcription of a reporter gene bearing the *GAL4* binding site, in transformed plant cells. Activation of transcription was dependent on the presence of the *VP1* domain in the fusion protein, demonstrating its potential to activate transcription. Second, introduction of the wild-type *VP1* gene into maize cells derived from a *vp1/vp1* mutant line restored the wild-type characteristic of activating the ABA-induced expression of a *Lea* gene. Thus it appears that the coordinating role played by the *VP1* gene is exercised through activating the transcription of other dormancy-related genes. Precisely how *VP1* is integrated into the ABA-perception and signal-transduction chain remains to be elucidated.

FRUIT DEVELOPMENT AND RIPENING

Introduction

In angiosperms (flowering plants), the seeds are enclosed by a structure known as the fruit. In strictly botanical terms the fruit is actually a ripened ovary which contains seeds, and thus the cereal 'grain' discussed in the previous section is actually a one-seeded fruit, with the pericarp being derived from the integuments of the ovary. By contrast, many of the structures we normally think of as fruit are actually formed from tissues other than the ovary, such as the receptacle, as in strawberry, and the peduncle (flower-stalk) as in fig. However, although such organs are technically not fruit, they certainly fulfil the same functions as fruit (i.e. protection and then dispersal of the seed), and their development is regulated in similar ways to 'genuine' fruit.

The development of fruit in most species depends on successful fertilization and thus proceeds in step with the development of the enclosed seeds. For fleshy fruits and other fruit-like structures, an initial phase of cell division is followed by a phase of massive cell expansion, leading to the formation of very large vacuolate cells (up to 700 μm in length). This cell expansion phase naturally enough leads to fruit enlargement, which is sometimes very spectacular: the volume of a mature water-melon may be as much as 30 litres, for example.

Climacteric and Non-climacteric Fruit

Following expansion, fruit ripening occurs. For the cereal grain this process parallels very closely the ripening of the enclosed seed, with the pericarp behaving like an extra seed coat and undergoing desiccation in parallel with the seed coat. For fleshy fruit, by contrast, the ripening of the fruit involves changes in colour, texture, flavour and aroma, these changes being associated with the attraction of suitable 'consumers' in order to facilitate the dispersal of the enclosed seeds. Such fruits have, on the basis particularly of their respiratory metabolism during ripening, been classified as either climacteric or non-climacteric. In climacteric fruit, the rate of respiration falls as the expansion/maturation phase finishes, then rises significantly during ripening before finally declining again. There is also in climacteric fruit a massive increase during ripening in the rate of synthesis of the gaseous hormone, ethylene. In non-climacteric fruit, there is no clear 'burst' of respiration during ripening and ethylene production is very limited.

Ripening in Tomato

In this section we concentrate on selected aspects of ripening in one climacteric-type fruit, the tomato, since research on this species is beginning to throw light on the way in which ethylene regulates ripening (or at least particular facets of the ripening process).

In tomato the cell division phase of fruit growth lasts about 2 weeks after pollination, following which cell expansion leads, after a further 3–4 weeks, to the formation of the mature, but unripe (i.e. green) fruit. The ripening process takes a further 1–2 weeks, depending on the particular cultivar of tomato and the ambient environmental conditions. Ripening is a very complex process, involving highly coordinated changes in gene expression, metabolism and ultrastructure throughout the fruit (Table 11.1). These changes are regulated by the hormone ethylene, as is shown by the failure to ripen tomatoes treated with inhibitors of ethylene synthesis or action (Figure 11.13) and of tomatoes carrying mutations affecting ethylene synthesis (Table 11.2). Ethylene synthesis is autocatalytic (i.e. is positively autoregulated) and although it is not clear what controls the initial low level of ethylene synthesis at the start of ripening, it is clear that low concentrations of ethylene then stimulate the massive burst of ethylene synthesis which characterizes these climacteric fruit. Interestingly, there is an indirect reference to this, relating to another fruit, dating back to Old Testament times. The prophet Amos referred to himself as a dresser of sycamore figs, which is almost certainly a reference to the process of slashing, cutting or punc-

Table 11.1. Major physiological and biochemical changes which occur during tomato fruit ripening.

Increased ethylene biosynthesis
Changes in gene expression
Increased respiration (the 'climacteric')
Loss of thylakoids and photosynthetic enzymes
Degradation of chlorophyll
Synthesis of pigments, especially lycopene
Changes in organic acid metabolism
Increases in activities of polysaccharide-
    hydrolysing enzymes, particularly
    polygalacturonase
Depolymerization of cell wall polyuronides
Softening
Increased susceptibility to pathogen attack

Table 11.2. Some mutations affecting tomato fruit ripening.

Name of mutation	Effects
rin (ripening inhibitor)	Fruit do not produce ethylene, nor do they exhibit a respiratory climacteric Fruit turn from green to yellow, but pigmentation changes proceed no further Polygalacturonase activity is greatly reduced (see Figure 11.14). Fruit fail to soften Normal ripening is not restored by exogenous ethylene
nor (not-ripening)	A separate gene from *rin*, but causing a very similar phenotype
nr (never ripe)	This is really a 'partially ripening' mutant The fruit turn orange: some lycopene (but much less than normal) is produced The fruit soften: some polygalacturonase is produced (but again much less than normal) Ripening may be restored by provision of exogenous ethylene

turing the figs in order to hasten ripening (as illustrated in carvings dating back to that era). The basis of this, as we now know, was to cause the production of wound-induced ethylene, which in turn led to the climacteric ripening of the figs. Harvesting took place 4–5 days after the initial wounding of the fruit.

So, how does ethylene control the ripening process? At present, it is very difficult to understand the overall coordination of ripening, but we

Figure 11.13. Inhibition of ripening in tomato. One half of a mature green tomato was injected with an inhibitor of ethylene action; the other half was not. The control half ripened normally, while the half containing the inhibitor failed to ripen. (Photograph supplied by Professor D. Grierson.)

do have information on some of the individual components of the process. At the molecular level, the preparation and differential screening of cDNA libraries from tomato has led to the isolation of several cDNAs that represent mRNAs which accumulate in response to ethylene as the fruit progresses from mature green to ripe. Here, we concentrate on a limited selection of these cDNAs/mRNAs and their corresponding genes.

The first of these is the mRNA for the enzyme polygalacturonase. This enzyme is under strict developmental control in the ripening tomato (Figure 11.14). In the mature green fruit it cannot be detected either by enzyme assay or by antibodies, but it appears about 24 hours after ripening commences and increases very significantly in both amount and activity as ripening proceeds, reaching maximum levels in red fruit. Its role in ripening is to digest the polygalacturonic acid in the cell wall, thereby contributing to fruit softening, although it must be emphasized that other hydrolytic enzymes are also involved in this process. Use of a polygalacturonase cDNA clone to assay polygalacturonase mRNA reveals that the changes in enzyme concentration and activity closely follow very similar changes in the amount of mRNA. Indeed, polygalacturonase mRNA accumulates to such high levels that at its peak it constitutes 2–4% of the total mRNA. Further,

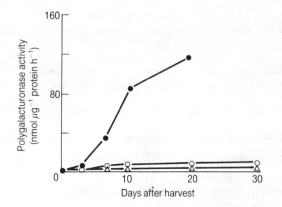

Figure 11.14. Production of polygalacturonase during ripening of tomato. ●, Control fruit; ○, fruit treated with an inhibitor of ethylene action; △, fruit of *rin* mutant plants. Fruit of *rin* plants neither produce ethylene, nor respond to it. Fruit were harvested at the mature green stage.

nuclear 'run-on' experiments indicate that this accumulation of mRNA represents a huge up-regulation of the rate of transcription of the poly-galacturonase gene. (Note: in nuclear 'run-on' experiments, transcripts made *in vitro* by isolated nuclei are assayed by hybridization with appropri-ate cDNA or genomic probes). By contrast, in the ripening mutants *rin* and *nr* (Table 11.2) the accumulation both of polygalacturonase mRNA and of the enzyme itself are reduced c. 100-fold compared to wild-type fruit, while in the *nor* mutant (Table 11.2) there is no detectable poly-galacturonase mRNA or enzyme. Thus, in these mutants there is a correlation between a failure to produce or perceive ethylene and a failure to up-regulate the polygalacturonase gene. Further con-firmation of this comes from the finding that treatment of wild-type fruits with norbornadiene, an inhibitor of ethylene action, also inhibits the transcriptional activation of the polygalactur-onase gene. The molecular mechanisms under-lying the regulation of gene activity by ethylene are discussed later.

In the meantime, we consider another facet of polygalacturonase molecular biology, viz. that this was one of the first plant genes to be 'artificially' down-regulated using 'anti-sense' technology.

Research groups in the UK and the USA have made gene constructs in which the polygalactur-onase cDNA is spliced *in reverse orientation*, to the promoter from cauliflower mosaic virus which normally regulates the synthesis of the virus 35S polycistronic mRNA (the CaMV 35S promoter); this promoter acts as constitutive promoter in transgenic (genetically engineered) plant cells. The constructs have been inserted into the T-region of the *Agrobacterium* Ti plasmid and intro-duced into tomato plants via *Agrobacterium* infection. The CaMV 35S promoter of course drives the synthesis of the reverse mRNA—i.e. anti-sense mRNA which is complementary to the 'real' polygalacturonase mRNA—in all cells of the transgenic tomato plants. However, only in ripening fruit is the real polygalacturonase mRNA also synthesized, and the anti-sense RNA is then able to base-pair with it (Figure 11.15). Such double-stranded RNA molecules are rapidly degraded, and when polygalacturonase mRNA is assayed in tomatoes carrying the anti-sense con-struct, very little can be detected. Furthermore, there is a concomitant inhibition of the rise in polygalacturonase enzyme concentration and activity (Figure 11.16). This is very important from a commercial viewpoint, since failure of the fruit to degrade the polygalacturonic acid of the cell wall (while all the other processes involved in ripening proceed normally) very much prolongs the post-harvest shelf-life of the fruit.

The precedent set by this use of anti-sense technology has led to its application in identifying a key gene in the ethylene response. Many of the cDNAs made from mRNAs which accumulate in ripening fruit are 'anonymous', i.e. the proteins for which these cDNAs/mRNAs code have not been identified; all that is known is that these mRNAs are ripening and/or ethylene response specific. However, Grierson's group at the Uni-versity of Nottingham has shown that one of these cDNA clones, known as pTOM13, encodes the elusive ethylene-forming enzyme. When pTOM13 is inserted into tomato plants in reverse orientation (as described earlier for the polygalac-turonase cDNA) ethylene production by the fruit is severely inhibited (Figure 11.17) and fruit

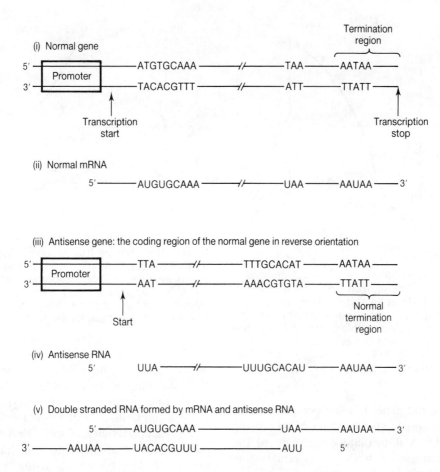

Figure 11.15. Diagram illustrating the use of anti-sense genes and the probable mechanism by which anti-sense mRNA inhibits normal gene expression. Reproduced from J.A. Bryant, 1988, *Plants Today,* **1**, 23–28, by permission of Blackwell Scientific Publications.

ripening proceeds very slowly. In the same transgenic plants, the pTOM13 anti-sense constructs also inhibit the wound-induced production of ethylene in leaves. Further, in the pTOM13 anti-sense fruits there is a build-up of the immediate precursor of ethylene, ACC (amino cyclo-propyl carboxylic acid), thus indicating that the conversion of ACC to ethylene is inhibited, and suggesting strongly that pTOM13 encodes the ethylene-forming enzyme (ACC oxidase). This has led to the identification and isolation of the ethylene-forming enzyme itself, a goal which had eluded biochemists for many years. Perhaps more significantly in the context of this chapter, these results

illustrate further the autocatalytic mode of ethylene synthesis: pTOM13 not only represents an ethylene-responsive gene, up-regulated by ethylene, but also actually encodes a key enzyme in the biosynthesis of the hormone.

So, as might have been suspected, the ethylene-regulated ripening process involves the up-regulation of several genes, including the auto-regulated genes controlling ethylene biosynthesis (and the down-regulation of several other genes). By comparison with other hormone or developmentally regulated genes, it would not be surprising to find that transcriptional regulation involves the tract of DNA immediately upstream (i.e. on

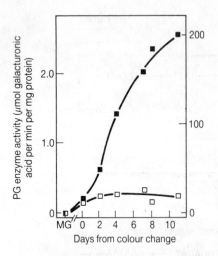

Figure 11.16. Production of polygalacturonase (PG) by control tomato fruit (■) and in fruit of tomato plants carrying an anti-sense version of the gene encoding polygalacturonase (□). The production of pigments during ripening was similar in the two sets of fruit. Reproduced with permission from Smith et al., Nature, 334, 724–726. © 1988 Macmillan Magazines Limited.

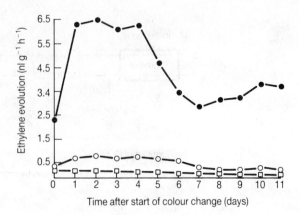

Figure 11.17. Ethylene production by control tomato fruit (●), by fruit of plants containing one anti-sense copy of 'pTom 13' (○) and by fruit of plants containing two anti-sense copies of 'pTom 13' (□). Reproduced with permission from A.J. Hamilton et al., Nature, 346, 284–287. © 1990 Macmillan Magazines Limited.

the 5' side) of the gene. For the polygalacturonase gene, constructs have been made where the 1500 bp of DNA immediately upstream of the gene have been spliced to a reporter gene coding for chloramphenicol acetyl transferase (CAT). In transgenic tomato plants carrying this construct, CAT was synthesized specifically in fruit, regulated by ethylene. 'Dissection' of the 1500 bp promoter region by making partial deletions has unfortunately not yet revealed any specific regions or sequence elements within the promoter region which are specifically essential for the ethylene-regulated ripening-specific response. However, for two other ethylene-responsive genes, 'E4' and 'E8', albeit anonymous (isolated from genomic DNA by using two anonymous ethylene-responsive cDNAs as probes), a sequence element common to both promoters has been identified (Figure 11.18). The spatial relationship between this sequence element and the transcription start site differs between the two genes, which for some researchers casts doubt on the idea that this element has a similar function

in the transcriptional regulation of these two genes. However, the element is recognized by a DNA-binding protein, which, as indicated by sequence competition experiments, is highly specific for the sequences illustrated in Figure 11.18. The presence/activity of this DNA-binding protein is developmentally regulated, further indicating that it may be a *trans*-acting transcriptional regulation factor involved in the regulation of these 'ripening genes'.

With results of the type just described, we are just beginning to understand the regulation of gene transcription by ethylene such as occurs in ripening fruit. However, it would be misleading to suggest that understanding the transcriptional regulation of a handful of genes actually provides us with a full understanding of fruit ripening. For a start, the ripening process itself, as emphasized earlier, is complex and highly coordinated,

$$ATT\frac{C}{T}C\frac{C}{T}A\frac{ACA}{TAT}A\frac{T}{A}AGAAA$$

Figure 11.18. Consensus sequence of a regulatory element from two ethylene-responsive genes in tomato. (Based on data in S. Cordes et al., 1989, The Plant Cell, I, 1025–1034, by permission of the American Society of Plant Physiologists.)

involving all cell compartments in all parts of the fruit and affecting many different facets of biochemistry, physiology and cell morphology. The spatial relationships within the fruit of the different processes rely on the pre-existing morphology of the fruit, established during fruit development before the ripening process begins. Our understanding of the control of fruit development is minimal, as is our understanding of the control of the initial low level of ethylene biosynthesis which apparently triggers the ripening process.

Within the ripening process itself there are clear indications that ethylene regulates different genes in different ways. In detached fruits supplied with ethylene, several genes, including that encoding the ethylene-forming enzyme, are up-regulated within about 2 hours, while for others up-regulation occurs much later; the accumulation of polygalacturonase mRNA, for example, does not start until 24 hours after ethylene treatment. Its up-regulation may represent a late phase in a cascade initiated during the short-term (say 2 hours) response to ethylene. This may explain why the putative 'ethylene-response element' present in the promoters of the rapidly responding genes 'E4' and 'E8' is apparently not present in the polygalacturonase promoter.

It should also be noted that the accumulation of one particular mRNA species in response to ethylene does not in fact involve transcriptional regulation. This mRNA is transcribed from another anonymous gene, E17, in mature green fruits in the absence of ethylene, as shown by nuclear run-on experiments. However, only after application of ethylene does the mRNA accumulate, clearly indicating the existence of a post-transcriptional regulatory mechanism.

Finally, some mention must be made of the overall signal transduction pathway involved in the ethylene response. At one end of the pathway there is the hormone. At the other end is the up-regulation of transcriptional activity (or, at least in one instance, some sort of post-transcriptional mechanism) leading to the appearance of a range of enzymes and other proteins which characterize the ethylene response in fruit ripening. For the genes involved in the early response, it is possible to take one step back up the transduction pathway to suggest that the early response may at least partially be coordinated by a *trans*-acting factor (DNA-binding protein) which recognizes an 'ethylene-response element' in the promoters of the early response genes.

But what of the hormone end of the pathway? Current thinking would suggest that the initial event in the pathway is likely to be the binding of the hormone to a receptor molecule, most likely a protein. To date, no ethylene receptors have been isolated from tomato fruit, although it is clear that sensitivity to ethylene increases very markedly in the later stages of fruit development, i.e. as the fruit reaches the mature green state. However, an extremely hydrophobic ethylene-binding protein, with binding and dissociation kinetics consistent with its being an ethylene receptor, has been purified from cotyledons of French bean. Its role in the cotyledons is not clear, since these organs do not exhibit an ethylene response! Nevertheless, purification, characterization and sequencing of the protein, coupled with the raising of antibodies, will facilitate a search for similar proteins in tissues and organs which are ethylene responsive, including tomato fruit.

GERMINATION

Introduction

Except in those seeds held in check by a specific dormancy mechanism, rehydration of seeds under conditions appropriate for growth will lead to germination, seedling growth and ultimately the establishment of the new plant. In the most simple terms, germination may be regarded as a resumption of growth which has been previously stalled by desiccation. However, the description of embryogenesis in the previous section has already indicated that the 'germinability' of developing seeds changes during development and there are actually clear metabolic and physiological differences between a developing embryo and a germinating embryo.

Hormones and Germination

The role of hormones in the germination process *per se* (i.e. the resumption of embryo growth leading to emergence of the radicle) is uncertain. It is certainly true that exogenous application of ABA inhibits germination in many species, while exogenous application of gibberellic acid (GA) and/or cytokinins promotes (i.e. increases the speed of) germination. However, the relationship between these experimental phenomena and the embryo's endogenous hormonal metabolism remains unknown.

Nevertheless there is clear evidence that hormones are involved in specific germination-related processes. The role of ABA in the prevention of precocious germination has been alluded to in a previous section. The prevention of precocious germination may be considered as the imposition of a period of dormancy during embryogenesis, and this is an indication of the wider role of ABA in the imposition and maintenance of the longer-term dormancy mechanisms, which are very widespread in wild plant species. There is also evidence, although so far from only

a handful of species, that the breakage of dormancy, e.g. by specific environmental factors such as low temperatures, is associated with a decline in ABA concentration in the embryo (Figure 11.19). Whether there is also a positive regulatory role for hormones in dormancy breaking is not so clear. Certainly, the application of GA (and in some instances, cytokinins) to dormant seeds leads, in many species, to breakage of dormancy in the absence of the usual natural dormancy-breaking mechanism. However, assays of the growth-promoting hormones in seeds undergoing natural dormancy breakage do not provide a consistent picture of increasing concentrations of GA or cytokinins.

The germination-related process where there is a well-established role for GA is the mobilization of the seeds' nutrient reserves in the family Gramineae (grasses and cereals). Two points must be emphasized before discussion of this system in detail. First, mobilization of reserves is not strictly part of the germination process. In the cereals, extension and emergence of the radicle occurs before mobilization of the endospermic nutrient reserves is initiated. However, except under experimental conditions in which isolated embryos are cultured in nutrient media, mobilization of reserves is essential for successful seedling establishment. Second, although the mobilization of nutrients in cereal endosperm is an excellent system for study, it is not a universal model for the mobilization of seed nutrient reserves. It is not justifiable to extrapolate from this system to those seeds in which reserves are laid down within the embryo (e.g. in cotyledons).

Mobilization of Reserves in Cereal Grains

In cereals the reserves are laid down in a tissue known as endosperm, which, as is apparent from the description of fertilization earlier in this chapter, is a tissue external to and separate from the embryo. Further, during the course of seed ripening, the cells of the endosperm die and are thus not capable of metabolism during germination. The endosperm is, however, surrounded by a

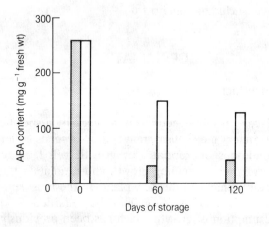

Figure 11.19. Dormancy in Norway maple (*Acer platanoides*) is broken by storage at 5 °C but not by storage at 17 °C. The diagram shows the abscisic acid (ABA) content of embryos in seeds stored at 5 °C (solid bars) or 17 °C (open bars). From J.A. Bryant, 1985, *Seed Physiology*, Edward Arnold, London, by permission of Cambridge University Press.

layer of living cells, the aleurone (Figure 11.20). Here then is a problem of coordination: the growing embryo requires the reserves which are located in a dead tissue that is external to the embryo. The key to this coordination is GA. GA is synthesized by the germinating embryo and secreted to the aleurone. In response to the GA, the aleurone cells synthesize hydrolytic enzymes which are secreted to the dead cells of the endosperm. These enzymes mediate the hydrolysis of the polymeric reserves; the hydrolysis products are taken up by the embryo, mainly via the transfer cells of the scutellum (which is actually a highly modified cotyledon; Figure 11.20).

This system is very amenable to experimental manipulation, in particular because in grains from which the embryo has been removed the aleurone is responsive to exogenously added GA. This response closely mimics the normal response to embryo-derived GA and, as in very many GA-mediated processes, is inhibited if ABA is added at the same time as the GA.

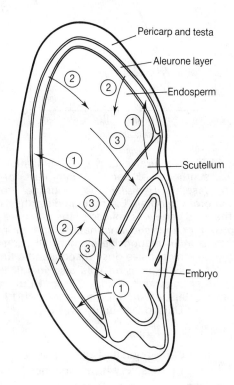

Figure 11.20. Diagram to illustrate control of the mobilization of reserves in germinating barley seeds. (1) The embryo secretes gibberellic acid (GA) to the aleurone layer. (2) In response to GA the aleurone cells synthesize hydrolytic enzymes and secrete them into the endosperm. (3) The hydrolysis products are taken up by the embryo. The diagram is a very simplified version of the real thing. In particular, the reader should note that the embryo is actively growing during this time and that germination (defined as emergence of the radicle) occurs well before the hydrolysis of the reserves. From J.A. Bryant, 1985, *Seed Physiology*, Edward Arnold, London, by permission of Cambridge University Press.

Regulation of Gene Expression by Gibberellic Acid

Differential screening of cDNA libraries reveals that addition of GA to the isolated aleurone–endosperm system leads to the accumulation of a small number (*c*. seven or eight) of new mRNAs. Several of these have been shown by sequencing and/or direct identification of the encoded protein to code for hydrolytic enzymes such as proteases and amylases, amylase mRNA being the most abundant class of mRNA in the aleurone. Nuclear run-on experiments (see p. 260) show that the accumulation of these new mRNAs reflects a very marked up-regulation of gene transcription, while comparison of the timing and kinetics of the accumulation of different mRNAs suggests that this up-regulation is a highly coordinated process.

Of the genes whose transcription is up-regulated by GA in the aleurone, those coding for α-amylase, the major starch degrading enzyme, have received the most attention. There are three families of α-amylase genes, which in wheat have been named α-*amy*-1, α-*amy*-2 and α-*amy*-3. Families α-*amy*-1 and α-*amy*-2 are expressed in the aleurone, although some members of the α-*amy*-2 family are also expressed in developing endosperm; α-*amy*-3 genes are only expressed in developing endosperm.

In order to understand this differential control of an α-amylase gene transcription, investigations have been carried out on different members of the wheat α-*amy*-2 family, which show some differences in transcription behaviour. Essentially, there are three patterns: (a) expressed at a low level in aleurone; (b) expressed at an enhanced level in aleurone (up-regulated by GA); and (c) expressed at an enhanced level in aleurone, but also expressed in developing endosperm. By comparing upstream (promoter) sequences of these genes (and of their equivalents in barley), it has been suggested that the first 90 base pairs upstream from the gene are required for the aleurone tissue-specific expression, while the next 190–200 base pairs upstream are involved in the regulation of transcription by GA.

In the earlier section on fruit ripening we described how the function of an upstream promoter region could be determined by splicing the promoter to a suitable 'reporter' gene and then using the promoter–reporter construct to make transgenic plants. Unfortunately, it is not possible to use this exact approach to test the α-*amy*-2 promoter sequences. This is, first, because there is no routinely workable genetic transformation technique for cereals and, second, because in the various dicot plants which are amenable to genetic manipulation there is no tissue which is of direct equivalence to aleurone. However, oat aleurone protoplasts can, under appropriate conditions, take up DNA from their incubation medium. The DNA is then expressed transiently in a promoter-dependent manner. Furthermore, the protoplasts also respond to GA in the normal way, including synthesis of α-amylase. Here then is a system for testing the α-*amy*-2 promoter. The whole promoter (1900 base pairs) was shown to drive the transcription of a reporter gene, β-glucuronidase (GUS), when the protoplasts were treated with GA, but not in the absence of GA. – Indeed, the pattern of GUS synthesis closely paralleled the pattern of synthesis of the protoplasts' own α-amylase (Figure 11.21). Use of partially deleted promoters showed that the sequences involved in the GA-regulated expression of GUS resided in the first 290 base pairs upstream of the gene, thus confirming the suggestion made on the basis of sequence data.

Although it appears that about 290 bp of DNA

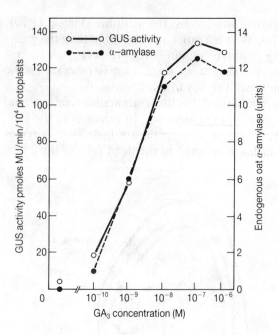

Figure 11.21. Gibberellic acid (GA) regulation of amylase synthesis (●) and of synthesis of β-glucuronidase (GUS) (○) in oat aleurone protoplasts. In order to obtain GA-regulated synthesis of GUS, the GUS gene was spliced to the α-amylase-2 gene promoter, thus showing that the latter promoter contains a GA-responsive element. In addition to giving an amylase-like dose response, the time-course for GUS expression was also identical to that of the endogenous amylase. From Huttly and Baulcombe, 1989, *EMBO Journal*, **8**, 1907–1913, by permission of Oxford University Press.

are needed for GA-regulated expression of the α-*amy*-2 genes, there is very little sequence homology between this α-*amy*-2 promoter and the promoters of the wheat α-*amy*-1 genes (and their equivalent in barley), which are also GA regulated. In fact specific sequence homology may be confined to a 7 bp sequence, the GA response element (TAACAG/AA), with the rest of the promoter providing more generalized features of promoter function, e.g. an appropriate molecular environment for binding of a *trans*-acting factor (regulatory protein) to the GA response element. The existence of a regulatory protein which binds to the promoters of GA-responsive genes has been demonstrated in aleurones of wheat, rice and barley. Further, the synthesis of this protein is itself GA regulated, i.e. the GA-dependent up-regulation of the α-amylase genes cannot occur until the GA-regulated synthesis of the DNA-binding protein has occurred. The antagonistic mode of action of ABA in preventing the GA-regulated response (noted earlier in this section) seems to involve the GA response element binding protein; ABA inhibits either the synthesis or the binding activity of the binding protein, thus preventing the GA-induced up-regulation of the activity of the genes coding for α-amylase and the other hydrolytic enzymes. This ability of ABA to prevent the response to GA does not need the synthesis of a new protein, unlike the GA response itself. This means that in the presence of appropriate concentrations of both ABA and GA, ABA will always override the GA response.

Concerning the early part of the transduction chain between the hormone and the gene product, we know very little. It is known that the ability of aleurone cells to respond to GA does not occur until relatively late in seed development. The most likely basis for this is in the initial perception of the hormone, but until a GA-binding factor has been detected and isolated it is not really possible to test this idea.

FLORAL DEVELOPMENT

The formation of flowers as reproductive structures is the characteristic feature of the angiosperms. These are formed by a change in the specification of cell fate at shoot apical meristems, which switch from the ordered production of leaf primordia in a regular phyllotactic array to the induction of the floral organs. The type, number and relative disposition of the floral organs within the developing inflorescence is stringently genetically defined. The shape, size, colour and pattern of the inflorescence is a diagnostic feature of most plant species—a consequence of the intensity of selection pressure acting to optimize the reproductive relationship between plant and pollinator throughout angiosperm evolution.

The induction of floral development may be either genetically preprogrammed, or prompted by environmental cues. Thus in many species the induction of flowering is triggered by the perception of light—the photoperiodic response. This defines the requirements of plants to undergo repeated cycles either of short days or long days before flowering is initiated. Investigations of the spectral quality of the light necessary to initiate a flowering response reveal the role of a red light receptor—*phytochrome*. This is a protein–pigment complex which undergoes reversible conformational changes upon absorption of light. Illumination of phytochrome with red light ($\lambda = 660$ nm) causes the pigment to alter its absorption properties, so that it preferentially absorbs light in the far-red range ($\lambda = 730$ nm). Upon absorption of far-red light, the pigment reverts to the red light absorbing form:

$$P_r \underset{730}{\overset{660}{\rightleftharpoons}} P_{fr}$$

Phytochrome is believed to mediate a wide range of light-dependent phenomena in plants in addition to photoperiodically related floral induction. It is a characteristic of these phenomena that they are activated by illumination with red light, but that this activation is reversed by a subsequent period of illumination with far-red light. The P_{fr} form of phytochrome is therefore regarded as the biologically active form of the molecule.

In plants which show a light-responsive pattern of floral induction, the perception of light, by phytochrome, occurs in the leaves rather than the apical meristem. This has been demonstrated by classical experiments in plant physiology in which the illumination of single leaves of a target plant has been sufficient to invoke a flowering response. Transfer experiments, in which a leaf from an illuminated plant is grafted onto an unilluminated plant, and causes the recipient plant to flower, indicate the passage of a transmissible signal from the site of light perception to the apex. However, years of research have failed to reveal the nature of this flowering inducer, often termed 'florigen'.

The events which occur within the apex upon receipt of the flowering stimulus have been much more extensively characterized. The recent combination of recombinant DNA techniques with classical genetic studies is enabling a dissection of the molecular basis of the genetic specification of floral morphogenesis. The approach which has proved most fruitful parallels that which has been so successful in unravelling the mysteries of animal embryogenesis: the analysis of homoeotic mutations.

Homoeotic mutations are defined as mutations which cause a change in the developmental specification of an organ or organ system. This type of mutation is exemplified most spectacularly in *Drosophila* development where, for example, mutations at the *antennipedia* locus cause flies to develop a pair of legs upon their heads, in place of the normal antennae. Such mutations, at individual genetic loci, must necessarily occur in genes whose function is to activate or coordinate the myriad of other genes whose combined expression is required to contribute to the formation of a specific organ. In plants a number of homoeotic mutations have been identified which alter the specification of floral organs. In two species these mutations are particularly amenable to molecular analysis.

Arabidopsis thaliana and *Antirrhinum majus* have a long history as subjects of genetic analysis. *Arabidopsis* is a member of the Crucifereae. It is a small weed, related to the mustards, which possesses a number of features recommending it as a subject of study. It has a rapid life cycle, passing from germination to flowering and the setting of seed in a period of about 5 weeks. It may be self or cross-pollinated, enabling rapid genetic analysis, and it is small in size, permitting large numbers of plants to be propagated within the laboratory. Following mutagenesis, by chemical or physical methods, several thousands of plants may be screened for genetic aberrations following germination on simple culture media in Petri dishes. Additionally, *Arabidopsis* is readily amenable to genetic transformation by the most efficient plant transformation vector—

the *Agrobacterium* Ti plasmid (chapter 12)—and the integration of T-DNA into the genome can also be used as a form of insertion mutagenesis permitting rescue of the mutated sequences.

Most importantly, the plant has a relatively small genome, consisting of c. 100 million bp of DNA, of which only c. 18 million bp exist as repetitive DNA sequence. It is the presence of large amounts of repetitive, non-coding DNA which contributes the bulk of the genome in most higher eukaryotic species, whereas in *Arabidopsis* the majority of the genetic material is present as single-copy sequences, presumably largely functional in nature. Such a small genome lends itself to molecular analysis, as the entire genome can be contained, as cloned DNA, within a relatively small number of clones in YAC vectors.

Sometimes referred to as 'the *Drosophila* of the plant kingdom', *Arabidopsis* has lately become the focus for a concerted international research programme which seeks to understand the molecular basis of plant development. This has resulted in the generation of a very well-marked genetic map, based both on classical features (mutations causing a recognizable phenotype) and on molecular features (restriction fragment length polymorphisms (RFLPs) detected by Southern blot analysis of genomic DNA using defined cloned probe sequences). Obtaining a close correspondence between the classical linkage map and the RFLP-based linkage map has been of prime importance, as it has enabled the isolation of genes which define a specific phenotype by chromosome walking. Interest in *Antirrhinum majus* has largely centred on the early discovery that this species contains transposable elements within its genome. These have been observed to cause a large number of mutations affecting floral morphology and flower colour, and the molecular analysis of this latter class of mutation has resulted in the isolation and characterization of a number of *Antirrhinum* transposons. These elements have thus been exploited as a means of tagging genes which cause homoeotic transformations in flower development.

Homoeotic Mutations in Floral Development

There are many similarities between floral development in *Antirrhinum* and *Arabidopsis*. In each species the inflorescence is complex, containing a number of individual flowers on an elongating stalk. Upon floral induction, the vegetative meristem undergoes a transition to become an inflorescence meristem, which retains an indeterminate growth habit. However, the primordia which arise on the flanks of the inflorescence meristem become not leaves but individual flowers. The inflorescence meristem thus gives rise to a series of floral meristems, each of which exhibits a determinate pattern of growth, i.e. it undergoes differentiation to produce the floral organs, and when these reach maturity further growth ceases. The floral meristems are produced sequentially around the inflorescence meristem with a spiral phyllotaxy.

The hermaphrodite flowers of both species develop from the floral meristems as concentric whorls of organs, illustrated diagrammatically in Figure 11.22. These organs are produced sequentially, the first whorl being the sepals (the outermost organs), followed by the petals, the stamens, which bear the male gametogenic tissue in the anthers, and finally the innermost, fourth whorl, the pistil, comprising the carpels and receptive tissue. In *Arabidopsis* the flower is actinomorphic: it has a radial symmetry with the organs evenly spaced around the centre. The pattern in *Antirrhinum* is more complex in that the flower is zygomorphic: the concentric array of organs is disposed with a bilateral symmetry, with some differentiation of shape between upper and lower petals.

When meristematic reprogramming occurs to initiate a flower, it is necessary that the cells within the meristem be able to interpret their position in order to develop the characteristic floral morphology. It is presumed that positional information is available to the cells within the meristem. A study of the mutations which cause homoeotic transformations illustrates how such a definition of developmental fate may be achieved.

These mutations occur in single genes, yet cause profound and specific changes in the identity and position of the floral organs. Two alternative roles may thus be envisaged for these genes. Either they are central in the establishment of positional information within the meristem, or they are primary regulators of families of organ-specific genes which respond to positional information. In this latter model these genes could be considered analogous to the homoeodomain-containing genes of *Drosophila*, which encode a range of transcriptional regulators active in morphogenesis.

Currently, a molecular description is emerging for two classes of homoeotic gene: those which regulate the growth habit of the meristem, and those which regulate organ identity. There is also an interesting third class, typified by the *cycloidea* mutant of *Antirrhinum*, in which the symmetry of the flower is altered from bilateral to radial. At the time of writing this last class of mutant has yet to be characterized at the molecular level.

Two mutations have been identified which cause a switch in the growth pattern of the floral meristem from determinate to indeterminate growth. In both the *apetala-1* mutants of *Arabidopsis* and the *floricaula* mutants of *Antirrhinum*, the floral meristems which arise from the flanks of the primary inflorescence meristem no longer develop to produce terminally differentiated

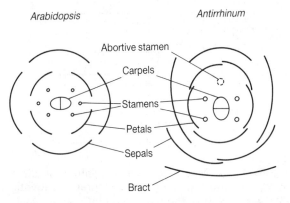

Figure 11.22. Diagram illustrating the different whorls of organs in the flowers of two dicot plants.

Table 11.3. Homoeotic mutants of *Arabidopsis* and *Antirrhinum*.

Site of action	*Arabidopsis*	*Antirrhinum*	Phenotype
Whorls 1 and 2	apetala-2	ovulata	Sepals become carpelloid Petals become stamenoid
Whorls 2 and 3	apetala-3 pistillata	deficiens sepaloidea globosa viridiflora	Petals become sepalloid Stamens become carpelloid

Whorls 3 and 4	agamous	petaloidea
		pleniflora
			Stamens become sepalloid or
			  petalloid
			Carpels revert to floral meristems,
			  giving an indeterminate
			  repetitive morphology

flowers. They have failed to undergo the switch from indeterminate to determinate growth. Instead these meristems themselves develop as indeterminate inflorescence meristems, giving rise to meristems on their flanks which similarly fail to give rise to determinate floral meristems, but which develop as inflorescence meristems. Thus a repetitive series of inflorescence meristems is continuously initiated in a potentially infinite branching pattern.

'True' homoeotic transformations, of one organ type into another, are also seen in both species, and are strikingly similar in their nature. It is a feature of these mutations that their effects are seen not on single classes of organ, but on organs in adjacent pairs of whorls of the flower. Thus three classes of such homoeotic mutation have been identified, and the characteristics of these are outlined in Table 11.3.

Consideration of the phenotypes of the individual mutants and those of doubly mutant plants has enabled a generally unifying model to be developed, explaining the possible function of these flower morphology genes. This model implies that the role of the flowering genes is to interpret positional information within the floral meristem, and to establish the developmental fate of the cells in which they are expressed.

The model proposes that organs develop in response to the presence within the developing meristem of three factors 'a', 'b' and 'c'. There

are three key features of this model (Figure 11.23):

1. These factors are spatially distributed within the floral meristem, so that factor *a* is present in the outermost pair of whorls, factor *b* is present within the second and third whorls and factor *c* is present within the innermost pair of whorls.

2. Factors *a* and *c* are mutually inhibitory: pre-

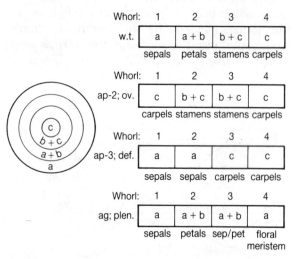

Figure 11.23. Model for the regulation of dicot flower structure by the activity of three homoeotic genes, *a*, *b* and *c*, and the molecular genetic basis of various types of mutant flowers; w.t. = wild type; ap = *apetala*; ov = *ovulate*; def = deficiens; ag = *agamous*; plen = *pleniflora*.

sence of factor *a* prevents the synthesis and accumulation of factor *c*, and vice versa.

3. Factor *c* provides a developmental 'stop signal', thus imposing the determinate nature of floral development.

According to this model, cells containing factor *a* alone develop to form sepals. Cells containing both factors *a* and *b* form petals, those containing factors *b* and *c* in combination form stamens, while those containing factor *c* alone produce the carpels. This model explains the nature of the homoeotic mutants which are observed. Loss-of-function mutants in factor *c* should lose normal specification of third and fourth whorls, and should also become indeterminate in developmental pattern. Such a phenotype is seen in the *pleniflora* and *agamous* mutants, and consequently these genes presumably establish the '*c*-function'. Similarly, loss of the *b* factor should result in loss of petals and stamens—a phenotype typical of the *deficiens* and *apetala-3* class of mutants, while loss of the *a* function predicts absence of sepals and petals, such as occurs in *ovulata* and *apetala-2* mutants.

That these homoeotic genes are instrumental in establishing these functions is further supported by observing the phenotypes of double mutants. Thus, *ap-2:ap-3* double mutants of *Arabidopsis* produce floral structures which are wholly carpelloid in appearance, as would be expected from the residual presence of the *c* factor alone, while *ap-3:ag* double mutants, lacking *b* and *c* functions, develop floral structures comprising repeating whorls of sepals—a consequence of organ development in response to the *a* factor, and of loss of the 'stop' function specified by the *c* factor.

It is important to emphasize that the homoeotic genes which establish the *a*, *b* and *c* functions appear to be expressed relatively late in floral morphogenesis, after the transition of the vegetative meristem to become a floral meristem, and the initiation of the individual floral meristems. This is apparent both from a molecular analysis of floral initiation and from observation of the developmental pattern of floral mutants. In particular, the expression of the organ-identity genes

must occur subsequent to the expression of genes which regulate the establishment of the determinate floral meristems. This is clear from the phenotypes of mutants of the *floricaula* and *apetala-1* class, in which no floral meristems are specified, and in consequence the organ-determination genes are not expressed. The isolation of these genes, by molecular cloning, has allowed the nature of the proteins which they encode to be inferred, following sequence analysis. It has also enabled the spatial pattern of expression of each gene to be determined.

Molecular Analysis of Homoeotic Genes

By using the techniques outlined earlier in this chapter, DNA sequences corresponding to the *deficiens* and *floricaula* genes of *Antirrhinum*, and the *agamous* gene of *Arabidopsis*, have been cloned. Sequence analysis of the protein-coding regions contained within the *deficiens* and *agamous* genes confirms their functional similarity, as they contain regions of striking homology. Not only are these sequences homologous with each other, but they are also homologous with proteins known to have regulatory roles in other, widely divergent species. Specifically, both the *deficiens* and *agamous* predicted protein sequences contain domains equivalent to those found in the mammalian serum response factor (SRF) and the yeast *MCM1* gene.

Both SRF and the *MCM1* product are transcriptional regulators. The former is responsible for controlling the proliferation of mammalian cells in response to serum proteins, the latter in establishing the mating type of yeast cells. The homology which exists between these genes of disparate origin lies in sequences which in the mammalian and yeast proteins are responsible for DNA binding and protein dimerization, respectively. Additionally, the protein sequences share homology at a target site for protein phosphorylation—a form of modification which is known to be instrumental in modifying the biological activity of the SRF protein, at least.

In this instance, therefore, two of the genes establishing organ identity—through the specification of the 'b function' (*def*) and the 'c function' (*ag*)—can be predicted to act by controlling the expression of other genes at the level of transcription, an activity which may be further 'fine-tuned' by interaction with protein kinases and phosphatases.

The molecular nature of *floricaula* is less certain, as although its sequence has been determined, there are no other previously determined sequences of known function with which it has obvious similarity. However, the *floricaula* gene product does have some sequence elements reminiscent of those which in other proteins have been postulated to interact with transcriptional regulators. For this reason it has been suggested that the *flo* gene product may, like the *VP1* protein in the developing seed, serve as a 'transcriptional *trans*-activator', which through protein–protein interaction with a DNA-binding protein could activate the transcription of specific sets of genes. This hypothesis awaits a functional test.

Sequence analysis of cloned genes allows prediction of gene function through the identification of known structural motifs within coding sequences. DNA–RNA hybridization techniques allow the temporal and spatial regulation of gene expression to be determined. Particularly valuable for this purpose is the detection of specific mRNA sequences *in situ* within tissue sections, by hybridization with a labelled, cloned hybridization probe sequence. This technique is illustrated diagrammatically in Figure 11.24. It is possible to detect relatively low quantities of specific mRNAs, located within individual cells or groups of cells, by this *in situ* hybridization.

Such analysis using the cloned *flo, def* and *ag* sequences as probes has proved to be particularly revealing. First, as predicted from the phenotype of *floricaula* mutants, it is clear that the *flo* gene is expressed early in floral meristem development, and prior to the expression of the organ-specifying genes. Second, it is expressed in a distinct temporal and spatial fashion, in a very limited number of cells. *Floricaula* mRNA is detected soon after the induction of meristematic

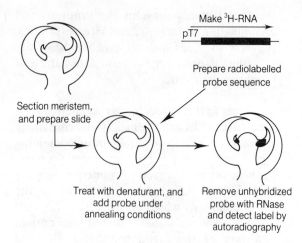

Figure 11.24. *In situ* hybridization to detect the presence of a particular mRNA.

transition—a process which can be promulgated by photoperiodic illumination in *Antirrhinum*. It accumulates first in the bracts, within whose axils the floral meristems arise, and within these meristems themselves. As the meristems develop, so the level of *flo* mRNA in the bract declines, and the expression of the *flo* gene is detectable only in those cells developing as sepal primordia. Subsequently, this sepal-related expression declines, and *flo* expression is detected within cells developing to form petals. Essentially, a pattern of serial transient expression of the *floricaula* gene is observed in the successively developing organs of the floral meristem. Notably, though, while a burst of transient *flo* expression is seen in the carpel primordia, none is detected within the stamen primordia. These observations within *Antirrhinum* can be synthesized with those made in *Arabidopsis* flower development to produce an overall hypothesis:

1. A *floricaula*-type gene is associated with the indeterminate reiteration of the inflorescence meristem, and it regulates the development from this of the determinate floral meristems. (*Loss of function in mutants causes loss of determinate status of floral meristems.*)
2. The *floricaula* gene product (or equivalent)

switches on the *a* function gene in sepal and petal primordia.

(Flo gene expression occurs early and transiently in these primordia.)

3. The *b* and *c* functions are switched on independently at this time—the timing of this switch determines the boundary between cells expressing *a* and cells expressing *c* whose products switch each other off. Expression of *b* inhibits expression of *flo*.

(in situ hybridization of floral meristem tissue sections using def and ag sequences as probes reveals their expression to be specific to 2nd and 3rd, and 3rd and 4th whorls, respectively. The timing of def expression relative to that of flo expression is consistent with this interpretation, as is the observation that flo is expressed in the 3rd whorl in deficiens mutants, which lack the b function.)

Thus, according to this model, a gene of the *floricaula* class instigates floral development, and the homoeotic genes subsequently coordinate the development of the flower through their interactions with each other, and with the multitude of genes which are required to lay down the essential features of the different organ primordia.

No doubt, as our knowledge of these genes, their products and their modes of regulation increases, this model will be subject to modification, and will become progressively more complex in its details. Nevertheless, the study of floral morphogenesis will remain of central importance in our future understanding of the processes involved in development and morphogenesis in higher plants.

FURTHER READING

This is not a comprehensive bibliography. Instead we have listed publications which either give a wider background than is possible in a chapter of this length or which are recent key references.

Coen, E.S. (1991) The role of homoeotic genes in floral development and evolution. *Annual Review of Plant Physiology and Molecular Biology,* **42**, 241–279.

Coen, E.S. and Meyerowitz, E.M. (1991) The war of the whorls: genetic interactions controlling flower development. *Nature,* **353**, 31–37.

Dickinson, H.G. (1990) Self-incompatibility in flowering plants. *BioEssays,* **12**, 155–161.

Dure, L.S., Crouch, M., Harrada, J., Ho, T.J.D., Mundy, J., Quatrano, R., Thomas, T. and Sung, Z. R. (1989) Common amino acid sequence domains among the LEA proteins of higher plants. *Plant Molecular Biology,* **12**, 475–486.

Fincher, G.B. (1989) Molecular and cellular biology associated with endosperm mobilization in germinating cereal grains. *Annual Review of Plant Physiology and Molecular Biology,* **40**, 305–346.

Grierson, D. (ed.) (1991) *Developmental Regulation of Plant Gene Expression.* Blackie, Glasgow and London/Chapman & Hall, New York.

John, P. (1991) How plant molecular biologists revealed a surprising relationship between two enzymes, which took an enzyme out of a membrane where it was not located, and put it into the soluble phase where it could be studied. *Plant Molecular Biology Reporter,* **9**, 192–194.

McCarty, D.R., Hattori, T., Carson, C.B., Vasil, V., Lazar, M. and Vasil, I.K. (1991). The *Viviparous-1* developmental gene of maize encodes a novel transcription factor. *Cell,* **66**, 895–905.

Nasrallah, J.B., Nishio, T. and Nasrallah, M.E. (1991) The self-incompatibility genes of *Brassica. Annual Review of Plant Physiology and Molecular Biology,* **42**, 393–442.

Old, R.W. and Primose, S.B. (1989) *Principles of Genetic Manipulation* (4th edn). Blackwell Scientific Publications, Oxford.

12

CELL CULTURE, TRANSFORMATION AND GENE TECHNOLOGY

R. Walden

Max-Planck-Institut für Züchtungsforschung, Köln, Germany

INTRODUCTION

The key element in the production of trans-formed, or transgenic, plant tissue is the unique ability of isolated plant cells to grow, divide and eventually regenerate into a plant. Protoplasts cultured *in vitro* with appropriate amounts of auxins and cytokinins grow and divide to form a mass of undifferentiated tissue known as callus. Callus, in turn, can be incubated under conditions which induce shoot formation or used for the initiation of a cell suspension which can be cul-tured further in liquid. Shoots can be excised from the callus and induced to form roots and

Plant Biochemistry and Molecular Biology. Edited by P.J. Lea and R.C. Leegood
© 1993 John Wiley & Sons Ltd

grow further to produce a normal plant. Hence, once a plant cell has taken up a sequence of foreign DNA, as long as the DNA is stable in the cell a transgenic plant can, in principle, be produced. Here I aim to introduce the basic concepts of plant cell and tissue culture, describe the experimental systems used to generate transgenic plants and discuss how transgenic plants may be used both in research and agriculture.

PLANT CELL AND TISSUE CULTURE

Protoplast Isolation

Protoplasts can be isolated from a variety of plant tissues, although most commonly leaf tissue is used. Their isolation involves the enzymatic removal of the cell wall by the incubation of tissue in the presence of cellulases, pectinases and hemicellulases contained in a solution of high osmotic potential in order to keep the protoplasts from bursting. Perhaps not unexpectedly, the exact conditions used in protoplast isolation varies not only from plant to plant but also from tissue to tissue. Hence protocols for protoplast isolation must be determined empirically. Moreover, it is important to bear in mind that protoplasts derived from different tissues—leaf, hypocotyl or root or suspension cells—are in different states of development and hence are likely to respond in a different manner not only to culture conditions but also to exogenously applied DNA.

Once released from the cell wall by enzymatic digestion, protoplasts are separated from the cell debris by a mixture of sieving and centrifugation. Depending on the origin of the cells, the population of protoplasts obtained can be uniform in size and developmental stage, particularly if the source of the cells has been leaf tissue or suspension cell culture.

Transient Expression Assays

Freshly prepared protoplasts can be used for the uptake of foreign DNA (DNA-mediated gene transfer, DMGT) and this has been used extensively to carry out transient assays of gene expression in order to investigate promoter function. This is because the expression of a large number of different gene constructs, for example those containing different deletions of a promoter, can be studied with relative ease using this approach.

Transient expression assays involve the treatment of a population of protoplasts with a stimulus which results in the protoplasts taking up exogenously applied DNA, generally in the form of a bacterial plasmid containing the desired gene construct, followed by measuring the level of expression of a marker gene. Routinely, two general methods have been used to introduce DNA into protoplasts for transient assays: treatment of protoplasts with polyvalent cations, or electroporation. Polyvalent cations, such as polyethylene glycol (PEG) or poly-L-ornithine (PLO), have been used extensively to induce protoplast fusion and they are thought to act to promote DNA uptake by precipitating the DNA, minimizing charge repulsion between the protoplasts and stimulating DNA uptake by endocytosis. With electroporation an electrical pulse is used to reversibly permeabilize the cell membrane, allowing the uptake of DNA. Both procedures rely on the use of a sensitive marker gene to assess levels of expression within the transformed cell. Routinely, chromogenic markers such as chloramphenicol acetyl transferase (CAT) or β-glucuronidase (GUS) can be used, with enzyme activity being detected 3–96 hours following DNA uptake, with a peak of activity occurring between 24 and 48 hours.

The functional analysis of the individual domains of a variety of promoters has been carried out by transient assays in protoplasts, and examples include promoters which direct expression following stress or infection by fungal pathogens. However, it is significant that not all promoters are active in transient assays, the most noteworthy examples being those which direct the expression of genes whose products are important in photosynthesis. This probably results from the developmental stage of the protoplasts or the

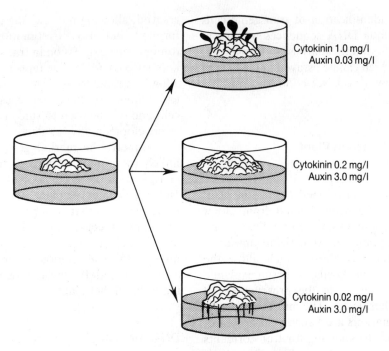

Cytokinin 1.0 mg/l
Auxin 0.03 mg/l

Cytokinin 0.2 mg/l
Auxin 3.0 mg/l

Cytokinin 0.02 mg/l
Auxin 3.0 mg/l

Figure 12.1. The developmental flexibility of plant cells. Plant cells will grow *in vitro* in the presence of a balanced mixture of auxins and cytokinins to produce callus. A high cytokinin : auxin ratio results in shoot initiation from the callus whereas a high auxin : cytokinin ratio results in root formation.

conditions under which the transient assays are carried out.

Plant Regeneration

Three to four days following isolation, protoplasts will begin to regenerate their cell walls and divide to form microcallus. Callus formation is dependent on the presence in the culture medium of auxins and cytokinins. The absolute requirement for these growth substances can only be circumvented by tumorous cells (see later) or attenuated cells which have, as a result of still unknown causes, become able to grow independently of plant growth substances. Callus is a mass of undifferentiated plant cells which, depending on the presence of different growth substances, can be induced to form shoots or roots. In what are now considered as a classic series of experiments in the 1950s, Skoog and coworkers[1] demonstrated, using callus derived from tobacco pith, that when

the ratio of auxin to cytokinin was high in the growth media callus would be induced to form roots, whereas when the ratio of auxin to cytokinin was low shoots were formed (Figure 12.1). Intermediate ratios of auxins to cytokinins favoured maintained callus growth. The developmental plasticity of the plant cell has probably arisen as a result of a sessile mode of life of the plant and the need to respond to predation as well as environmental stress. In the long term it allows the experimenter a means to investigate how a growth substance can determine a developmental pathway and in the short term it provides a means of producing transgenic plants.

THE GENERATION OF STABLE PLANT TRANSFORMANTS

The development of plant transformation strategies was coupled with the construction of genetic markers functional in plant cells. Such marker genes allow the identification of transgenic cells by providing unique DNA sequences which can be detected by Southern analysis and novel enzyme activities which can be assayed and which in some cases allow direct selection of transformants.

Genetic Markers used in Plant Transformation

The genetic markers developed for use in plant cells in general have been derived from either bacterial or plant sources and can be divided into two types: selectable or screenable markers. Selectable markers are those which allow the selection of transformed cells, or tissue explants, by their ability to grow in the presence of an antibiotic or a herbicide. The most frequently used selectable markers are kanamycin or hygromycin. In addition to selecting for transformants such markers can be used to follow the inheritance of a foreign gene in a segregating population of plants (for example, see Figure 12.2). Screenable markers, such as CAT, encode gene products whose enzyme activity can be easily

assayed, allowing not only the detection of transformants but also an estimation of the levels of foreign gene expression in transgenic tissue. Markers such as GUS, luciferase or β-galactosidase allow screening for enzyme activity by histochemical staining or fluorimetric assay of individual cells and can be used to study cell-specific as well as developmentally regulated gene expression.

Routinely, the marker genes used have been linked to promoters derived from plant pathogens, such as the T-DNA of *Agrobacterium tumefaciens* (see later) or plant viruses. In this way it is to be expected that such promoters function well in all plant cells, although increasingly it has become clear that the levels of expression directed by such promoters may be modulated to a certain extent by developmental factors operating in the intact plant.

DNA-mediated Gene Transfer

As discussed above, protoplasts can be induced to take up exogenously applied DNA and, by virtue of their ability, under appropriate conditions to regenerate into plants, a transgenic plant can be

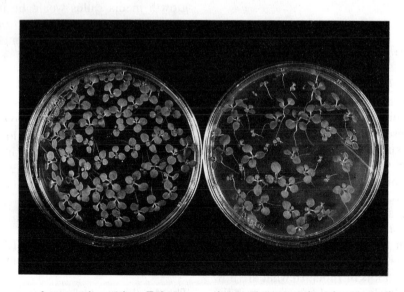

Figure 12.2. The use of a genetic marker. Tobacco engineered to contain a hygromycin resistance gene. Shown is a segregating F1 population of developing seedlings grown in the absence (left) or the presence (right) of selective levels of hygromycin in agar on Petri dishes. In this example the marker is segregating 3 : 1 (normal large) resistant plants to (shortened) sensitive plants.

obtained if the DNA is stably maintained within the cell. In addition to polyvalent cations and electroporation, fusion of protoplasts with liposomes containing foreign DNA, and microinjection, have been used to introduce DNA into cells which have then been regenerated into transgenic plants. In these cases, the mechanism by which foreign DNA integrates into the genome of the plant cell remains unclear, but it appears to be most efficient when the plant genome is undergoing replication. Once integrated into the genome the DNA can display a complex organizational array, suggesting that it may have undergone replication, ligation and recombination prior to integration.

Homologous recombination between the plant genome and the introduced foreign DNA raises the possibility of using DNA-mediated gene transfer (DMGT) in gene targeting. This was first demonstrated[2] by generating transgenic plants containing inserts of DNA comprising, amongst other sequences, a portion of a kanamycin resistance gene which alone, following insertion into the plant genome, was unable to confer resistance to kanamycin. Protoplasts were isolated from this plant and a DNA uptake experiment was performed using DNA containing the remaining portion of the kanamycin resistance gene and a region of the gene shared by the fragment integrated into the genome. Once again the second fragment of the kanamycin resistance gene alone was unable to confer resistance to the antibiotic. Following DMGT, protoplasts were selected for their ability to grow in the presence of kanamycin. Kanamycin-resistant individuals, appearing at a frequency of approximately 0.5 to 4.2×10^{-4} of those protoplasts which had taken up the complete kanamycin resistance gene in a parallel experiment, were subjected to Southern analysis. It was found that indeed the kanamycin resistance gene was reconstructed as a result of homologous recombination between the portion of the gene located in the genome and the fragment of the gene carried into the cell during the DNA uptake experiment.

The transformation protocols described above rely on the ability to regenerate plants from isolated protoplasts. However, this is not feasible with protoplasts from all plants, including many of the world's important cereal and leguminous species. In order to overcome this limitation, protocols have been devised for the transformation of embryogenic explants which are themselves able to develop into plants. The most important aspect of this approach is the introduction of DNA into somatic cells which are destined to form the germ line, so that even if the primary transformant is a genetic chimera progeny will be completely transgenic. In order to achieve this the transforming DNA may have to travel through several cell layers, and this requirement has resulted in the development and use of particle acceleration, or more colloquially 'the gun', to introduce DNA into embryogenic tissue. Transformation mediated by particle acceleration involves an explosive force, provided by either gunpowder or an electric discharge, propelling particles of

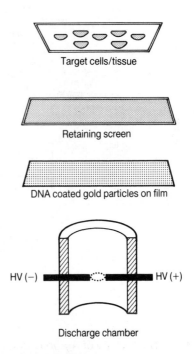

Figure 12.3. The particle accelerator. The motive force in the discharge chamber, in this example an electrical discharge, propels gold particles coated with DNA towards the target cells. The retaining screen allows passage of the particles while retaining the remains of the mounting film.

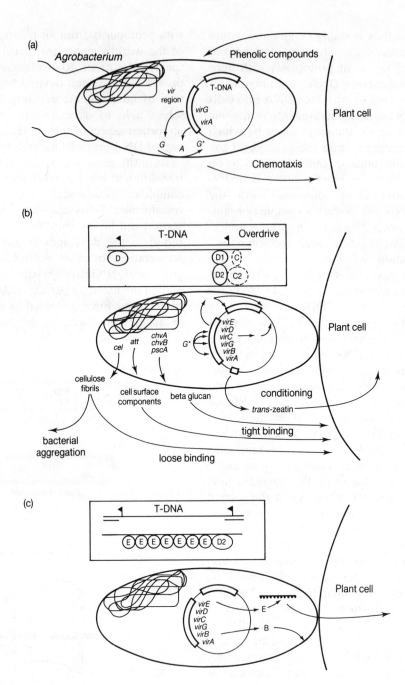

Figure 12.4. The biology of *Agrobacterium*-mediated gene transfer. (a) The release of phenolic compounds from wounded plant cells is sensed by the transmembrane sensing protein encoded by *virA* of the Ti plasmid. The activated *virA* in turn phosphorylates the product of *virG*, which results in the chemotaxis of the bacteria to the plant cell. (b) Expression of the bacterial chromosomal genes *att, chvA, chvB* and *pscA* result in the binding of the bacteria to the plant cell, whereas the product of *cel* is involved in the formation of cellulose fibrils which assist in bacterial aggregation. The activated *virG* product serves as a transcrip-

CELL CULTURE, TRANSFORMATION AND GENE TECHNOLOGY   281

heavy metal, such as tungsten or gold, coated in DNA at the tissue explant (see Figure 12.3). The particles enter the tissue, occasionally passing through several cell layers, and carry the DNA into the cell, where it will eventually integrate into the genome. While at first such an approach may appear a little cumbersome or crude, if not destructive, it has proven to be remarkably successful, being used recently to produce transgenic soya bean and maize.

Agrobacterium-mediated Transformation

While DMGT has been used extensively in plant transformation, one of the most reliable and versatile means of introducing foreign DNA into plant cells harnesses the normal functions of an unlikely biological system—bacteria.

Biology of the *Agrobacterium*–Plant Interaction

The study of the interaction of the soil bacterium *Agrobacterium tumefaciens* and the closely related *A. rhizogenes* with wounded dicotyledonous plants is one of the most fascinating areas of plant biology, and the results obtained, in addition to being utilized in developing vectors for use in plant transformation, have shed important light on plant/bacterial interaction, hormone biosynthesis, plant development and the control of gene expression.

 A. tumefaciens and *A. rhizogenes* are the causative agents of the plant neoplasias known respec-tively as crown gall and hairy root disease. As with cancer in mammalian systems, plant cells suffering from either crown gall or hairy root grow *in vitro* in the absence of external growth substances. Both bacteria contain a large plasmid which is required for initiation of the neoplasia. In the case of *A. tumefaciens* the plasmid is called the tumour-inducing or Ti plasmid, whereas it is called the root-inducing or Ri plasmid in the case of *A. rhizogenes*. During the establishment of infection a region of either the Ti or Ri plasmid, termed the T-DNA, is transferred from the bacteria to the plant cell and stably integrated into the genome. The process of this transfer appears to be analogous in both crown gall and hairy root formation.

 The natural process of gene transfer carried out by *Agrobacterium* is initiated by the release of phenolic compounds, such as acetosyringone, from the wounded plant cell. These serve to trigger a cascade of events, including chemitaxis of the bacteria to the plant cell, as well as binding of the bacteria to the plant cell wall (Figure 12.4). Moreover, acetosyringone induces expression of the virulence or *vir* genes encoded by the Ti (or Ri) plasmid. The *virA* gene is expressed constitutively in the bacteria and its product is a cell wall protein which senses the presence of acetosyringone. When acetosyringone is present the *virA* product becomes autophosphorylated, which in turn phosphorylates the *virG* product, which is also constitutively expressed, and this activates transcription of the remaining *vir* loci. The *vir* gene products serve to recognize the 25 bp imperfect direct repeat sequences that border the T-

tional activator for the *vir* genes. The products of the *virC* operon bind to overdrive—a sequence located to the right of the T-DNA (see box) and in association with the products of the *virD* operon, one of which, D2, acts as a strand-specific single-strand endonuclease—results in the appearance of single-stranded nicks at the border sequences of the T-DNA. The border sequences are denoted by flags. *Trans*-zeatin, encoded by the *tzs* gene of the Ti plasmid, is thought to 'condition' the plant cell for transformation by inducing cell division. (c) A single-stranded DNA, the T-strand, representing the bottom strand of the T-DNA (see box), accumulates in the bacterial cell and is protected from nucleases by the *virE2* product, which serves as a single-stranded DNA-binding protein, and the *virD2* protein binds at the 5' end. The products of the *virB* operon are associated with the bacterial membrane and are thought to be involved in the passage of the T-strand from the bacteria to the plant cell, a process thought to be analogous to conjugational transfer of DNA between bacteria.

DNA, induce single-strand nicks at the border sites and produce the T-strand, which is a single-stranded DNA corresponding to the bottom strand of the T-DNA. The T-strand is protected against nuclease attack by a single-stranded DNA-binding protein and 5′ DNA-binding protein, which are both encoded by the *vir* region. Transfer of the T-strand to the plant cell is thought to take place by a process analogous to transfer of plasmids during bacterial conjugation. The process of integration into the plant genome is little understood, although it appears that the T-DNA preferentially integrates into regions of the DNA that have the potential to be transcribed and is mediated by a recombinational event. Single, or low copy number inserts can be obtained in the genome, although tandem integration events can be obtained. The process of integration is likely to involve both host encoded enzymes as well as those that are bound to the T-strand. Once integrated into the plant genome, the T-DNA is stably maintained.

Following integration into the plant genome the genes encoded by the T-DNA are expressed. The promoters directing the expression of the T-DNA encoded genes, although of a bacterial origin, function in plant cells but levels of expression directed by them appear to be modulated by subtle changes in plant growth substances. T-DNA genes of the Ti plasmid encode products involved in the synthesis of auxin (gene 1 or *iaaM* and gene 2 or *iaaH*) and cytokinin (gene 4 or *ipt*), whereas genes 5 and 6b appear to deregulate auxin and cytokinin levels. The *rolA*, *B* and *C* genes of the Ri plasmid T-DNA modify the cellular response to growth substances such as auxin and induce root differentiation. Thus the T-DNA genes encode proteins which cause major alterations in the differentiation and development of the transformed plant cell and hence are responsible for the neoplastic phenotype.

In addition to enzymes that serve to disrupt the hormone balance of the plant cell the T-DNA also encodes enzymes that both synthesize and aid in the secretion from the plant cell of novel amino acid and sugar derivates that are collectively known as opines. The type of opine—for example, nopaline, octopine or agrocinopine—is dependent on the strain of *Agrobacterium* that has carried out the infection of the plant cell. The Ti plasmid contained within the bacteria encodes enzymes able to catabolize the appropriate opine. For example, nopaline-type Ti plasmids contain, at the right border of the T-DNA, a gene encoding nopaline synthase (*nos*), while gene 6a encodes a protein allowing its secretion. The *Noc* region of nopaline Ti plasmids encodes nopaline oxidase, arginase and ornithine cyclodeaminase, which serve to convert nopaline to proline. It is thought that the catabolism of opines provides the bacteria with a source of carbon and nitrogen and hence the *Agrobacterium*–plant interaction can be considered as a novel type of parasitism.

Transformation Using *Agrobacterium*

The use of *Agrobacterium* to mediate the production of transgenic plants capitalizes on the observations that the DNA transferred to the genome of the plant cell is defined by the 25 bp border repeat sequences, the products encoded by the T-DNA play no role in the transfer of DNA to the plant cell and integration into the genome, and that the *vir* region functions in *trans*. This has allowed the construction of a wide variety of plant transformation vectors and generally they can be classified as being one of two types: cointegrative or binary vectors.

Cointegrative vectors are deletion derivatives of the Ti plasmid from which the majority of the T-DNA between the border repeats has been replaced by a defined sequence of DNA. Foreign DNA to be inserted into the plant genome is cloned into an intermediate vector able to replicate in *E. coli* but not *Agrobacterium*, and contains appropriate selectable marker genes as well as a sequence homologous to that located between the border repeats of the cointegrative vector. The intermediate vector is introduced into *Agrobacterium* containing the cointegrative vector by conjugation. Lack of a functional origin of replication results in the eventual loss of the intermediate vector from *Agrobacterium* but selection for a bacterium containing a marker carried on the intermediate vector will select those bacteria in which homologous recombination has resulted in the foreign DNA being integrated between the border sequences of the cointegrative vector.

Binary vectors are plasmids that contain origins of replication that are active in both *E. coli* and *Agrobacterium*, as well as selectable marker genes functional in both types of bacteria. Flanking a region that allows the insertion of foreign DNA are the T-DNA border repeat sequences. Binary vectors containing foreign DNA to be inserted into the plant genome can be introduced into *Agrobacterium* containing a Ti plasmid from which the T-DNA has been deleted but retains a functional *vir* region by either transformation, electroporation or conjugation.

Production of transgenic plants by *Agrobacterium*-mediated transformation depends on the ability of the bacterium to carry out its normal process of natural gene transfer to a single cell, which can then subsequently regenerate into a plant. Routinely, two general protocols which allow this are used. Protoplast co-cultivation involves incubating protoplasts which are regenerating their cell walls with the bacterium carrying the appropriate transformation vector. After the incubation period the bacteria are removed by centrifugation and the protoplasts cultured to form callus in the presence of antibiotics, which select for the growth of transgenic tissue and suppress the growth of any remaining bacteria. Callus can be cultured further and induced to form shoots which can then be placed on root-inducing media. Tissue explant inoculation involves the incubation of explants, for example leaf discs, with *Agrobacterium*, followed by placing the explants on media which allow the induction of callus growth from the cells at the periphery of the explant. Once again antibiotics can be used to select for the growth of transgenic callus cells and interfere with the growth of remaining bacteria. Transfer of callus to the appropriate media results in the subsequent growth of shoots and these can be removed and placed on root-inducing media.

A large number of plants from dicotyledonous species have been transformed using *Agrobacterium* and it is now clear that *Agrobacterium* can also transfer DNA to the cells of monocotyledonous plants. However, the lack of success in generating transgenic plants from these species results largely from difficulties in devising regeneration protocols for these plants.

As we have seen with DMGT, an alternative to the limitations of tissue culture is to attempt to transform organized plant tissue. One method using this approach in *Agrobacterium*-mediated transformation has been developed for *Arabidopsis*. In this case seeds are imbibed in a solution of *Agrobacterium* and the resulting plants are self-fertilized. The resulting seeds are germinated in the presence of a selective antibiotic to demonstrate the presence of foreign DNA. Currently, the precise mechanism by which transformation takes place is not known, but the resultant transgenic plants can inherit the T-DNA in a Mendelian manner.

USING TRANSGENIC PLANTS

Plant transformation is now a routine tool for the plant biologist, and a large number of genes, viral genomes and plant transposable elements have been transferred to the genomes of new host plants. As such, plant transformation has proven to be vital in attempts to define DNA sequence elements important for directing tissue-specific and developmentally regulated expression of promoters in addition to achieving the expression of foreign genes and determining what novel phenotype they might impose on the plant. Moreover, plants engineered to contain novel traits have been proposed as being of future agricultural benefit.

Promoter Analysis

As described earlier, transient assays in protoplasts following DMGT has proven to be a rapid and convenient method for analysing the levels of gene expression directed by promoters and promo-

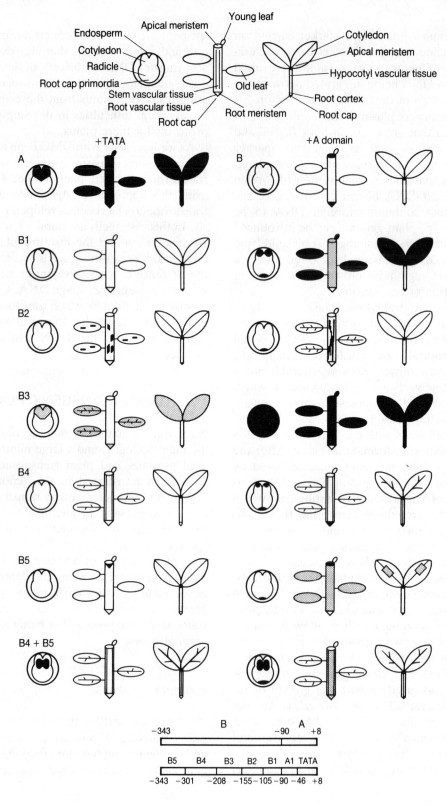

ter deletion derivatives. Promoter analysis can be carried a step further by generating transgenic plants containing different promoter constructs linked to an appropriate marker gene. In this way, when a sensitive marker gene such as GUS, which allows histochemical detection of enzyme activity, is used an investigation of the developmental and tissue-specific expression directed by a particular promoter can be carried out. One of the best examples of this has been provided by the analysis of the 35S RNA promoter of cauliflower mosaic virus (CaMV). This virus normally infects *Brassica* plants and the 35S RNA promoter in its natural state directs high levels of expression of a genome length RNA which is considered to act as a replicative intermediate. The promoter has been found to be active in the cells of many dicotyledonous plants and has been considered as being constitutive in its activity, directing high levels of expression of foreign genes. Exhaustive studies of the 35S RNA promoter have indicated that it comprises a collection of sequences which individually direct a defined pattern of tissue-specific and developmentally regulated gene expression and that these, once combined in the intact promoter, result in the constitutive pattern of expression. These studies were carried out by linking subdomains of the promoter to GUS and introducing them into the genome of tobacco, with the resultant GUS activity being assayed in regenerated plant tissue[3] (see Figure 12.5). This type of study indicated that the individual domains of the promoter interact differentially with each other during development and that these in turn are likely to interact with different factors, the so called *trans*-acting factors, which serve to control levels of expression by interacting directly, or indirectly, with RNA polymerase.

Biochemical Function of Gene Products

While the biochemical activity of a particular gene product may be well established, the precise role that it may play during growth and development often may not be so clear. Hence the analysis of the phenotype of transgenic plants engineered to express a specific gene product can be used to assess this. This approach was initially used to assist in the identification of the hormone biosynthetic genes encoded by the T-DNA. Analysis of the ability of *Agrobacteria* containing different deletions within the T-DNA to produce tumours allowed the identification of genes important in inducing hormone-independent growth[4]. Direct demonstration of the function of the individual genes was provided by subcloning them from the T-DNA and reintroducing them, using a cointegrative transformation vector, into plant cells. In this manner it was demonstrated that gene 4 was able to induce small tumours which could subsequently spontaneously produce shoots. This suggested that the activity of the gene 4 product resulted in a high cytokinin to auxin ratio. Similarly it was shown that genes 1 and 2 individually were unable to induce tumours; however when introduced into cells together they were able to do so. Subsequently, by applying intermediates of auxin biosynthesis to cells containing the individual gene, it was shown that the product of gene 1 is involved in the synthesis of α-naphthalene acetamide, which is converted by the product of gene 2 to produce auxin. The precise enzyme activity encoded by the individual gene could be identified by subcloning the gene into *E. coli* expression vectors and assessing any novel enzyme activity present in the bacteria containing the clone. Hence it was found that gene 4

Figure 12.5. Patterns of gene expression directed by the subdomains of the 35S RNA promoter. Schematic representation of the patterns of expression conferred by the B domain and individual B subdomains linked to a minimal TATA region (left) and by the A domain alone or linked to the B subdomains (right). The domains of the promoter are shown at the bottom of the figure with bases numbered with regard to the transcriptional start site. Expression is depicted in (left to right) seeds, mature plants and seedlings. Hatched areas indicate low levels of expression, with maximal levels of expression shown by filled regions. Modified from Benfey and Chua, 1990.

encoded an isopentylenyl transferase activity functioning in cytokinin biosynthesis and that genes 1 and 2 encode a tryptophan monooxygenase and indole acetamide hydrolase, respectively, which are involved in auxin biosynthesis.

More recently, this approach has been used to investigate the effect of expression of an anionic peroxidase in plants. In this case[5], the peroxidase gene, linked to the 35S RNA promoter, was introduced into the tobacco genome. The resulting plants developed chronic wilting throughout the plant at the onset of flower bud formation, which was more pronounced when the plants were in direct sunlight. The wilted leaves were able to recover turgor in darkness or subdued light; however, this ability to recover declined at later stages of development. As yet, the precise reason for the wilting is unclear, but this example illustrates the power of this approach and indicates that the tissue-specific expression of peroxidases is likely to provide insight into the regulation and function of peroxidase activity in different developmental processes.

ENGINEERING OF POTENTIALLY USEFUL AGRONOMIC TRAITS

Not unnaturally, the development of protocols for plant genetic transformation has aroused much interest in the possibility of engineering useful agronomic traits into crop plants. This is largely because recombinant DNA technology may allow the transfer into crop plants of traits, such as tolerance to herbicides, viral infection and insect predation, which are currently not available to the plant breeder. Moreover, this technology may allow these traits to be transferred directly to breeding lines which are currently in use, potentially reducing the time required to produce an elite line. Nevertheless, it needs to be borne in mind that gene transfer will not necessarily remove the need for extensive testing of transgenic progeny for phenotypic stability, impact on yield and testing of possible toxicological effects which may result from the expression of the foreign genes. Several traits, which may be transfer-

red into plants using recombinant DNA technology, have been identified as having potential agronomic benefit. However, these traits remain relatively limited to those resulting from the expression of a single gene. Those traits involving complex characters, such as yield or time of flowering, have yet to be sufficiently understood at the molecular and biochemical level before they can become the target of the genetic engineer. Hence it is safe to say that plant transformation is unlikely to replace the established methods of plant breeding and selection of advantageous traits, but rather become another important tool available to the breeder.

Herbicide Tolerance

Tolerance to herbicides has been one of the oft-cited goals of the plant genetic engineer for several reasons. First, there are sound economic advantages for farmers to have crops engineered to tolerate herbicides. Second, the biochemical pathways inhibited by herbicides have been well characterized and strategies to achieve resistance are relatively simple to devise. Third, resistance to herbicides can often be achieved by the transfer and expression of a single gene in a new host plant. Finally, resistance to certain herbicides may allow the farmer to adopt the use of some of the newer herbicides which are considered as being environmentally safe. Many plants are naturally resistant to herbicides, or have been produced by conventional plant breeding. These resistances generally arise from the plant either not taking up the herbicide or being able to detoxify it metabolically. However, the biochemistry of these resistances is poorly understood. Moreover, these resistances are localized in relatively few crop plants. If strategies can be devised to provide tolerance to herbicides, genetic engineering raises the possibility of extending the host range of resistance which is currently available.

Several experimental strategies are available for engineering herbicide resistance and include reduction of herbicide uptake, overproduction of the target site of the inhibitor or its mutation or

Table 12.1. Strategies for engineering herbicide-tolerant plants.

Herbicide	Pathway inhibiter	Primary target	Strategy adopted	
			Target modification	Detoxification
Glyphosate	Amino acid biosynthesis	EPSPS	EPSPS amplification/ overexpression *aroA* mutant	
Sulphonylurea/ imidazolinone	Amino acid biosynthesis	Acetolactate synthase (ALS)	ALS mutant	
Phosphinothricin	Amino acid biosynthesis	Glutamine synthetase (GS)	GS overexpression	*bar* gene
Atrazine	Photosynthesis	Q_B protein	Mutant *psbA*	GST gene
Bromoxynil	Photosynthesis	Q_B protein		*bxn* gene
2,4-D	Growth regulation	Unknown		*tfdA* gene

modification so that its affinity for the herbicide is reduced, or alternatively engineering the conjugation, modification or metabolism of the herbicide itself (see Table 12.1). The first strategy is difficult to realize owing to a lack of knowledge concerning how herbicides are taken up into the cell; however, each of the later strategies have proven to be successful.

Several herbicides inhibit amino acid biosynthetic pathways and as such can be considered as advantageous, as approximately half of the amino acids used in protein biosynthesis are synthesized uniquely in plants or bacteria. This can ease the isolation of the enzymes that either detoxify a herbicide or that have become resistant to its action because bacteria can simply be used to screen for those able to grow on high levels of herbicides as a result of mutagenesis. As an example of how both target protein amplification, as well as modification, can be used in engineering tolerance to a herbicide the case of glyphosate will be described here. Glyphosate is a wide-spectrum, post-emergence herbicide which is considered as being ideal because it is non-toxic to animals and is rapidly degraded by soil microorganisms. The target site of the herbicide is 5-enolpyruvylshikimic acid 3-phosphate synthase (EPSPS), an enzyme localized predominantly in the chloroplast. Initially, it was found that plant cell lines resistant to glyphosate contained an amplified EPSPS gene, suggesting that overexpression of EPSPS could result in tolerance. This notion was experimentally confirmed by linking a petunia EPSPS gene to the 35S RNA promoter and transferring the construct to tobacco. Resultant transgenic plants were able to grow in the presence of four times the dose of glyphosate required to kill wild-type plants[6]. The alternative approach involved isolating a mutated EPSPS gene (*aroA*) from *Salmonella typhimurium* resistant to glyphosate, linking it to a plant-specific promoter and transferring it to plants. In this example, the bacterial protein did not contain sequences required to target the protein to the chloroplast, so the resultant EPSPS was presumably localized in the cytoplasm. Nevertheless, plants expressing the bacterial gene showed herbicide tolerance[7]. Hence there may be a cytoplasmic shikimate pathway or the substrates and products of the reactions involved are able to pass through the chloroplast membrane. Both strategies have been combined in an attempt to optimize tolerance. In this case a mutant *aroA* gene isolated from a glyphosate-resistant *E. coli* was engineered to contain the chloroplast transit sequence of the native plant protein and linked to the 35S RNA promoter and used to transform plants[8]. The resultant plants were more tolerant of higher levels of glyphosate than those overexpressing the normal plant protein.

Other herbicides which inhibit amino acid biosynthesis include phosphinothricin (PPT), and the sulphonylurea and imidazolinone herbicides. PPT, an analogue of glutamate, inhibits glutamine synthetase. Uniquely in plants, glutamine synthetase plays a vital role not only in ammonium assimilation but also in general nitrogen metabolism. Plants have two forms of glutamine synthetase: one associated with the chloroplast, the other located in the cytosol. As with EPSPS, it was found that a cell line resistant to PPT had an amplified glutamine synthetase gene which resulted in enhanced levels of enzyme activity. Hence it was to be expected that engineering the overexpression of glutamate synthetase in tobacco could result in resistance to PPT and this was indeed the case. In addition to this approach, engineering the metabolic inactivation of PPT has also been successful in conferring resistance to PPT in plants. The bacterium which produces PPT, *Streptomyces hygroscopicus*, also synthesizes phosphinothricin acetyl transferase (PAT), encoded by the *bar* gene which converts PPT to a non-toxic acetylated form acting to protect the bacteria against PPT. Plants engineered to express the PAT gene were found to be able to grow in PPT at levels four to ten times normal field application rate[9]. Sulphonylurea and imidazolinone herbicides are notable because of their low toxicity to animals and low rate of application required for effect. Both groups of herbicide inhibit acetolactate synthase (ALS), a chloroplast protein. Here tolerance to these herbicides has been achieved by engineering the expression of a mutant herbicide-tolerant ALS gene derived from plants.

Similarly to the case of PPT and PAT above, other bacterial products which serve to detoxify herbicides have been used successfully to confer resistance in transgenic plants. These include the product of the *bxn* gene of *Klebsiella ozaenae*, which is a nitrilase detoxifying bromoxynil, and 2,4-dichlorophenoxyacetate monooxygenase encoded by the *tfdA* gene of *Alcaligenes eutrophus*, which degrades 2,4-D.

The target sites of herbicides inhibiting photosynthesis are generally chloroplast-located proteins that are also encoded by the chloroplast genome. Resistance to these herbicides has proven to be more difficult to engineer because of the technical inability to reproducibly introduce foreign DNA into the chloroplast using current transformation protocols. However, in the case of engineering resistance to atrazine this has been partially overcome by mutating the chloroplast target gene, *psbA*, so that it was insensitive to the herbicide, fusing it to a transit peptide coding sequence which would ensure transport into the chloroplast and transferring this, fused to a nuclear transcription signal, to the plant nuclear genome. The protein product in the transgenic tissue was synthesized in the cytoplasm and post-translationally transported into the chloroplasts, where it appeared able to function normally in photosynthesis and resulted in tolerance to the herbicide.

Engineering Virus Resistance

Traditionally, several approaches are available to counter the effects of virus infection on crop yield and quality. These include breeding for resistance, propagation of virus-free material, application of insecticides to control insect vectors as well as modifications in agronomic practice such as changing the timing of planting. Nevertheless, the effects of these measures can at best be limited and by themselves cannot be completely effective. In some cases (for example, in the infection of cereal viruses such as barley yellow dwarf) there are few effective controls and crop losses can be substantial. Gene transfer has been effectively used to study both the replication and expression of viral genomes and has also provided a means of developing several novel strategies which may be used to reduce the effects of viral infection.

Coat Protein-mediated Protection

For some time it has been known that the prior inoculation of a host plant with a virus producing a mild infection would reduce the symptoms produced by the subsequent inoculation of a more virulent strain of the same virus. The extent of this cross-protection is dependent on how closely related the two virus strains are to each other and is characterized by a delay in symptom development and reductions in both lesion number and accumulation of the challenge virus. While the precise mechanism of cross-protection remains unknown, it was found using tobacco plants previously inoculated with tobacco mosaic virus (TMV) that plants were more susceptible to superinfection if the viral RNA was inoculated as opposed to virions. Moreover, cross-protection could be overcome altogether if the challenge TMV RNA was encapsidated in brome mosaic virus coat protein. These results implicated the viral coat protein as playing an important role in cross-protection and prompted experiments involving the transfer of viral coat protein genes, linked to a constitutive promoter, to plants and investigating whether these transgenics responded in a different manner to viral infection. This approach first proved successful using a cDNA representing the coat protein gene of the U_1 strain of TMV linked to the 35S RNA promoter in transgenic tobacco plants (Figure 12.6a). These plants accumulated the viral coat protein to a level 0.1% of the total protein of the plant, and when these plants were subsequently inoculated with TMV they displayed delayed or no TMV symptoms[10]. Subsequent analysis showed that the inoculated plants displayed both reduced numbers of lesions as well as reduced levels of virus accumulation. This approach has been extended to other viruses such as alfalfa mosaic virus, tobacco streak virus, and the transfer of the coat protein genes of both potato virus x and potato virus y to potato has been shown to mediate resistance to infection by mixtures of both viruses. To date, reports of success in this method of engineering resistance to viral infection have been limited to viruses with a messenger sense single-stranded RNA genome which is encapsidated by a single type of capsid protein. This would suggest that a common point in the infection process of these viruses is susceptible to disruption as a result of the expression in the plant of a previously engineered coat protein. The coat protein may act to interfere with the uncoating of the virus particle in the initial stages of viral infection but it is clear that, because there are also reductions in the rate of systemic passage of the virus as well as symptom expression, other as yet unidentified effects may be playing a role in determining this mode of resistance.

Resistance Mediated by the Expression of Satellite RNAs

An alternative approach for engineering resistance to a virus has been demonstrated by engineering plants to express virus satellite RNA from cucumber mosaic virus (CMV)[11]. Satellite RNAs are molecules which show little, if any, sequence homologies with the virus to which they are associated, yet are replicated by the virus polymerases and appear to affect the severity of infection produced by the virus, in some cases attenuating symptoms but in others increasing their severity.

Plants expressing CMV satellite RNA (Figure 12.6b) when inoculated with CMV show normal symptoms of infection on the inoculated leaves as well as the first leaves showing systemic infection. However, no subsequent symptoms of infection developed. Screening the levels of RNA accumulating in the test plants, it was found that upon infection the accumulation of the satellite RNA encoded by the test plant was substantially increased, whereas replication of CMV RNA decreased compared to inoculated control plants. In addition accumulation of the viral coat protein was reduced and interestingly the infectivity of the virus isolated from the infected plant was greatly reduced. Similar results were obtained when plants expressing the tobacco ringspot virus (TobRV) satellite RNA were inoculated with TobRV. The mechanism by which symptom

(a)

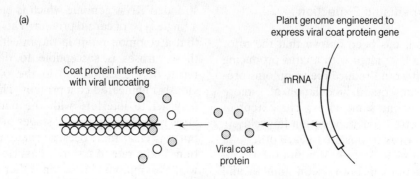

(b)

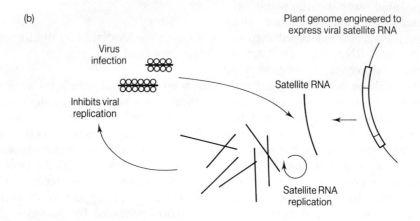

(c)

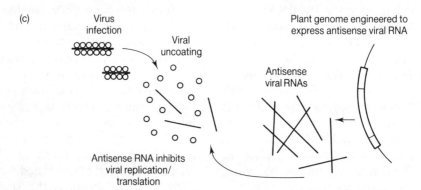

Figure 12.6. Strategies for engineering resistance to viral infection. (a) Expression of a coat protein gene engineered into the target plant is thought to interfere with uncoating of infecting virions and interferes with the later spread of the virus. (b) Following infection of plants engineered to express a satellite RNA, the satellite RNA becomes actively replicated and appears to interfere with viral replication and suppresses symptom expression. (c) Plants engineered to express anti-sense viral RNA can inhibit viral infection by interfering with either replication or translation of the viral RNA.

reduction takes places in these examples remains unknown.

Anti-sense-mediated Protection

It is now well established in a variety of experimental systems that gene expression can be controlled by anti-sense RNA. It has been proposed that anti-sense RNA may play a role in cross-protection, and this has prompted several workers to investigate the effect of the expression of anti-sense RNA on viral infection. In order to do this, cDNAs representing viral RNA genomes were cloned in an anti-sense orientation to a promoter and transferred to plants which were then subjected to test inoculations (Figure 12.6c). This approach has been shown to be effective against infection of TMV, although the level of protection did not appear to be as effective as that achieved as a result of the expression of the coat protein gene. The mechanism by which protection takes place has not been resolved, but it has been shown that an anti-sense RNA containing a sequence representing the 3′ end of the viral RNA is more effective in producing resistance than an anti-sense RNA lacking this region[12]. It is thought that viral replicases bind to this region of the RNA during normal infection, suggesting that resistance in this case results from hybridization of the sense to anti-sense RNA in the 3′ region of the RNA and interferes with the interaction of the replicase in this region.

Engineering Resistance to Insect Predation

Two general approaches have been used to engineer resistance to insect predation in plants. One utilizes the insecticidal protein produced by the bacterium *Bacillus thuringiensis*[13], whereas the other exploits the natural defence mechanisms developed by plants[14].

Bacillus thuringiensis synthesizes an insecticidal crystal protein which resides in inclusion bodies produced during sporulation of the bacterium. The crystal protein, when ingested by insect lar-vae, is solubilized in the alkaline conditions of the insect midgut and processed by midgut proteases to produce a protease-resistant polypeptide which is toxic to the insect. Hence the activity of the protein is limited to insects and this specificity has encouraged the use of the crystal protein as a biological insecticide. Although the polypeptide is able to induce the lysis of insect cells *in vitro*, the exact mechanism of the toxicity is unknown but it is specific for a certain type of insect. Hence, the protein produced by *Bacillus thuringiensis* varieties *berliner* and *kurstaki* is active against Lepidoptera, whereas the protein from the variety *israelensis* is active against Diptera. Extensive work has been carried out on the isolation of the genes encoding the insecticidal protein from the bacteria as well as investigating the importance of different domains of the protein in the insecticidal activity. In the case of the *berliner* variety the protein consists of 1155 amino acids, but the amino terminal region between amino acids 29 and 607 is essential for toxin activity. When this region of the gene was fused to the kanamycin resistance-encoding NPTII gene as a translational fusion and in turn linked to a constitutive plant promoter and transferred to the tobacco genome, the resulting plants which were resistant to high levels of kanamycin were also resistant to insect predation. The insect bioassay involved cutting discs from the transgenic tissue and feeding these to young larvae of *Manducta sexta* (tobacco hookworm)—a pest of tobacco—with insect mortality and weight loss being scored. It was found that in 75% of the plants which were resistant to high levels of kanamycin there was 75–100% mortality of the larvae. By using proteins specific for other insects, the range of protection can be increased to include engineering of resistance to insect predators that have a major economic impact on yield. A recent case in point has been the engineering of cotton to be resistant to *Heliothis zea* (cotton bollworm)[15]. Here the insecticidal protein was engineered so that the codon usage in the structural gene was more similar to that used in plants and this, allied with engineering to allow maximal levels of transcription, resulted in the insecticidal protein accumulating to a level of between 0.05% and 0.1% of the total leaf protein. These plants proved to be resistant to lepidopteran pests which are considered to be agronomically important.

The alternative approach to engineering resistance to insect predation involves the transfer and overexpression of plant genes that are considered to play a role in resistance to insects. One such example involves protease inhibitors, the expression of which can be induced in the plant following insect predation or microbial infection or which accumulate to high levels in seeds[14]. Using this strategy cowpea trypsin inhibitor has been cloned and inserted into tobacco, where its expression resulted in the resistance of the plants to predation by *Heliothis virescens* (tobacco budworm)— a normal pest of tobacco—as well as several other coleopteran pests.

Engineering Plants to Accumulate Novel Proteins

The advent of gene transfer to plants has raised the possibility of engineering plants to produce novel polypeptides which may be of pharmaceutical importance. The idea here is that plants may provide a variable alternative to protein production by microbial or eukaryotic cell expression systems because the production of plant biomass is inexpensive. Moreover, plant cells may be able to correctly process the translation product, thus producing a functional protein in the correct conformation—something that is not always possible to achieve with microbial expression systems. Naturally, to be a successful proposition it is important that the product itself can be produced in high and pure enough levels to make its isolation or utilization economic. For instance, it might not be worthwhile considering the production of a low-yield, high-cost pharmaceutical in a plant if it requires extensive purification. On the other hand, such an undertaking may be worthwhile if the product is simple to purify and might increase the value of a fraction of the plant, or a by-product that is normally discarded. An example of this is provided by engineering potato to

express human serum albumin[16]. In humans this is synthesized as a prepro-protein which is processed to the mature product by sequential protease digestion as the protein passes through the endoplasmic reticulum and Golgi complex. In microbial expression systems the mature protein is either incompletely processed, not secreted efficiently or is difficult to purify from the host cells in culture. Various constructs containing the human serum albumin gene were fused to both the 35S RNA promoter and the translational enhancer sequence derived from alfalfa mosaic virus so as to maximize gene expression and used to transform potato. It was found that the normal prepro-processing signal was able to function in plant cells and target the protein across the plasma membrane. However, normal processing was incomplete and plant-specific processing signals were required for this to be carried out. The advantage of this system is that in the production of starch from potatoes the liquid fraction resulting from the wet milling process is essentially a waste product and could be utilized in the further purification of any protein synthesized in and secreted from the potato cells.

Engineering Plants to Display Agronomically Useful Phenotypes

As has been discussed previously, many of the agronomically useful traits that are of interest to plant breeders are likely to be encoded by a variety of genes, many of which are yet to be identified, let alone cloned and characterized. Moreover, apart from the auxin biosynthetic genes of the T-DNA, there have been no reports of the insertion of more that one gene into a plant which results in the introduction of a novel biochemical pathway in the transgenic. While this naturally limits the useful traits that can be transferred, it does not preclude all which can be agronomically useful and a case in point is provided by male sterility. Male sterility is important in crops such as maize, where there can be an improvement in yield as a result of hybrid vigour. Hybrid seed is produced by crossing parental plants, and if self-fertilization during this process is to be prevented mechanical emasculation, which can be both time consuming and costly, must be carried out. This can be overcome by the use of genetic male sterility, preventing self-fertilization. There are a variety of genetic methods available to produce male sterility but these can be unstable and associated with deleterious effects. Hence there has been great interest in the production of a dominant male sterility gene which could be introduced into plants by transformation. With this in mind, recently a gene encoding a ribonuclease (*barnase*) derived from *Bacillus amlyloliquefaciens* has been fused to a plant promoter which directs expression uniquely in the tapetum cells of the anther and was transferred to both tobacco and oilseed rape[17]. The tapetum surrounds the pollen sac during the initial phases of pollen development and produces a number of proteins and other substances which are thought either to aid in pollen development or become pollen cell wall components. Transgenic plants bearing the *barnase* linked to the tapetum-specific promoter were found to be male sterile and upon closer examination it was found that in these plants the tapetum was destroyed, presumably as a result of the expression of the *barnase*, and no pollen produced. Transfer of this gene into plants where fruit is not the harvested product can immediately be used to produce hybrid seeds by crossing with a pollinator line. On the other hand, where fruit (i.e. seed) is the harvested product a restorer gene which counteracts the activity of the *barnase* is required. However, as part of the strategy of the bacterium to protect itself against the activity of *barnase*, *Bacillus amyloliquefaciens* produces a very potent inhibitor of *barnase* called *barstar*, and engineering this into the plant could be used as a method of inducing fertility restoration.

CONCLUSIONS AND FUTURE PROSPECTS

As we have seen, it is the developmental plasticity of isolated plant cells that has allowed the development of a large array of tissue culture techniques and has culminated in recent years in our ability to transfer DNA to cells and ultimately regenerate a transgenic plant. Gene transfer technology has provided an essential tool in our understanding of how a particular sequence of DNA functions in the plant cell. As illustrated by the case of the T-DNA genes and the anionic peroxidase gene, transfer now can be used as an aid to the biochemist in understanding the role that a particular gene product might play in the life cycle of the plant. With this in mind, many are now working to isolate genes that are thought to play key roles in development, such as those that are important in linking the recognition of plant growth substances to the induction of specific patterns of gene expression or development of a particular organ. The ultimate aim is that once the gene is isolated and characterized it can be reintroduced into plants, linked to a specific promoter, and the effect of the expression of the gene can be investigated. However, while the results of this type of work can be of great interest, the potential results may not always be simple to interpret. The prime factor for these difficulties lies in the limitations of our current understanding of the pattern of expression directed by a particular promoter and our inability to control it experimentally. This limitation must be taken into account when the effect of a particular gene product on a particular cell type or tissue is being investigated.

While the initial work on plant transformation was carried out exclusively on dicotyledonous plants which can be manipulated with relative ease in tissue culture, in recent years much progress has been made in transformation of both cereals and legumes. It is likely that in the near future protocols will become available for the transformation of all of the world's important crop plants. The question that remains is which genes it will be worthwhile to transfer to crop plants. Obviously, in the short term the transfer of single genes encoding agronomically relevant traits is feasible. It remains to be seen whether further research will reveal ways in which multigenic traits may be engineered into crops.

The advances in plant genetic transformation have been followed with much excitement both in the scientific and lay press. The question remains whether, in the current world situation, genetically engineered crops will become generally accepted. On the long path to economic success a new plant has to satisfy many diverse interests. Farmers require a crop which can be grown consistently to make maximum (or at least acceptable) profit and which satisfies the market requirement of governments, commodity dealers and eventually the consumers. These interests are interwoven and can be difficult to predict. Whether a genetically engineered crop will be commonly accepted is likely only to become clear once the first engineered crop approaches marketing. One thing that is certain is that genetically engineered crops will not solve the problems that currently face the world, but they may, in the right circumstances, help to alleviate them.

REFERENCES

1. Skoog, F. and Miller, C.O. (1957) Chemical regulation of growth and organ formation in plant tissues cultured *in vitro*. *Symposia of the Society for Experimental Biology*, **11**, 118–130.
2. Paskowski, J., Baur, M., Bogucki, A. and Potrykus, 1. (1988) Gene targeting in plants. *EMBO Journal*, **7**, 4021–4026.
3. Benfey, P.N. and Chua, N.-H. (1990) The cauliflower mosaic virus 35S promoter: combinatorial regulation of transcription in plants. *Science*, **250**, 959–966.
4. Inze, D., Follin, A., Van Lijsebettens, M., Simoens, C., Genetello, C., Van Montagu, M. and Schell, J. (1984) Genetic analysis of the individual T-DNA genes of *Agrobacterium tumefaciens*: the further evidence that two genes are involved in indole-3-acetic acid synthesis. *Molecular and General Genetics*, **194**, 265–274.
5. Lagrimini, L.M., Bradford, S. and Rothstein, S. (1990) Peroxidase induces wilting in transgenic tobacco plants. *Plant Cell*, **2**, 7–18.
6. Shah, D., Horsch, R.B., Klee, H.J., Kisherer, G.M., Winter, J.A., Turner, N.E., Hironaka, C.M., Sanders, P.R., Gasser, C.S., Aykent, S., Siegel, N.R., Rogers, S.G. and Fraley, R.T. (1986) Engineering herbicide tolerance in transgenic plants. *Science*, **233**, 478–481.
7. Comai, L., Facciotti, D., Hiatt, W.R., Thompson, G., Rose, R.E. and Stalker, D. (1985) Expression in plants of a mutant aroA gene from *Salmonella typhimurium* confers tolerance to glyphosate. *Nature*, **317**, 741–744.
8. della-Chioppa, G., Baier, S.C., Taylor, M.L., Rochester, D.E., Klein, B.K., Shah, D.M., Fraley, R.T and Kishore, G.M. (1987) Targeting of herbicide-resistant enzyme from *Escherichia coli* to chloroplasts of higher plants. *Bio/Technology*, **5**, 579–588.
9. Deblock, M., Botterman, J., Vandewiele, M., Dockx, J., Thowen, C., Gossell, V., Rao Movva, N., Thompson, C., Van Montagu, M. and Leemans, J. (1987) Engineering herbicide resistance into plants by expression of a detoxifying enzyme. *EMBO Journal*, **6**, 6873–6877.
10. Powell Abel, P., Nelson, R.S., De, B., Hoffman, N., Fraley, R.T. and Beachey, R.N. (1986) Delay of disease development in transgenic plants that express the tobacco mosaic virus coat protein gene. *Science*, **232**, 738–743.
11. Harrison, B.D., Mayo, M.A. and Baulcombe, D.C. (1987) Virus resistance in transgenic plants that express cucumber mosaic virus satellite RNA. *Nature*, **328**, 799–805.
12. Hemenway, C., Fang, R.-X., Kaniewski, W.K., Chua, N.-H. and Tumer, N.E. (1988) Analysis of the mechanism of protection in transgenic plants expressing the potato virus x coat protein or its antisense RNA. *EMBO Journal*, **7**, 1273–1280.
13. Vaeck, M., Reynaets, A., Hofte, H., Jansens, S., DeBeukeleer, M., Dean, C., Zabeau, M., Van Montagu, M. and Leemans, J. (1987) Transgenic plants protected from insect attack. *Nature*, **328**, 33–37.
14. Hilder, V.A., Gatehouse, A.M.R. and Boulter, D. (1990) Genetic engineering of crops for insect resistance using genes of a plant origin. In *Genetic Engineering of Crop Plants* (eds G. Lycett and D. Grierson), pp. 51–67. Butterworths, London.
15. Perlak, F.J., Deaton, R., Armstrong, T.A., Fuchs, R.L., Sims, S.R., Greenplate, J.T. and Fischhoff, D.A. (1990) Insect resistant cotton plants. *Bio/Technology*, **8**, 939–943.
16. Sijmons, P.C., DeBeuckeleer, M., Truettner, J., Leemans, J. and Goldberg, R.B. (1990) Production of correctly processed human serum albumin in transgenic plants. *Bio/Technology*, **8**, 217–221.
17. Mariani, C., De Beuckeleer, M., Truettner, J., Leemans, J. and Goldberg, R.B. (1990) Induction of male sterility in plants by chimeric ribonuclease gene. *Nature*, **347**, 737–741.

FURTHER READING

Draper J., Scott, R., Armitage, P. and Walden, R. (1988) *Plant Genetic Transformation and Gene Expression: A Laboratory Manual*. Blackwell Scientific, Oxford.

Lindsey, K. and Jones, M.G.K. (1989) *Plant Biotechnology in Agriculture*. Open University Press, Milton Keynes.

Ream, W. (1989) *Agrobacterium tumefaciens* and interkingdom genetic exchange. *Annual Review of Phytopathology*, **27**, 583–618.

Schell, J. and Vasil, K. (1989) *Cell Culture and Somatic Cell Genetics of Plants: Molecular Biology of Plant Nuclear Genes*. Academic Press, San Diego.

Walden, R. (1988) *Genetic Transformation in Plants*. Open University Press, Milton Keynes.

Walden R., Koncz, C. and Schell, J. (1990) The use of gene vectors in plant molecular biology. *Methods in Molecular and Cellular Biology*, **1**, 175–194.

Weising, K., Schell, J. and Kahl, G. (1988) Foreign genes in plants: transfer, structure and expression. *Annual Review of Genetics*, **22**, 421–477.

Index

δ-aminolaevulinic acid 188–9
aminotransferases 162–3
ammonia (NH₃) 130, 156, 157
 assimilation 159, 160, 162
  diazotrophs 137, 142, 143, 146–8
 metabolic role 165–7
 photorespiratory, reassimilation 41, 166–7, 178
AMP 14
α-amylase genes 265–7
α-amylases 92, 93, 95
β-amylases 92, 95
amylopectin 90, 91, 93
amylopectin 6-glycanohydrolase (R enzyme) 92, 95
amyloplasts 93–5, 210
amylose 90, 93
Anabaena 132, 142, 144
anaerobes 2, 138
anaerobic response element 226
Ananas comosus (pineapple) 60, 64
Andria 147
anf genes 141
animals, pigments 195
anomers 77–8
anthocyanidin 191–2
anthocyanins 191–3, 195
anthraquinones 193
anti-sense RNA 236, 260, 261
 engineering virus resistance 290, 291
 ethylene-forming enzyme 260–1, 262
antibodies, screening cDNA libraries 238, 240
antimycin A 23–4, 25
Antirrhinum majus 236, 268, 269–73
apetala-1 mutants 269–70, 271
apetala-2 mutants 270, 271
apetala-3 mutants 270, 271
apoplastic transport 88
aquatic plants
 CO₂ availability 66–7
 light absorption 8
Arabidopsis thaliana 32, 92, 158, 217
 DNA content 222, 223
 floral development 268, 269–73
 photorespiration 39, 40–1, 43, 44
 transformation 268, 283
arabinans 105, 106, 107
arabinogalactans 105, 106, 107
arabinose 76
arginine synthesis 174–5
Arum maculatum 16, 17–18, 19
ascorbate 177
ASF-1 234
asparaginase 164
asparagine 147, 163–4
asparagine synthetase 163–4
Asparagus officinalis (asparagus) 102, 103
aspartate 147, 164

C₄ plants 48, 49–50, 53
 transport 19, 55
aspartate aminotransferase 162, 163
aspartate-derived amino acids 167–73
 regulation of biosynthesis 167, 171–3
aspartate kinase (AK) 167, 168, 169, 172, 173
aspartate:2-oxoglutarate aminotransferase 179
AT-1 231
ATP
 Calvin cycle 28
 control of glycolysis 16
 hydrolysis 2–3
 nitrogen fixation 134, 135, 138
 synthesis 3–11, 22–3
 transport 19, 37
ATP:fructose 6-phosphate 1-phosphotransferase, *see* phosphofructokinase
ATP sulphurylase 176
ATP synthetase complex 9, 203
ATPase, F₀F₁ 3–5, 19, 20
atrazine 287, 288
Atriplex rosea 70
autocatalysis, Calvin cycle 28
autotrophs 2
auxins 277, 282, 285–6
Avena sativa (oats) 223, 266
Azolla 132, 144, 148
Azorhizobium 132
Azorhizobium caulinodans 131
Azospirillum 140, 148
Azospirillum lipoferum 132
Azotobacter 140
Azotobacter chroococcum 137
Azotobacter vinelandii 132, 136–7, 141

Bacillus amyloliquefaciens 293
Bacillus thuringiensis 291
bacteria
 genome organization 204–5, 206
 nitrogen-fixing 131–2
 pigments 188, 195
bacteriochlorophylls 187, 188
barley (*Hordeum*)
 amino acid metabolism 169, 170, 172, 173–4
 nitrate uptake/reduction 156–7, 158
 photorespiration 43, 44
 seed germination 265, 267
barnase gene 293
barstar 293
bean (*Phaseolus vulgaris*) 33, 160–1, 164, 235
Benson–Calvin cycle, *see* Calvin cycle
Beta vulgaris (beetroot) 194, 195
betalains 194, 195
bicarbonate ions (HCO₃⁻) 50, 66, 67
bile pigments 186, 187, 188
bilirubin 187

Index compiled by Liza Weinkove